THE
BUSY
NARROW
SEA

THE BUSY NARROW SEA

A SOCIAL HISTORY OF THE ENGLISH CHANNEL

ROBIN LAURANCE

First published 2024

The History Press
97 St George's Place, Cheltenham,
Gloucestershire, GL50 3QB
www.thehistorypress.co.uk

© Robin Laurance, 2024

The right of Robin Laurance to be identified as the Author
of this work has been asserted in accordance with the
Copyright, Designs and Patents Act 1988.

All rights reserved. No part of this book may be reprinted
or reproduced or utilised in any form or by any electronic,
mechanical or other means, now known or hereafter invented,
including photocopying and recording, or in any information
storage or retrieval system, without the permission in writing
from the Publishers.

British Library Cataloguing in Publication Data.
A catalogue record for this book is available from the British Library.

ISBN 978 1 80399 682 0

Typesetting and origination by The History Press
Printed and bound in Great Britain by TJ Books Limited, Padstow, Cornwall.

 Trees for LYfe

Contents

Introduction   7

1	When Britain First Left Europe	11
2	The Long Crawl	27
3	A Tale of Two Cities	35
4	Cawsanders and Smugglers	57
5	The Channel's Islands	77
6	The Channel at War	129
7	Packets, Balloons and *Mal de Mer*	149
8	Writers, Poets, Painters and Debussy	169
9	Just Another Day at the Office	189

Acknowledgements   213
Bibliography   215
Index   217

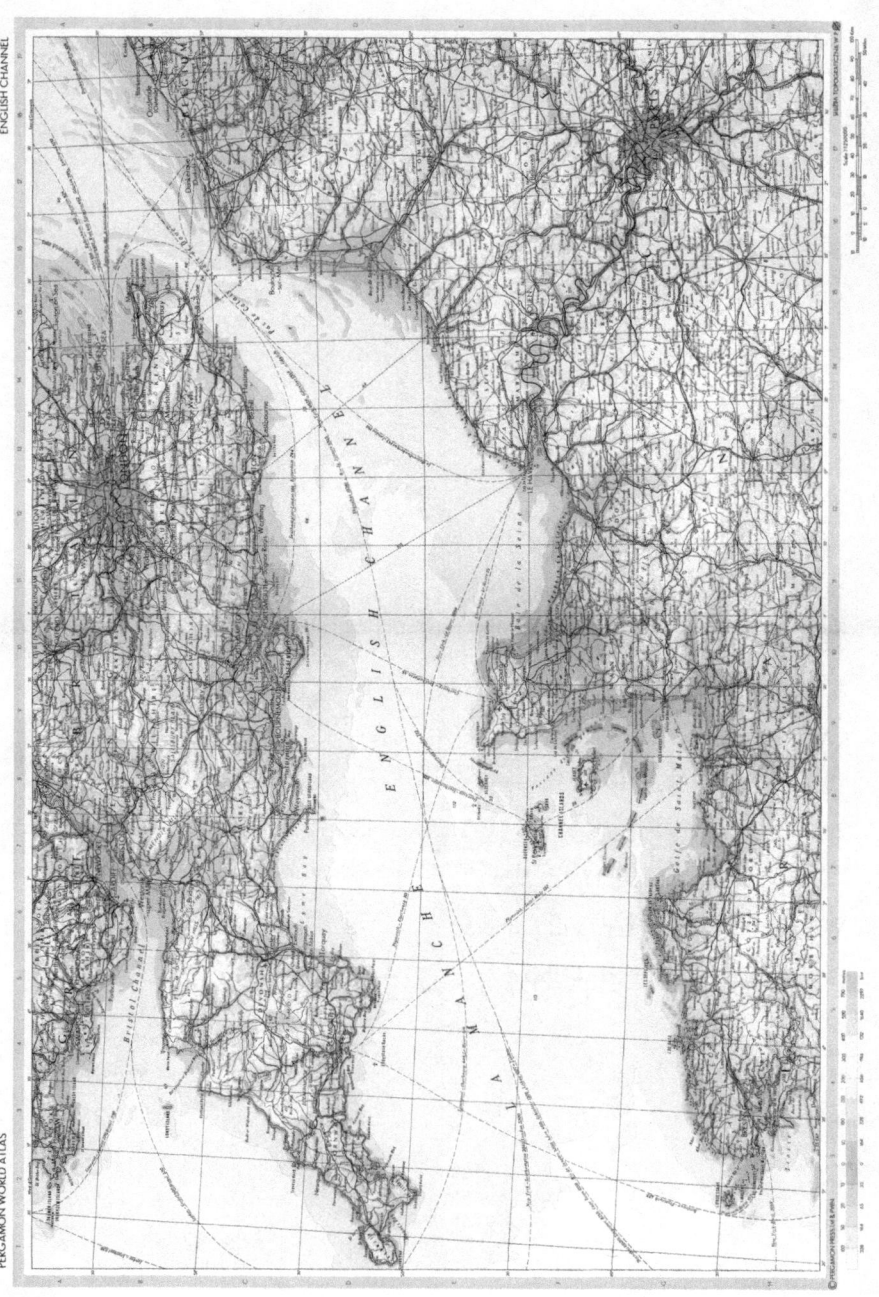

The English Channel, or *La Manche* as it is known in France, from *Pergamon World Atlas*, 1968. (The Rumsey Collection)

Introduction

Britain first left Europe half a million years ago when a mass of icy water cascaded across the narrow piece of tundra that joined this small island to the continental landmass of Europe. Since then, the English Channel – *La Manche* to the French – has played host to humankind in all its various guises. For potential invaders, from Julius Caesar and the Normans to the Nazis, it has been the final hurdle, crossed successfully by the Romans and William the Conqueror but defeating Hitler in the twentieth century. Boulogne was where invasions were planned, yet neighbouring Calais was ruled by the British for 200 years from 1360. Dick Whittington was the town's mayor in 1407.

The Channel's waters were charted by Edmond Halley after he had tracked his eponymous comet. Henry Winstanley, a no less extraordinary pioneer, built the first Eddystone Lighthouse of wood on a small, slimy piece of rock, 9 miles from the shore.

Brighton and Deauville mirrored each other by growing from small fishing villages into seaside resorts that set the pattern for beach resorts the world over as they entertained the great and the good and not so good. The ever-changing Channel light inspired the birth of Impressionism, with Monet, Boudin and Pissarro leading the way from the Channel coast of France. Writers and poets similarly found inspiration in this narrow seaway – Wordsworth, Shakespeare, Charlotte Smith, John Masefield, Dickens and Matthew Arnold all had cause to turn to the Channel. Victor Hugo wrote *Les Misérables* while living on

Guernsey, and on a tiny islet off Saint-Malo lies the body of the French romanticist, Chateaubriand.

Fishing remains an important industry along both sides of the Channel. In the seventeenth century, pilchard fishing off the Devon coast went a long way to finance the famous Bodleian Library at Oxford University. Today, herring fishing sustains the fishermen of Boulogne, and the trawlermen of Brixham put Dover sole on the high tables in London.

The Channel's islands have, over the years, enjoyed a mixture of fame, fortune and fascism. The Isle of Wight became the summer home of Queen Victoria and Prince Albert; the meeting place for the world's finest yachts and yachtsmen; home to the Poet Laureate, Alfred Lord Tennyson, and to the outstanding portrait photographer of the Victorian era, Julia Margaret Cameron (Cameron was snubbed by a visiting Giuseppe Garibaldi, who took her for a vagrant). The island's prison, Parkhurst, held many of the country's most notorious criminals and suffered one of the British Prison Service's most shameful breakouts. In 1968, the island became the site of a pop festival that crowned the counter-culture of the sixties and set the trend for festivals of music, mud, sex and drugs that became a rite of passage for teens and twenties-somethings the world over.

Jersey farmers discovered an exceptional and early potato, while their dairy colleagues found a little brown cow that produced an inordinate amount of milk. And the islanders discovered just how attractive the local tax breaks could be. But both Jersey and Guernsey suffered horribly at the hands of the Nazi occupying forces during the Second World War.

That war saw terrible fighting in and over the Channel. And it witnessed acts of extraordinary courage. But there were some awful and unnecessary disasters too, not least at Slapton Sands in the days before the D-Day Landings. Pigeons flew the Channel in the service of MI6, and the State Rooms of the Royal Pavilion at Brighton became chandeliered hospital wards for wounded Indian combatants.

Throughout its existence, and beyond the days of war, crossing the English Channel has proved a challenge for sailors, engineers and not least for adventurers. Jean-Pierre Blanchard was the first to fly the 21 miles across the Channel in his hot-air balloon. In 1909, Louis Blériot was the first to fly a plane across.

Introduction

Its engine was made by a company that produced lawnmowers, its propeller was of varnished walnut, its fuselage and wings used a combination of ash, oak and poplar. And by converting an old coastal collier into the Channel's first car ferry, in 1928 Stuart Townsend met the challenge set by British tourists who wanted to drive their own cars on the Continent.

Matthew Webb, an officer in the Merchant Navy, was the first to swim the Channel. In 2010, Philippe Croizon crossed in thirteen hours – he had no arms or legs – and Hilary Lister sailed across, controlling her dinghy by blowing through straws – she was paralysed from the neck down.

Today, the English Channel is the world's busiest seaway – a super maritime highway for tankers, container ships, bulk carriers, cruise liners, ferries, fishing boats and yachts. Deep-sea pilots guide the freighters through the crowded waters, first climbing rope ladders up the sides of huge vessels in what are often rough seas. Trawlermen squabble over fishing grounds in a stretch of water that both links and separates two nations who find it so hard to get along. And it separates an island state, not only from its immediate neighbour but from the wider continent of Europe. A narrow sea that is deep in history but too often mired in conflict.

1

When Britain First Left Europe

When Britain first left Europe half a million years ago there were no politics involved: it did so by the force of nature. Before the parting, you could take a Sunday afternoon stroll from England, across Siberian-like tundra, to the European continent and then walk back again before dark. It would not have been a particularly interesting trek across a land bridge with little vegetation. You would not have met anybody on the way, but you might have encountered mammoths and the occasional elk, and you would have got your feet wet crossing the river that ran through the valley.

And then there was a catastrophic flood – a mega flood created by much greater forces than prolonged and heavy rain. Earth scientists believe that the land bridge that joined Britain to the Continent at what is now the Channel's eastern end had formed a dam holding back the waters of a huge lake fed by the waters of the rivers Thames, Rhine, Meuse and Scheldt, with ice from the mountains of Scandinavia and Scotland forming the lake's eastern and northern boundary. Evidence for the lake comes from glacial-like sediments found in the coastal areas of south-east England and on the continental coast where Holland meets Germany. The Cambridge Professor Philip Gibbard, an expert on the Channel's birth, suggests that the water level in the lake was about 30m above current sea levels and was held back by the land bridge that joined England to the Continent.

It was about 450,000 years ago that the waters of the lake began to spill over the land bridge, filling the valleys on its western side and creating what Gibbard calls a 'channel river', and a very narrow beginning of the Dover Straits. Erosion caused the river to widen gradually over the next 300,000 years, until the next mega flood occurred when a second ice sheet in the southern North Sea created a new lake which was dammed to the south by a land ridge north of the Dover Straits.

When this weaker dam burst, the flood was catastrophic. Sanjeev Gupta and his colleagues at Imperial College London studied data from the examination of the Channel floor, which showed grooves indicating that it was once exposed to the air and then subjected to an enormous deluge of water. This evidence of vast quantities of water cascading from the southern limits of the North Sea was what scientists had been waiting for.

Land on both the northern and southern sides of this new torrent of water was eroded, and Britain became an island. Its history had been defined by the forces of nature; not until the opening of the Channel Tunnel could you once again walk (in theory) to the continent of Europe from England's green and pleasant land.

'Nature has placed England and France in a geographical location which must necessarily set up an eternal rivalry between them,' wrote Napoleon's spin doctor, Jean-Louis Dubroca, in 1802. How right he was. The two kingdoms kept apart physically by the waters of the Channel have been kept apart by long-established cultural differences, peppered by outbursts of pique. And when the French decided they had had enough of kings, the Republic proved no better friend to their cross-Channel neighbours.

Robert Tombs, a Professor of French History at Cambridge University, writes:

> Geography comes before history. Islands cannot have the same history as continental plains. The United Kingdom is a European country, but not the same kind of European country as Germany, Poland or Hungary. For most of the 150 centuries during which Britain has been inhabited it has been on the edge, culturally and literally, of mainland Europe.

There are no marker buoys defining the extent of the Channel, but by general consent the Straits of Dover mark the eastern end, and a line between Land's End and the Île Vierge, off the Brittany coast, defines the point where the Channel meets the Atlantic Ocean. Altogether, this is a stretch of about 350 miles. The massive granite lighthouse on Vierge rises nearly 100m above the ground and is the tallest in Europe. The first Longships Lighthouse, on the rocks off Land's End, was also built of granite in 1795. The modernised automatic light now has a helipad on the top.

Known originally as the 'Narrow Sea', the British now call this stretch of water the English Channel, although nobody is entirely sure why. Early Dutch sea charts refer to it as the *Engelse Kanaal*, and English statesmen would, at an early stage, have wanted the French to be in no doubt about who had sovereignty over the water. The French call it *La Manche* (the sleeve). On days when the sea is angry and giant freighters are crowding the space as they fight to meet ever-more demanding schedules, ferry captains crossing the freighters' path call it something else altogether. Separation lanes keep the eastbound and westbound traffic apart, with the eastbound, fully laden ships taking the French side. But in the Straits of Dover, where the ferry crossings are at their most prolific, there are just under 22 miles between the chalk cliffs of Dover and Cap Gris-Nez, squeezing the traffic lanes and demanding the unflinching concentration of helmsmen.

That unflinching concentration required of helmsmen was clearly absent when the cargo ship *Mont-Louis* sank in the Channel in 1984 after colliding with a German vessel in thick mist. The collision raised worrying questions about the cargoes carried through the narrow seaway. The French authorities claimed the ship had left Le Havre with a cargo of medical supplies, but, on persistent questioning from the environmental protection campaigners Greenpeace, admitted the cargo was not medical supplies but barrels of uranium hexafluoride. The *Mont-Louis* had been heading for Riga in the Soviet Union, where the uranium was to be enriched. There were no fatalities as a result of the collision, and the cylinders of uranium hexafluoride were recovered intact.

The Channel is classed as international waters. The 12-mile Territorial Sea Convention applies until the Dover Straits, where there are less than 24 miles between the French and English coasts. The two countries share responsibility and management for this stretch of the Channel, and for the 500-plus commercial ships that pass through the straits every day. (There was a time when the range of territorial waters was measured by the distance travelled by a shot fired from a cannon on the shore.)

The Channel weather is notoriously fickle. Fog and heavy rain that reduce visibility and winds that regularly reach gale force test the best of seamen. But the shipping forecast, broadcast four times a day by the BBC, does at least provide warnings of what is to come. This forecasting for shipping around the British Isles was conceived after a fatal storm off the Welsh coast in 1859 claimed 130 ships and the lives of 800 sailors. As a result, Robert Fitzroy, who had captained Charles Darwin's *Beagle*, was moved to pioneer a system of telegraphed warnings of storms to harbours around the British Isles – a system that grew into today's Meteorological Office. The Met Office divides the Channel into four regions – Dover, Wight, Portland and Plymouth – advising ships' captains of predicted wind directions and force, and likely visibility. And BBC Radio 4 broadcasts the information at four precise times during the day on behalf of the Maritime and Coastguard Agency.

Tides are strong in the Channel. High tide at Plymouth comes some seven hours after high tide at Dover. At Calais, high tide comes five hours before Saint-Malo. And as the tide rises and falls it creates tidal currents. Tidal ranges – the difference in height between low and high tide – differ greatly. At Dover it is about 17ft; at Brighton it is nearer 19ft 6in. The Bristol Channel experiences a rise in water levels of some 30ft, while at Saint-Malo it can reach 26ft. (It climbs to just under 40ft in the Minas Basin of the Bay of Fundy on Canada's Nova Scotia coast.) Just beyond Saint-Malo, the tidal range at the mouth of the Rance river reaches a full 35ft, which prompted the French electricity company EDF to build the world's first tidal barrage across the estuary. Its twenty-four turbines generate enough tidal power in a year to satisfy the needs of a town the size of Rennes with its population of 217,000.

Spring tides, which have nothing to do with the seasons, occur when the earth, sun and moon are in alignment and the gravitational pull of the sun adds to the pull of the moon. The oceans bulge more than usual and tidal heights increase. Seven days later, when the three are out of alignment, the high tides are a little lower and the low tides a little higher. These are referred to as neap tides. In the Channel during spring tides, tidal currents flow at terrifying speeds through the Alderney Race – the funnel between the island and Cap de la Hague on the north-western tip of the Cherbourg Peninsula. Yachtsmen have logged the current at 12 knots (22km/hour), which is fine if you are going with the tide but to be avoided at all costs if you are coming the other way.

The Channel tides proved too much for Julius Caesar. Coming from a part of the world that sees very little tidal movement, the tides in the Channel confused him. After his failed first crossing, he made a base for himself at Boulogne, where he ordered the building of hundreds of landing craft. In all, he had gathered 800 ships, which were made ready at their moorings on the River Liane, just upstream from Boulogne. With some 17,000 men and 2,000 horses, he set sail at sunset one summer evening in 54 BCE. But the tidal currents in the Dover Straits drove him way off course up the Kent coast towards the Thames Estuary. He had failed to add those tidal currents into his calculations. His own account claims he made landfall 7 miles along the coast but he fails to say in which direction.

In Rome, there was already apprehension about the seas around Britain, not least about the tides. Marcus Cicero wrote about his misgivings, 'I was afraid of the ocean, afraid of the coast of the island.' And the historian Cornelius Tacitus, writing about the exploits of his father-in-law Agricola, noted:

> Nowhere has the sea a wider dominion, that it has many currents running in every direction, that it does not flow and ebb within the limits of the shore but penetrates and winds far inland and finds a home among hills and mountains as though in its own domain.

The tidal problems that faced Caesar in the Channel were recorded by the Greek historian Strabo:

> The deified Caesar crossed over to the island [Britain] twice, although he came back in haste, without accomplishing anything great or proceeding far into the island, not only on account of the quarrels that took place in the land

of the Celti among the barbarians and his own soldiers as well, but also on account of the fact that many of his ships had been lost at the time of the full moon, since the ebb-tides and the flood-tides got their increase at that time.

Strabo also remarked that while the Britons were taller than the Celts, their bodies were of 'looser build'. He also recalled that it rained a lot and that fog often hung over the Channel.

Ten years previously, the Chinese had produced tide tables that research suggests were extraordinarily accurate. 'Time and tide,' observed Geoffrey Chaucer, 'wait for no man.' (See Edmond Halley below.)

Unlike Julius Caesar, William I – William the Conqueror – was provided with a good tidal understanding from his seamen, many of whom had been working in cross-Channel trade. Tidal conditions were decisive in the choice of his departure date and time from Saint-Valery. The optimum time for departure was sunset on 27 September 1066. The crossing to Pevensey was 127km. William had amassed 10,000 men and 3,000 horses at Cabourg, and had gathered 300 ships to put them all in. The Norman Conquest was under way, recalled in pictures created with wool thread on 70m of linen cloth.

Who commissioned the Bayeux Tapestry is unclear, but recent research at the University of York adds weight to the long-held assumption that it was intended to decorate the nave of the cathedral at Bayeux. The city's Bishop Odo was alongside William at the Battle of Hastings (and is portrayed in the tapestry), prompting some historians to suggest that the tapestry was commissioned by the bishop himself. For a time, it appeared in the cathedral on an annual basis, but the tapestry now rests permanently in a purpose-built museum in the city. There is a copy in the museum at Reading.

Six centuries later, in the first decade of the eighteenth century, it was an astronomer who set about charting the tides in the English Channel. The fact that it took an astronomer to do the work is not as surprising as it might seem. After all, the tides were all about the pull of the sun and the moon, so when Their Lordships at the Admiralty – not least Josiah Burchett, the Secretary to the Admiralty – sanctioned astronomer Edmond Halley's trip into the English Channel, Halley, who already had a comet to his name, was eager to be on his way.

Halley was a man of immense learning. The son of a wealthy soap-maker, he went first to St Paul's school and then to Oxford, where he gained entry to Queen's College when he was just 16. But rather than complete his degree, he took off for St Helena with a friend, a sextant and some telescopes, and spent a year looking at the stars. He recorded the celestial longitude and latitude of 341 of them.

Back home, he was something of a star himself. But there was little time to bask in any glory for he was soon off again, this time into the Atlantic, charged with 'improving the knowledge of the Longitude and variations of the compass, with all the accuracy you can'. His masters at the Admiralty had no idea whether his findings were accurate but appeared delighted with the work.

When Halley wrote his next Admiralty assignment for himself, Their Lordships were perfectly happy to give him the go-ahead, but to save any loss of face, published the instructions under their name.

A pink (after the Dutch *pinke*, meaning pinched) was a type of sailing vessel, flat-bottomed with a narrow pinched stern, and the *Paramour* was the one chosen for this research:

> Whereas his Majesties Pink the Paramour is particularly fitted out and Putt under your Command that you may proceed with her, and observe the Course of the Tydes in the Channell of England, and other things remarkable.
>
> You are therefore hereby required and directed to proceed with the Said Vessel, and use your utmost care and Diligence in observing the Course of the Tydes accordingly, as well in the Midsea as on both shores; As alsoe the Precise times of High and Low Water of the Sett and Strength of the Flood and Ebb, and how many feet it flows, in as many, and at such certaine places, as may Suffice to describe the whole. And whereas in many places in the Channell there are irregular and halfe Tydes you are in a particular Manner to be very carefull in observing them.
>
> And you are alsoe to take the true bearings of the Principal head Lands on the English Coast one from another and to continue the Meridian as often as conveniently you can from side to side of the Channell in order to lay downe both Coast truly against one another.

And in case dureing your being employed on this Service, any other Matters may Occur unto you, the observing and Publishing whereof may tend towards the Security of the Navigation of the Subjects of his Majestie or other Princes trading into the Channell You are to be very careful in the takeing notice thereof: And when you Shall have performed what Service you can, with relation to the particulars before mentioned you are to returne with the Ship you Command into the River of Thames, giving Us from time to time an Account of your Proceedings. Dated this June 12th 1701.

He sailed the three-masted, six-gun naval pink for five months in the summer and early autumn of 1701, cruising the Channel shores of both England and France, observing the tides' rise and fall. To feed themselves, the crew caught fish using lines over the side of the boat, providing Halley with the only soft food he could manage – he was only 45 but had lost all his teeth.

From the Channel, he dutifully informed his masters in the Admiralty of his progress, writing from various ports of call as he made his way along the British coast before crossing from the Lizard to Ushant, a tiny island off the Brittany coast. He would address his letters to the Admiralty, 'Honoured Sr' and close with 'Your Honrs most obedt: servt'.

From Spit Head (*sic*) on 29 July 1701, he wrote:

I have ... endeavoured to get as exact an account of the Tides in the Channell as possible, and have ankered all over it ... and have been particularly curious in this part between the Isle of Wight and Portland ... where I find the Course of the Tides very extraordinary.

He had witnessed the notorious tidal race at Portland Bill.

He got into the habit of referring to Their Lordships at the Admiralty as 'Lopps' – as in, 'In obedience to their Lopps orders ...', 'If their Lopps have any further orders for me ...' or 'As their Lopps shall direct...'

With the job done, he was back at Deptford by the second week in October. He had already sent a note to Josiah Burchett, the Admiralty Secretary, saying he had 'discovered beyond my expectation the general rule of the Tydes in the Channell; and in many things corrected the charts thereof'. He would publish his findings in the form of a chart, which he first presented to a meeting of the Royal Society and then had published under the title 'A new and Correct Chart of the Channell between England and France: with considerable improvements not extant in any Draughts hitherto Publish'd; shewing the Sands, shoals, depths of water and Anchorage, with ye flowing of the Tydes, and the setting of the Current'.

The chart records depths in fathoms. 'Ye Hour of High-water, or rather ye End of the stream that setts to ye Eastward on ye Day of ye New & Full Moon' was expressed in plain Roman numerals. The tide tables produced more than three centuries later do not differ greatly from Halley's findings.

His seafaring days ended with his appointment to the Chair of Geometry at the University of Oxford. He lived in a small house in New College Lane, where he had an observatory built into the roof. He entertained his friend Isaac Newton here and was alone responsible for encouraging Newton to publish his *Principa* (some would say his encouragement of Newton was the most valuable thing he ever did). Yet Halley's career was crowned with his appointment as Astronomer Royal in 1720, when he already held the post of Secretary to the Royal Society. He died in 1740 aged 86.

Newton and Halley were working in an age of increasing scientific knowledge into which stepped a mildly eccentric engineer/inventor from Saffron Walden in Essex called Henry Winstanley, the son of the estate manager at the Earl of Suffolk's Audley End House. Henry went to Saffron Walden Grammar School, and then briefly into service at Audley End. When still in his late twenties, he took himself off on a tour of Europe, and on his return gathered together some of his whimsical engineering apparatus and put it on show in Piccadilly as a visitor attraction. He called it the Water Theatre, and much to everyone's surprise, not least his own, it proved an outstanding success. Children were entranced and their parents enthralled.

His prosperity grew and Henry and his parents enjoyed a comfortable lifestyle. Alison Barnes, in her book about Henry, says the family 'ate roast swan at New Year, pancakes on Shrove Tuesday, leeks and fish during Lent, tansies [a yellow-flowered perennial] in June, pheasants, partridge and oysters in September and turkey, geese, ducks, frumenty [wheat boiled in milk and flavoured with cinnamon], plum pudding and marzipan sweetmeats at Christmas'.

Clearly a man of substance, Henry took for his bride Elizabeth Taylor, from Little Munden in Hertfordshire, shortly before his 40th birthday. He then did something remarkable: he built a lighthouse on a small slippery piece of rock, 9 miles out in the middle of the Channel.

Winstanley's business had expanded, and he invested in a cargo ship called the *Constant*. Shortly after he added this new addition to his business, the ship was wrecked on rocks off Plymouth on Christmas Eve in 1695, with the loss of all on board. The *Constant* wasn't the only casualty of what became known as the Eddystone Rocks. This reef was just visible above the surface when the sea was calm but disappeared under even moderate waves. Trinity House had given the go-ahead for a lighthouse on the rocks in 1692 to one Walter Whitfield, a Plymouth businessman, but it proved too much for him, and after more wrecks and more loss of life, Mr Winstanley took up the challenge.

As a starting base, he had a small, uneven, slippery piece of rock on which he could work when the sea was calm. It was hard to get a foothold at the best of times, yet in two years he had a solid stone base 12ft high and 14ft in diameter, bound with copper and iron. Waves would unexpectedly and periodically wash over the rock, taking with them his tools and his lunch. Twelve iron stanchions anchored the structure to the rock, each stanchion fixed in a man-made hole with molten lead.

He enlarged the base and on it constructed a tower of wood, which he bound with iron straps, the work being interrupted when French corsairs (privateers) kidnapped him and shipped him off to France demanding a ransom. The Admiralty made urgent representations to the French authorities, and the king himself, Louis XIV, ordered the lighthouse builder's immediate release, saying that while France was at war with England, she was not at war with humanity.

Undaunted by his French sojourn, Winstanley went back to work on the tower, adapting it from its original octagonal shape to an improved and strengthened twelve-sided design and taking it to 115ft. In the base there was a windowless storeroom, above which was a bedroom. On top of that was a grand visitor's room with sash windows and two closets.

Pipes from the roof channelled rainwater into a cistern. The kitchen had a domed roof and access to an alfresco platform. The eight-sided glass lantern with sixty candles was first lit on 14 November 1698. For three years, not a single vessel came to grief on the Eddystone Rocks, with Winstanley boasting that his lighthouse would withstand anything that nature could throw at it. He was an optimist – and anyway, he had faith in the work his team had done.

Yet, bad weather engulfed the Channel in November 1703, with the vicious tail of an American hurricane bringing ever-increasing winds as the month progressed. By the night of 26–27 November the wind had reached gale force, and the western Channel was experiencing the worst storm anyone could remember. Winstanley and some of his builders were in the lighthouse.

Daniel Defoe, the author of *Robinson Crusoe* and a very able reporter, relates what happened that night in his *Tour Thro' the Whole Island of Great Britain*. He begins at Plymouth:

> In the entrance to this bay, lyes a large and most dangerous rock, which at high-water is covered, but at low-tide lyes bare, where many a good ship has been lost, even in the view of safety, and many a ships crew drown'd in the night, before help could be had for them.
>
> Upon this rock, which was call'd the Edystone, from its situation, the famous Mr. Winstanley undertook to build a light-house for the direction of sailors, and with great art, and expedition finished it; which work considering its height, the magnitude of its building, and the little hold there was, by which it was possible to fasten it to the rock, stood to admiration, and bore out many a bitter storm.
>
> Mr. Winstanly [sic] often visited, and frequently strengthened the building, by new works, and was so confident of its firmness, and stability, that he usually said, he only desir'd to be in it when a storm should happen, for many people had told him, it would certainly fall, if it came to blow a little harder than ordinary.
>
> But he happen'd at last to be in it once too often; Namely, when that dreadful tempest blew, Nov. the 27, 1703. This tempest began on the Wednesday before, and blew with such violence, and shook the light-house so much, that as they told me there, Mr. Winstanly [sic] would fain have been on shoar, and made signals for help, but no boats durst go off to him; and to finish the tragedy, on the Friday, Nov. 26, when the tempest was so redoubled, that it became a terror to the whole nation; the first sight there seaward, that the

people of Plymouth, were presented with in the morning after the storm, was the bare Eddystone, the light-house being gone; in which Mr. Winstanly [sic], and all that were with him perish'd, and were never seen, or heard of since: But that which was a worse loss still, was, that a few days after a merchant's ship call'd the Winchelsea homeward bound from Virginia, not knowing the Eddystone lighthouse was down; for want of the light that should have been seen run foul of the rock itself, and was lost with all her lading, and most of her men, but there is now another light-house built on the same rock.

What other disasters happen'd at the same time, in the Sound, and in the roads about Plymouth, is not my business: They are also publish'd in other books, to which I refer.

One thing, which I was a witness too, on a former journey to this place, I cannot omit: It was the next year after that great storm, and but a little sooner in the year, being in August, I was at Plymouth, and walking on the Hoo, which is a plain on the edge of the sea, looking to the road, I observed the evening so serene, so calm, so bright, and the sea so smooth, that a finer sight, I think, I never saw; there was very little wind, but what was, seem'd to be westerly; and, about an hour after, it blew a little breeze at south west, with which wind there came into the Sound, that night, and the next morning, a fleet of fourteen sail of ships, from Barbadoes; richly loaden, for London: Having been long at sea, most of the captains and passengers came on shore to refresh themselves, as is usual, after such tedious voyages, and the ships rode all in the Sound on that side next to Catwater: As is customary, upon safe arriving to their native country, there was a general joy and melanchol, both on board and on shore.

The next day the wind began to freshen, especially in the afternoon, and the sea to be disturbed, and very hard it blew at night, but all was well for that time; but the night after it blew a dreadful storm, not much inferior, for the time it lasted, to the storm mentioned above, which blew down the light-house on the Eddy Stone; about midnight the noise indeed was very dreadful, what with the roaring of the sea, and of the wind, intermixed with the firing of guns for help from the ships, the cries of the seamen and people on shore, and, which was worse, the cries of those, which were driven on shore by the tempest, and dash'd in pieces.

In a word, all the fleet, except three, or thereabouts, were dash'd to pieces against the rocks, and sunk in the sea, most of the men being drowned: Those three, who were sav'd, received so much damage, that their lading was almost all spoil'd: One ship in the dark of the night, the men not knowing where they were, run into Catwater, and run on shore there, by which she was however sav'd from shipwreck, and the lives of her crew were saved also.

This was a melancholy morning indeed; nothing was to be seen but wrecks of the ships, and a foaming furious sea, hi that very place where they rode all in joy and triumph, but the evening before: The captains, passengers and officers who were, as I have said, gone on shoar, between the joy of saving their lives, and the affliction of having lost their ships, their cargoes, and their friends, were objects indeed worth our compassion and observation; and there was a great variety of the passions to be observed in them: Now lamenting their losses, then giving thanks for their deliverance, many of the passengers had lost their all, and were, as they expressed themselves, utterly undone; they were, I say, now lamenting their losses, with violent excesses of grief; then giving thanks for their lives, and that they should be brought on shore, as it were, on purpose to be sav'd from death; then again in tears for such as were drowned; the various cases were indeed very affecting, and, in many things, very instructing.

Winstanley's lighthouse was replaced by one designed by John Rudyard. It had a base of solid wood, laid on flattened pieces of the rocks. Wood and stone were used to form a conical tower that rose to 92ft and was topped with an octagonal lantern with twenty-four candles.

The lighthouse lasted almost fifty years before a fire in the lantern burnt it down on a winter's night in 1755. As the fire took hold, the keeper, Henry Hall, who was reportedly in his nineties at the time, was joined by his two colleagues who threw buckets of water up into the lantern. But their efforts were to no avail and the fire spread down through the building, forcing the three men to scramble for their lives out onto the rocks. There they huddled together, helplessly watching the flames work their way down through their circular living quarters and their circular kitchen until the whole tower had been consumed.

The blaze had been seen from the shore and a dozen fishermen jumped into small boats and rowed furiously out to the reef. All three men were rescued, but Mr Hall complained of terrible stomach pains. He died some days later, having somehow ingested molten lead from the lantern's roof. Speculation suggests the poor man gazed upwards open-mouthed at the flames at the very moment the molten lead fell from the burning roof.

Just two months later, in February 1756, Robert Weston, who held the lease to the rocky reef, was introduced to one John Smeaton, a civil engineer and instrument-maker, who suggested the new lighthouse be made of stone. He used granite brought from Cornwall and lined the inside with limestone from nearby Portland. And in the process, Smeaton pioneered the use of 'hydraulic lime', a mortar that cured under water and led to the development of today's concrete. He also cut his stone so that the pieces dove-tailed into each other, and he made sure that when the pieces were assembled, the gaps between them did not form a continuous straight line – a technique that made the walls stronger.

His overall design for the tower reflected the shape of an oak tree, broad at the base and gradually narrowing towards the top. It had a diameter of 28ft at the base and 17ft at the top. When heavy seas pounded the structure, the waves rolled up the column and were then deflected back into the angry sea. The lighthouse rose to just 59ft and was topped with a lantern consisting of a chandelier with twenty-four very large tallow candles. It was lit for the first time on a moonless night in 1759.

The candles were later replaced with oil lamps and a series of reflectors. Smeaton's lighthouse served Channel sailors for more than 100 years, until the rock on which it was built began to erode and Trinity House built the present lighthouse to replace it. Smeaton's contribution to maritime engineering had equalled that of Henry Winstanley, making a very significant improvement to the safety of seafarers in the Channel.

It was the renowned Smeaton who Charles Rashleigh turned to for designs for a harbour at West Polmear on Cornwall's Channel coast. Charles was the tenth child of Jonathan and Mary Rashleigh, Cornish landowners, who lived in gracious style at Menabilly, the grand house that Daphne du Maurier was to live in a century and a half later and on which she modelled Manderley in her novel *Rebecca*. Jonathan Rashleigh was the Member of Parliament for Fowey, but his son Charles chose law rather than politics. When Rashleigh senior died, Charles returned to Cornwall to manage the family estate. He bought an elegant townhouse in St Austell and two years later, married Grace Tremayne, a young woman from Heligan. He also bought land at West Polmear.

It was here on the beach that ships collected the copper extracted from local mines. But running ships up onto the sand was not safe nor an efficient way to load the minerals onto the freighters. So in 1791, Charles Rashleigh took it upon himself to build West Polmear a port. The man he engaged to design and construct it was lighthouse builder John Smeaton.

The key to the plans was a dock that would remain full of water when the tide went out. Cornish miners, who had made an extra penny working on Smeaton's lighthouse, now did some moonlighting digging Smeaton's new harbour. Lock gates were constructed and fitted to the inner dock. To keep the water within the lock topped up, a leat – a narrow, open waterway – was dug from the pleasant, wooded valley of the Par river, 7 miles away.

As he did for the Eddystone Light, Smeaton used local granite. A larger, outer tidal dock was created and protected by a breakwater. The new harbour served the export of copper, and when the copper mines ran dry, it was the turn of china clay. And the local pilchard fishermen rejoiced in having a safe and convenient quayside on which to land their catch. Shipbuilding and rope-making, as well as pilchard curing, all took advantage of the new facility, which finally opened in 1800.

With trade through the port growing fast, the tiny fishing community of West Polmear, which had a population of just nine when Charles Rashleigh set Smeaton to work, had grown to some 3,000 by the turn of the century. So grateful were the people of Polmear to Rashleigh that they changed the name of their village to Charlestown.

Smeaton was not to be so publicly feted, and he died before the harbour works were finished. Historians, however, rightly credit him with outstanding advances in engineering that are still in evidence today. His harbour may no longer be a trading port, but tall ships tied up in the inner dock create a scene popular with filmmakers looking for ready-made sets in which to spin their eighteenth- and nineteenth-century yarns of swashbuckling, seafaring folk. Charlestown is the Channel's gift to the silver screen, still making money for the local community 200 years after it first opened for business.

2

The Long Crawl

If you were minded to swim across the Channel from England to France, you would, after months or even years of preparation, likely set off from Shakespeare Beach. This long shingle beach on England's Kent coast, just a mile west of Dover, is followed by the railway line, which soon disappears into the tunnel beneath Shakespeare Cliff en route for Folkestone and London.

William Shakespeare was a regular visitor to these parts: in *King Lear*, the blinded Earl of Gloucester threatens to throw himself from the cliff top. The bard was here again with *Henry VI*, as his protagonists go to and fro across the Channel: 'My thoughts are whirled like a potter's wheel; I know not where I am, nor what I do!' cries Lord Talbot.

Train passengers witnessing your preparations on the beach will wave a silent 'good luck' and offer a thumbs up. Gulls will be circling above looking for scraps of your friends' sandwiches, calling excitedly to each other as they wheel and dive. You will be anxious, of course you will. From the beach, it is 19 nautical miles to France, the equivalent of just under 22 land miles. If you are crossing at the end of August – the most popular time to take on the Channel – you will dip your toes into water at around 65°F and then, after a few steps, you will be out of your depth. The Channel drops to an average depth of just over 200ft. (At Hurd's Deep, just off Alderney, it drops to 571ft, but you would be off course by a long way if you strayed into the Deep.) Salt comprises 3.5 per cent of the sea water, which will give you a crusty mouth by the time you finish.

As Edmond Halley had discovered, there are devilish tides and currents out there. You will have studied these and planned your swim accordingly. Ships of unimaginable size will be crossing your path. Their cargoes will be oil, gas, motor cars, exotic fruits, televisions from Japan, white goods from South Korea, frozen meats from the Antipodes, and almost anything you care to mention from China via Hong Kong. And these giant ships won't be able to see you – if they could, it would take them a couple of miles to stop.

Your limbs are taking the strain, but it is your eyes that must keep you safe. You are crossing what has become the world's busiest seaway. The fully laden ships bound for Rotterdam, Hamburg, London and the ports of Scandinavia keep to the French side. They return on the English side.

However, you will not be alone, and if these giants of the ocean can't see you, they should be able to see your support boat, which will appear on their radar as a very small dot. No rational swimmer attempts a Channel crossing without a pilot and a support boat. Neither come cheap. A Channel swim will set you back a minimum of £3,000. You will need to register your intended swim with one of two organisations that regulate and authenticate Channel swims. You may grease your body – duck or goose fat do the job well – but may not wear a wetsuit, and your swimming costume must be a simple one- or two-piece affair. Goggles, a hat, nose clips and ear plugs are allowed. Courage and determination are a must.

You will be sharing the seductive waters of the Channel (hundreds of swimmers have been drawn to the crossing before you and hundreds more will surely follow) with herring, mackerel, right-eyed flat fish and left-eyed flounders. Dover sole, so called because early nineteenth-century stagecoaches would race to London's fancy restaurants with sole caught by the port's fishermen, are making a comeback. Shoals of flat fish will be well below you as they forage on the seabed. Scallops, the object of many a squabble between French and British fishermen, will be on the seabed too – plenty of them as you approach the French coast. Dolphins might break the surface in welcome; grey seals and harbour porpoise, less intent on human company, will take less notice of you.

On the seabed, interrupting the food trails of foraging flat fish lie the ships lost in battle. The rotting galleons of the Spanish Armada that survived the British fire ships are beyond recognition. The German battleships of two world wars, and the ships of the British Navy that succumbed to enemy action rot alongside the fishing boats that lost their battle with the cruel sea. And here too are the rusting hulls of cargo ships caught in the melee of what is the world's busiest shipping lane.

You will swim above some of them, but you will be far from the worst British naval disaster in the Channel when in 1744, HMS *Victory*, the predecessor of Admiral Lord Nelson's ship of the same name, went down off the Devon coast with the loss of 1,100 lives. She was heading home after relieving a British convoy trapped by French ships in the mouth of Portugal's Tagus river. It became separated from the rest of the fleet, but why it foundered remains a mystery.

Further beneath you, trains, some with foot passengers, some with passengers in their cars and some with the freight of continental trade, use the world's second-longest underwater tunnel to shuttle between England and France in rather less time than it will take you to swim. The chalk substrate that stretches from one coast to the other proved ideal for underwater tunnelling – the chalk marl soft but impermeable.

Along with the tunnel, various cables lie hidden in the seabed. There is a high-voltage electricity cable (the UK imports French electricity) and a communications cable pioneered by Samuel Morse, who insulated his wire with tarred hemp. Charles Walker refined the cable, wrapping his copper wire in gutta-percha from the *Palaquium gutta* tree, grown in Malaysia. In 1850, John and Jacob Brett laid a telegraph cable across the Channel, which was severed the next day by a French fisherman who believed he had come across some stubborn seaweed. The language which greeted the discovery on board the brothers' cable-laying paddle steamer *Goliath* is best left below decks. Their anger and frustration set aside, the brothers repaired the damage, wrapping their cable in a form of armoured coating, enough to deter a whole fleet of French fishermen.

In the skies above you, the gulls, gannets and guillemots will weave and dive, hoping for an easy meal from your support boat. But there will be no bluebirds over those white cliffs of Dover. The American songwriters of the Vera Lynn wartime favourite were blissfully unaware that the native American bluebird had never made it across the Atlantic. No bluebirds, but if you are lucky there will be yellow butterflies on the last stage of their migration from north Africa. They create low-flying clouds of brilliant yellow stretching for miles – hundreds of thousands of Clouded Yellows (*Colias Crocea*) heading for the Kent and Sussex coasts, where they feed off the nectar of cat's ear, mouse-ear, marjoram and Devil's-bit scabious. Then, well nourished, they breed in the fields of clover. If the cloud is down, you may also be witness to a flypast of homing pigeons flying low across the water on one of their summer cross-Channel races.

The light will change in the twelve or so hours it will take you to reach France and collect your souvenir pebble. (Pebbles are what the Roman Emperor Caligula claimed he used to fight Neptune in the Battle of the English Channel.) Never mind the pebbles and the fantasies of a mad emperor, it is that ever-changing light that brought the French impressionists to the coast that you now proclaim as 'yours'. Forsaking the Mediterranean, with its unchanging and unchallenging blue skies, the future impressionists congregated to the west of where you stand – Honfleur, at the mouth of the Seine, was a favourite haunt. Deauville too. Monet, Sisley, Pissarro and local boy Eugène Boudin were all drawn to the light of the Channel coast. And it was here they gave birth to impressionism.

You will be forgiven, as you plough your way through the dark, heavy waters, for believing this is *your* Channel. But you will not be the first to enjoy the swim. Matthew Webb, a captain in the Merchant Navy, was the first to do so in 1875. Covered in porpoise grease and sustained by brandy, beef tea and sandwiches from his escort boats, it took him nearly twenty-four hours after the currents had driven him off course. He died tragically, eight years later, trying to swim through the rapids at the foot of the Niagara Falls.

A young Frenchman, Philippe Croizon, was met by dolphins on his thirteen-hour crossing. The swim, while not a record, was the result of quite extraordinary human endeavour. Croizon has no arms or legs. He lost his limbs as the result of a horrific accident while fixing a television aerial on the roof of his house. He used prosthetic flippers on his leg stumps and stabilisers on his arm stumps.

Tom Gregory was a few days short of his 12th birthday when he made the crossing, and Sarah Thomas, a 37-year-old from Colorado, went there and back twice without stopping.

However, for sheer dogged determination, nothing compares with the efforts of Jabez Wolffe. Jabez was the son of German–Polish immigrants who settled in Glasgow at the end of the nineteenth century to make and sell jewellery. Wolffe was a well-built young man who quickly developed a passion for swimming, and distance swimming, in particular. He did an apprenticeship in cabinet-making, and to earn a living making furniture, moved to London, where he went into serious training for long-distance swims.

He made his first attempt to swim the Channel in July 1906 at the age of 30. Covered in grease, he set off from the beach at South Foreland, about 3 miles north-east of Dover. In his support boat there were two pipers whose playing provided the rhythm he needed to modulate his strokes and keep them steady. On later attempts, he used a gramophone in the boat to save money – its horn throwing scratchy rhythms at him over the water.

He was fed every half an hour with broth, Oxo, weak tea and cocoa, all lowered to him on the end of a rope. But after nearly ten hours in the water, his strength failed him and he had to be pulled into his boat. Twelve days later, he was back, but managed only three hours and forty-two minutes.

Nor was it third time lucky when he tried again in August, this time managing nearly eight hours. He made two attempts the following year, four in 1908, two in 1909, two in 1910, two in 1912, two in 1913 and one in 1914 before he was called up to fight the Germans – not a comfortable calling being half German himself. Not once did he complete the journey, although in 1911 he did get within 100 yards of his target but could not find the strength for the final few yards. On three occasions, he got within a mile before tidal currents got the better of him.

To be fair, Sarah-Jane Stirling was equally determined. She stepped into the water in August 2005 and promptly felt sick. And in time, she was sick – very sick. But she had spent £2,000 getting this far, so this lady was not for turning. She refused to look at the French coast because it seemed so far away, and when her tongue became swollen, her diet of bananas, Jaffa cakes and peach slices had to be reduced to the peach slices.

A young gull travelled with her and settled on the water next to her every time she stopped to be sick or deal with diarrhoea. She was stung by jellyfish and had fits of nerves when her irrational mind conjured all manner of disasters, from being attacked by sharks to being dragged into impossible swirling depths. When she reached France, she climbed onto the rocks only to be caught by a wave and thrown on her back onto a bed of barnacles. Back on her feet, she was all smiles, thumping the air with her fists. Her crossing had taken eleven hours and twenty-three minutes, heaps of guts and buckets of determination.

Equally determined, but thwarted, were eight women who planned a relay crossing in 1964 led by three sisters whose enthusiasm for swimming had begun when they were students at Cheltenham Ladies' College. After school, they joined the oil company Shell and found, to their delight, that the company's London HQ boasted a swimming pool in the basement. Here they began serious training, found a Channel-swim organiser and advertised for five women to join them.

The new recruits included a long-distance swimming champion and the 45-year-old Women's Editor of the *Financial Times*. By September in 1964 they were ready to go. At 5.30 on a chilly morning, Brenda Bourne, the long-distance swimmer, led them off, while the remaining seven climbed into a cramped rowing boat for their short journey to the dilapidated fishing boat that was to serve as their escort vessel. The boat stank of dead fish and diesel oil and had nowhere for the women to change. The water was a chilling 17°C, but covering the body in grease, an oft-used deterrent against the cold, was discouraged by the organisers due to the difficulty it posed when hauling oneself back on to the boat.

In the water, Brenda faced a heavy swell, while those on board, waiting their turn, were throwing up and growing increasingly weak. The wind increased to a Force 5, and after nine hours at sea, the team found they had swum 40 miles in a Channel that was just 21 miles wide. The attempt was called off, but it nevertheless hit the headlines. In the planning stage, someone had joked about doing the swim in the nude. In a glorious precursor to twenty-first-century fake news, two newspapers – the *New York Sunday News* and the *Sydney Morning Herald* – ran stories about 'nude women in a long UK swim'.

Now, dear reader, clothed and with your feet on a French beach, exhausted and exhilarated in equal measure, you stoop to collect your souvenir French pebble. Having recovered and enjoyed the adulation of family and friends, you may ponder on how this strip of water you conquered has played its part in the history of the two peoples it separates. They are the people of two nations: variously enemies, rivals, partners and, at a stretch, friends.

They have much in common, but centuries of wars, distrust and grievances, petty and lasting, have kept them emotionally, culturally and psychologically apart. For some, the Channel is a barrier or a defensive moat; for others, it is a separation of political ideals; while many more see the Channel as access to a new land, offering a journey that satisfies a longing for foreignness. Its waters are the graveyard of the soldiers, sailors and airmen of two world wars. And those same waters, touched by ever-changing light, gave rise to the impressionist movement in France, which was to challenge the canons of the art world establishment with a technique that would eventually define painting in the nineteenth century.

For centuries, the Channel was productive territory for smugglers who dealt in wool, tea, lace, tobacco, brandy and all manner of luxury goods; and, latterly, in hapless, desperate, exploited migrants. (Small-time amateur smugglers tried their luck on the 'booze cruise' and may be encouraged to do so again, now that Britain has left the EU.)

For fishermen and ferry operators, the Channel is a place of work; for divers, a place of discovery; for leisure seekers, a pleasure ground of beaches and seaside towns with grand hotels, casinos, horse racing and promenading. For the French navy, it is home to their fleet of nuclear submarines, just round the corner at Brest on the Atlantic coast. For the four north-coast French nuclear power stations, *La Manche* is a source of cooling water. Plymouth and Portsmouth host the warships of Britain's navy, and for yachtsmen of both coasts, the Channel waters provide leisure cruising and competitive racing.

3

A Tale of Two Cities

Brighton and Deauville

With the twenty-first century's preference for holidays in the sun, whether it be on the beaches of the Spanish Riviera or among the tropical palms of more distant lands, it seems unlikely that the taste for seaside tourism began on the often-chilly Channel coasts of England and France. Yet the Channel towns of Brighton and Deauville launched aristocratic tourism at the beginning of the nineteenth century.

The two towns had much in common. They were both within reach of their respective capital cities and when the railways arrived, they provided easy, quick and cheap connections between the capitals and the coast. Both towns enjoyed the patronage of celebrities, both royal and independently monied. Both towns soon became attractive settings for horse racing and gambling. And crucially, both offered safe sea bathing, which was hailed as the cure for all manner of complaints. And they fast became towns where you went not just to bathe, but to be seen.

It was in the mid-1700s that Dr Richard Russell brought his theories about the medical properties of seawater to Brighthelmstone (as Brighton was then). Having trained at the university in Leiden, he had published a paper in Latin on the treatment of glandular ailments with seawater. The treatments involved not just bathing in it but drinking it too. He failed to obtain a licence to practise medicine in Britain but nonetheless gained the support of Dr Anthony Addington, a physician and psychiatrist who had attended George III in the early stages of the king's illness. Addington, like Russell, was another advocate of the use of seawater for certain medical conditions and his backing gave added impetus to his colleague's endeavours. Russell moved to Brighton and built a handsome house, backing onto the beach, where he lived and treated his patients.

Dr Russell found Brighton a depressed and impoverished place. It had, in times past, prospered on the back of its fishermen, who formed one of the most successful fleets operating in the Channel. But war with France and successive storms had wrecked their boats, claimed the lives of their men and reduced dwellings to uninhabitable ruins.

No one of any standing lived there, the population was in decline and Russell's new home became an attraction. Once installed, he wrote about his sea-based treatments in *The Oeconomy of Nature in Acute and Chronical Diseases of the Glands*, and with unexpected speed, word of the doctor's cures and the benefits of sea bathing spread. Brighton became the destination for bathers who believed the Channel waters would do wonders for their health.

⛵

On the other side of the Channel, it was art rather than sea bathing that first turned France's north coast into a nineteenth-century bourgeois playground. The movement had its beginnings in Paris in 1820 when the second son of the Mozin family, all talented musicians, decided the competition at home would be stiff for yet another musician, so apprenticed himself instead at the workshop of the painter Xavier Leprince. Charles Louis Mozin proved as talented an artist as his family were talented musicians and collaborated with Leprince on a painting of Honfleur, an artist's delight of a town at the estuary of the River Seine.

The work prompted the young Mozin (he was still in his teens) to travel north himself. But rather than linger in Honfleur, he explored the Channel coast to the west and, he proclaimed, 'discovered' Trouville, then a tiny fishing village on the north of the estuary of the River Touques. Mozin was enchanted. The light over the water was ever changing, and the small fishing boats which decorated the gentle sea when the tide was in lay prettily in the mud when the tide went out.

Satisfied he had captured the essence of his new find, Mozin left Trouville and headed back to Paris with his paintings in his bag. Back in the capital, he exhibited his new work and set about encouraging his artist and writer friends to join him when he returned to 'his' Trouville.

Monet was one of the first to follow his lead, as captivated as Mozin was by the light over the estuary. Eugène Boudin moved to Trouville from Honfleur. Then came Alexandre Dumas, to stay with his family at the Auberge de la Mère Oserais, where he would eat and drink with Mozin, the pair paying their bills with poems and pictures. Dumas' eloquent word paintings, together with the canvasses of Mozin, not only encouraged more visitors but also prompted an increasing number of families from the aristocracy to build homes here.

The popularity of sea bathing soon reached the French side of the Channel, and when the illustrated Paris weekly *L'Illustration* described the beach at Trouville as 'a velvet carpet strewn with gold glitter and silver', the Channel resort, discovered by artists, fast developed into a playground for the upper middle classes. They built grand houses and brought their titles and their fashionable aristocratic ways to what, just thirty years earlier, had been a modest village of fishermen and smugglers.

Trouville prospered until competition arrived on the other side of the Touques Estuary. Duc de Morny, the illegitimate son of Hortense de Beauharnais, the estranged wife of Napoleon's brother Louis, was a man addicted to the pleasures of life. He financed his extravagant lifestyle by farming sugar beet and by putting his fingers in a variety of highly profitable commercial pies.

He was not a duke, not yet anyway, but figured the title would help smooth the journey on which he had set out. The circumstances of his birth had been carefully concealed on a forged birth certificate. And the fact that his mistress was the strikingly beautiful wife of the Belgian Ambassador only enhanced his reputation.

Morny's doctor, Joseph Olliffe, invited his patient to holiday with him in Trouville. Together, they strolled along the banks of the Touques, from which the attention of the self-proclaimed duke was drawn to the acres of barren marshland and sand across the estuary. A small village stood on the hill overlooking the sea. Here, he thought, he could create a resort that would equal, if not better, the facilities on offer in Trouville. Back in Paris, Duc de Morny used his contacts to put together a team of bankers and architects, whose plans soon became known to speculators keen to be part of the growing potential of this part of the country's Channel coast. Deauville was in gestation.

⛵

Then came the railways. The line from Paris to Trouville opened in 1863, twenty years after the London–Brighton line had opened. For both towns, their rail links to their respective capitals were a further fillip to their growth and prestige. The station at Trouville served both sides of the Touques Estuary. Deauville passengers would be met by horse-drawn carriages that would whisk them across the river to their chosen hostelry, while the hotels and guest houses of Trouville were all within an easy and comfortable walk – especially when porters sent from the hotels led the way, carrying the luggage.

Brighton was already a fashionable resort due, in no uncertain measure, to the growing popularity of sea bathing. The opening of the railway line from London in the autumn of 1841 made the journey easy, reliable and cheap. But its arrival followed a long and tortuous journey. Early schemes failed to gain parliamentary approval, and by 1836 there were no fewer than six routes under consideration.

Robert Stephenson, the designer of the *Rocket* and son of George Stephenson, proposed linking a Brighton line with the London–Southampton line at Wimbledon and then proceeding via Epsom, Dorking and Horsham. John Rennie, a less illustrious but much respected engineer, proposed a more direct but costly and challenging route. The final chosen route was something of a compromise, following Rennie's route after Redhill but linking with the yet to be completed London and Croydon line. The journey was first class only to start with, a ticket costing 5s; and when third class was introduced at 3s 50d, eighteen months later, passenger numbers soared. In the thirty years between 1841 and 1871, Brighton's population doubled from 46,661 to 96,011.

The railway from Paris to Trouville had the same effect on the pair of French Channel towns. It arrived in 1863, with a journey that took about six hours. Its gestation had been as long and tortuous as the London–Brighton line, with various interested parties proposing routes that would favour their various interests. On top of which, the French government considered the canals their transport priority, seeing them as a more financially efficient means of moving goods and freight.

However, a railway, built by British engineers no less, did already exist between Paris and the Channel towns of Le Havre and Dieppe (where Winston Churchill would choose to play golf, and to which Oscar Wilde and the Prince of Wales became regular visitors). The Gare Saint-Lazare, the capital's first railway terminus and a favourite painting subject for both Monet and Manet, served the north of the country and was the departure point for the Dieppe line and later for the Trouville link, which arrived in 1863 with a handsome station built by de Morny. On both sides of the Channel, the railways were giving new impetus to the growth of the burgeoning Channel resorts.

⛵

What finally clinched the future of the Channel coasts as holiday destinations was the patronage of the good and the great – and the great and not so good. In Brighton, Dr Russell's sea-bathing initiatives were in full swing when he died in 1759. But in his case, the good that men do did not die with him. Not only did the pleasures of sea bathing continue to attract more enthusiasts, his grand and now vacant house caught the eye of King George III's brother, the Duke of Cumberland, who became a regular summer visitor. And when his nephew, the future George IV, came to visit in the late summer of 1783, the town had its royal seal of approval.

George, the 'Prince of Pleasure', liked the place so much he set about finding suitable accommodation for his mistress, Mrs Maria Fitzherbert, before employing Henry Holland to upgrade the modest lodgings he rented into a handsome villa. When he became Prince Regent, after his father's mental health had rendered him incapable of carrying out his royal duties, he commissioned the pre-eminent Regency architect John Nash to turn the villa into something more regal. With money no object, he turned what had become a playboy's pavilion – the Brighton Pavilion – into an opulent oriental-style palace with minarets and domes and uninhibited arabesque flourishes.

The people of Brighton, many of whom were gainfully employed building the palace, were pleased to see their new king in town, even if his appearances were rather fewer than those of his father. They were fewer because George, still Prince Regent, grossly overweight and suffering from the swollen limbs of gout, built a tunnel under the palace so he could move unseen by ordinary people between the living quarters and the stables. His still unfettered appetite was reflected in a banquet he threw for Tsar Nicholas of Russia in 1817 to celebrate the defeat of Napoleon. There were eight fish dishes, preceded by eight soups, forty entrées, eight roasts, thirty-two desserts and savouries, and a pastry model of the Pavilion itself. The entrées included venison, lamb, chicken, teal, grouse, pheasant, fried brains, foie gras truffles in warm linen and a terrine of larks and thrushes au gratin. For dessert, there was an upside-down lemon jelly, rose ice cream, a model of a Swiss hermitage fashioned in fondant and marzipan, and a tower of profiteroles with aniseed, all creations of the Regent's favourite celebrity chef, Antonin Carême. The Pavilion's servants feasted for days afterwards.

George IV's successor, his younger brother William, enlarged the palace still further to provide room for his Queen Adelaide and her very extensive household. And when Victoria became queen, day trippers, using the new rail service from London, would flock to Brighton in the hope of catching a glimpse of the royal family. The more affluent visitors hired wheeled bathing machines and stepped modestly into the shallow water, hoping, although they would never admit it, that a royal head might bob above the waves next to them.

Across the Channel, Duc de Morny was using his contacts at the French court to entice Parisian aristocrats to visit Deauville, his own town of pleasure by the sea. They did not need much encouraging. Speculators were putting up grand mansions, and the Parisian bourgeoisie were soon tripping over their crinolines and canes to get a foot on the riviera of the north. Napoleon III had already made visits to the coast, the word about Dr Russell's benefits of sea bathing having reached the Continent. And there was a smart train direct from Paris. A casino and a racecourse only added to the charms of the place. The surrounding country was ideal for the breeding and training of thoroughbreds. The Aga Khan established stables here during the 1920s, in company with leading owners and trainers from England and Ireland.

Deauville's mile-long boardwalk, Les Planches, was the place to promenade. Gaily decorated parasols or wide-brimmed floppy hats sheltered delicate skins from the afternoon sun, while for men, black top hats were de rigueur. As the years passed, the top hats were followed by blazers and boaters, until white vests and baseball caps, worn back to front, replaced any sense of smart casual. Today's female promenaders cover as little as possible, shading their eyes with huge sunglasses while flaunting their carefully tanned bodies.

Back in the day, the Normandy Hotel was *the* address in town, soon to have competition from L'Hermitage and L'Hôtel du Golf. These were grand hotels matching the best to be found in Paris, with guest lists recording the comings and goings of well-heeled famous and infamous Parisian socialites, not to mention equally infamous foreign royalty.

Two kings, Farouk of Egypt and Alfonso XIII of Spain, were great playboys who were habitually lost in clouds of extravagance and indecent behaviour that hung heavily about their every move. Farouk, portly and bespectacled and with a mustachio that did nothing for his looks, had been described in a diplomatic cable sent to the Foreign Office in London as 'uneducated, lazy, untruthful, capricious, irresponsible and vain'.

This unattractive king travelled to Deauville with a thirty-strong entourage that included Albanian bodyguards, secretaries, doctors and Sudanese food tasters. He spent his nights at the casino and his days philandering.

Philandering at home had come to a temporary halt after he set his eyes on the Cairo teenager Narriman Sadek, a young beauty who was engaged to a Harvard-educated lawyer. Royal pressure put an end to the engagement and when the young woman was rumoured to be on her way to join the king, who had travelled 'incognito' to Deauville in a cavalcade of black limousines with police outriders, horns blaring and lights flashing, the town went into a frenzy. In the expectation of a glitzy royal wedding, the Egyptian flag was hoisted at the Hôtel du Golf, where Farouk had installed himself in a suite on the fifteenth floor, and on every other flagpole in town.

In the end, and to howls of local disappointment, the celebrity couple left the coast and got married instead in Cairo – Sadek in a dress embroidered with 20,000 diamonds. The marriage lasted less than three years, but it gave Farouk the son he had been denied in his first marriage.

Alfonso XIII, Spain's king since 1886, preceded Farouk by a couple of decades and proved a hard act to follow. The Associated Press reported:

> King Alfonso of Spain is leading the social world at Deauville, the world-famous French seaside resort, where a frenzy of gambling enlivened by constant thrills is running its maddest pace. Wherever the King goes scores of merrymakers of noble birth gather. The Shah of Persia, the crown prince of the Balkans, Prince Christopher of Greece, the Maharajah of Kapurthala, queens of movieland and captains of industry cluster about him.

So did a host of ambitious and beautiful women. In addition to his six legitimate children with his wife Victoria Eugenie, Queen Victoria's granddaughter, he sired another six children illegitimately.

The antics of the two kings did Deauville's reputation no harm at all. The veteran British actor Tom Courtenay mused once, 'When you see who's considered to be in fashion, it is a relief to be out of it.' For high society in France at the beginning of the twentieth century, the opposite was true. And Deauville was very much in fashion.

⛵

Brighton had always been a step ahead of its French counterpart, and so it was with horse racing. Having settled comfortably into Dr Russell's grand house on the Steine, the Duke of Cumberland organised horse racing on land a mile inland from the coast. The Prince of Wales came down from London and he and his aristocrat friends raced their horses on the duke's new course. The sport of kings enhanced Brighton's reputation, and when the prince was no longer an attraction for the aristocratic racegoers, the railway brought commoners down from London in sufficient numbers to put horse racing at Brighton on a firm footing.

The town then extended its reach into the Channel with a pier, lined with stalls selling cylindrical sticks of colour-patterned, peppermint-flavoured boiled sugar and glucose, which they called Brighton Rock (nothing quite so vulgar would appear at Deauville). The early pier served as landing stage for passenger boats making the crossing from France, but storms, neglect and arson finished off those early piers. The present version has become one of the country's top visitor attractions with its space invaders, fish and chips, dodgem cars, horror hotel and free deckchairs. At night, it is lit by 67,000 lightbulbs, which the pier's owners are quick to point out are of the long-life, low-energy variety.

But Brighton's evolution had its bleak side. By the 1930s, the town had developed a distinctly dark underbelly, with opposing gangs running protection rackets on the racetrack and spreading their criminal activities into town. All this formed the backdrop for Graham Greene's 1938 novel *Brighton Rock* and a 1947 film recreating the dark years of the 1930s had the *Daily Mirror* newspaper proclaiming, 'No woman will want to see this film. No parents will want their children to see it. The razor-slashing scenes are horrific.'

In time, the town's reputation was revived and then enhanced with the opening in 1961 of the University of Sussex, which quickly gained a reputation for innovation within academic circles, not least through its programme combining related subject areas in a new curriculum. But the town's repute faced a setback a year later when a local nightclub owner shot his young wife in a jealous rage.

Harvey Leo Holford, the handsome, self-styled King of Brighton, lording it over the town from the wheel of his scarlet open-top Pontiac, was the man Brighton loved to hate. But he was not good enough for his young wife, who was unfaithful soon after they married, and when she revealed their child was not his, he shot her, took an overdose of barbiturates and cradled her bloody, bullet-holed body until police arrived. He was convicted of manslaughter on the grounds of diminished responsibility and provocation. Sentenced to just three years in prison, he served less than two.

⛵

The town which began as a Saxon village, developed into an active fishing community, attracted Georgian aristocrats and became a holiday home for kings and queens finally found its place in twenty-first-century Britain with a wind farm off its coast, a football team in the Premier League, trendy shopping in the Lanes and a liberal culture that welcomed a growing gay community. However, it struggles to shrug off the memory of a terrible night in 1984 when the Provisional Irish Republican Army attempted to assassinate members of the British government, including the prime minister, Margaret Thatcher.

In the September of that year, Patrick Magee, a member of the IRA, had checked into the seafront Grand Hotel, where the prime minister and her Cabinet were to stay during the Conservative Party Conference a month later. Under the bath in his room, which was above the VIP suite in which Thatcher would stay, he placed a time bomb. It duly detonated just before 3 a.m. on the morning of 12 October. Mrs Thatcher was still at work, preparing for the conference, and neither she nor her husband Denis were hurt. But one Member of Parliament and four Tory Party members died and more than thirty were hurt, some sustaining terrible injuries.

Deauville has so far evaded the bombers. It does have trendy boutiques catering for the well-heeled, fashion-conscious Paris crowd – Coco Chanel, the doyenne of France's fashion designers, opened her first boutique here in 1912. But there is no pleasure pier and no sticks of sickly rock, and the town still boasts belle époque architecture, although the grande dame of early twentieth-century hotels, the Hôtel du Golf, no longer attracts the playboy kings it once entertained. The hotel still has grand suites, but it also has special offers – and a kids' club.

In summer, the beach is a splash of multicoloured umbrellas, but promenading is not what it was. The dainty parasols, the profusion of chiffon and silk, the spats and top hats are no longer part of the glamorous parade. For glamour, the town must wait until September and the annual film festival devoted to American cinema. But that's celebrity movie-makers' glamour, not the real-life glamour that personified 1930s bourgeois France. However, the horses still gallop here, and when the humans have left the summer beach, well-bred mares put on a fine show as they exercise, splashing through the shallows. But the glitterati have moved south, leaving the prototype of French seaside holiday-making clinging to its past.

Bournemouth and Boulogne

When Brighton and Deauville were setting the trend for trips to the beach, Bournemouth was a few cottages gathered around the mouth of the River Bourne in Dorset. It was a remote area, favoured by smugglers like Isaac Gulliver, whose convoys of packhorses loaded with smuggled tea, silks, lace and alcohol stretched further than the eye could see. He built his smuggling headquarters at Kinson, today just ten minutes from Bournemouth city centre. When his former smuggling base was finally demolished, several secret passages were revealed, access to one of them through a door which was 10ft up the chimney. Arrested, charged and imprisoned, Gulliver was eventually pardoned by George III before the king went mad. Mr Gulliver went on to enjoy a career in banking.

Then, along came Lewis Tregonwell, a captain in the Dorset Yeomanry. The eighteenth century was drawing to a close, and the well-heeled army officer brought his wife to the Dorset coast to help rouse her from the grief she was suffering from the loss of their child. They both flourished in their new surroundings and decided to build a second home, with a view of the Channel, on the banks of the Bourne. They acquired more land and built several cottages, including one for their butler, Mr Symes. (When the Symes' cottage was destroyed by fire, it too revealed hidden passages and storage facilities, suggesting his boss, the respected army officer and landowner, indulged in smuggling himself.)

By the 1820s, Bournemouth was taking shape. Much of the land was owned by the Tapps-Gervises, a long-established, well-to-do Dorset family from whom Tregonwell had bought his land. Come the 1830s, and

Sir George Tapps-Gervis began to see the potential for creating a resort around the mouth of the River Bourne in much the same way as Duc de Morny saw the potential at Deauville at the mouth of the Torques in Normandy. In 1836, Sir George appointed a local architect, Ben Ferrey, to draw up plans that would include villas for rent during the summer months.

By 1840, the stagecoach from Southampton to Weymouth was calling at Bournemouth, and families from Victorian London were taking up summer residence here. When the railway arrived in 1880, those same wealthy Victorians complained that it would ruin their peaceful retreat. To appease the most loquacious of the objectors, the railway company built the station in a cutting to reduce the disturbance. A pier was opened in the same year.

By the early years of the twentieth century, Bournemouth had lost little of its charm, prompting the poet John Betjeman to describe the town as 'a garden with houses in it'. One of the grander houses was an Edwardian manor built in 1905 as a holiday villa for the Austro-Hungarian Ambassador, Count Albert Joseph Michael von Mensdorff-Pouilly-Dietrichstein, a second cousin of Edward VII. Along with Robert Louis Stevenson and the actor-manager Sir Henry Irving, the count was a regular guest at performances in the private theatre built by Percy Bysshe Shelley's son, Charles, at Boscombe. The ambassador's palatial retreat eventually became a hotel – the Miramar – to which J.R.R. Tolkien retired, taking treasured personal copies of *The Hobbit* and *The Lord of the Rings* with him. The Shelley theatre has yet to reopen after closing during the Covid pandemic.

Many of Betjeman's 'houses in a garden' have morphed into guest houses, while couples in retirement take tea in the bay-windowed lounges of grand hotels. The beach, cleaned by the council and washed by the Channel's tides, is today one of the most popular in Britain, but there are fewer deckchairs, these days, for gentlemen in braces with knotted white handkerchiefs on their heads. Their places are taken by much younger people, who wear eye shades fashionably on their heads, avoid getting wet and pose endlessly for the necessary selfies.

The town has its own symphony orchestra and boasts a large collection of stuffed animals, which dwell in a large Victorian house that has been turned into a museum. It has a host of English language schools, and a large quantity of ice cream. Its two current Members of Parliament are Conservatives with thumping great majorities, ensuring the town's conservative Victorian founders can lie easily in their Dorset graves. Would they turn in their graves in the knowledge that the town has two mosques, Hindu and Buddhist temples and a synagogue by the sea? A synagogue would be no cause for surprise, but the domes and minarets of temples and mosques might be.

While John Betjeman had described Bournemouth as 'a garden with houses in it', Thomas Hardy thought of it as a 'Mediterranean lounging place on the English Channel'. Hardy began his working life as an architect, and his early attempts at writing earned him neither critical acclaim nor financial success. But with *Far from the Madding Crowd*, published in 1874, he gained both the literary recognition and financial reward that had previously eluded him. For his novels, he recreated Wessex, the Anglo-Saxon kingdom in the south-west of England during the first millennium. He has Tess, she of the D'Urbervilles, living in the Sandbourne district of Bournemouth and he uses the Channel coast in subsequent novels. Portland, a 'peninsular carved by Time out of a single stone', becomes Hardy's 'Isle of Slingers'. Yet Hardy, who could read even before he went to school, thanks to his mother's tutelage and encouragement, thought of himself more as a poet than a novelist, and in 1914, before the outbreak of war, wrote 'Channel Firing' about the futility of war in which he imagines firing practice over the Channel waking the dead:

> That night your great guns, unawares,
> Shook all our coffins as we lay,
> And broke the chancel window-squares,
> We thought it was the Judgment-day
>
> And sat upright. While drearisome
> Arose the howl of wakened hounds:
> The mouse let fall the altar-crumb,
> The worms drew back into the mounds,
>
> The glebe cow drooled. Till God called, 'No;
> It's gunnery practice out at sea

Just as before you went below;
The world is as it used to be:

'All nations striving strong to make
Red war yet redder. Mad as hatters
They do no more for Christe's sake
Than you who are helpless in such matters.

Hardy wrote to John Galsworthy that 'the exchange of international thought is the only possible salvation for the world'. Hardy died in 1928. His remains joined those of Charles Dickens and Rudyard Kipling in Poets Corner at Westminster Abbey.

Boulogne has little glamour in its history. It was where invasions of England were planned.

The town barely existed before Julius Caesar arrived in 55 BCE, with his centurions and his aspirations to establish a Roman Britain. The emperor Caligula arrived here in a state of near madness nearly a century later. He claimed that he battled with Neptune in the Channel using nothing more than pebbles as weapons. As spoils of war, he ordered his soldiers to collect seashells with which they packed their helmets and filled the folds of their tunics.

Napoleon set up camp here in 1803 while he contemplated crossing the Channel. 'Let us,' he said, 'be masters of the Channel for six hours, and we are masters of the world.' And Adolf Hitler, thinking along much the same lines, prepared to launch his invasion from here in 1940.

If Caligula arrived here in a state of near madness, his mental state was well matched by an Englishman, some nine centuries later. Lord Northcliffe, he who created the *Daily Mail*, paid a visit to Boulogne in the summer of 1922. Northcliffe had, by then, become seriously deranged and was habitually in possession of a pistol because he thought the Germans were out to get him. On arriving at the station in Boulogne, the ennobled press baron loudly proclaimed to anyone who cared to listen that God had an enthusiasm for sodomy.

Today, the ships in the port of Boulogne are not warships preparing for an invasion. Boulogne is France's leading fishing port, where its trawlermen land a full third of their country's fish and seafood. At times, the smell of fish that hangs over the place can challenge the sensitivities of urban visitors. Herring is fished here in November, when huge shoals come to spawn, and scallops are found in the winter months from October till May.

Mullet, whiting, mackerel, cod, sole, squid, crab, lobster and oysters in season are auctioned in the early morning in a room that looks and feels like a modern university lecture theatre. Buyers sit in tiered rows, scanning screens with the day's prices. Buying starts at five o'clock and is over before the bells of the town's Basilica Notre-Dame strike six.

When all is done, buyers and sellers migrate to Le Chatillon for breakfast. Patrice Baude, a one-time fishmonger, runs the brasserie with his daughter in an unprepossessing building among the fish warehouses. At six, the brasserie has already been open for an hour. Baude offers an interesting fishy breakfast of mackerel with pepper, marinated *lisette* (baby mackerel), soft herring fillets, smoked salmon, a glass of white wine, baguettes and a mug of hot chocolate.

There was one fisherman who was conspicuous by his absence at M. Baude's, one November morning in 2018. Strong winds had prevented Pascal Deborgher taking his trawler *L'Épervier* out for a night's fishing. When he came to check on his boat in the morning, it had gone. The Channel controllers in their operations centre atop the cliffs at Cap Gris-Nez had reported a vessel without lights behaving erratically as it headed towards the English coast. By dawn, British authorities found the trawler, with its distinct green hull, in Dover Harbour. On board were seventeen very cold Iranian refugees – three of them children. While Home Office officials in Britain decided what to do with the asylum seekers, police in Boulogne were wondering how best to pursue charges of theft.

⛵

There are no longer trawlers at Fécamp – but not because someone has gone off with them. This small town, further south down the coast from

Boulogne, is spread across two sides of the Valmont River, where it empties into the sea. Freed from its British occupiers in 1449, it became the capital for the dukes of Normandy. The writer Guy de Maupassant lived here and, as a teenager, he saved the English poet Charles Swinburne from drowning when he got into trouble off the coast at Étretat.

The monk Remigius, who was William the Conqueror's first Bishop of Lincoln, was a son of Fécamp. He built Lincoln's great cathedral with local oolitic limestone (a particularly hard form of the stone was used in New York's Empire State Building and at the Pentagon in Washington). Building began in 1072 and took twenty years to complete. But God did not look kindly on the monk from Fécamp, for he died just two days before his cathedral was consecrated. For many years, the Benedictine monks of Fécamp lived quietly but productively in the abbey, mixing spices and herbs to produce a potent intoxicant.

Dom Bernardo Vincelli was the monk who led a team of robed colleagues in pursuit of their new brandy-based liqueur. They gave their name to their subtle alchemy, which is still produced here in a very grand palace built by one Alexandre Le Grand.

A wealthy industrialist, Le Grand, who included wine in his trading portfolio, rekindled the production of Benedictine in 1863. He claimed to have found some of the monks' recipes in the remains of the abbey, which, if not entirely true, was a great story for his marketing team. He built the next best thing to a palace to house the distillery, designed a bottle with a distinctive shape, and gained permission from Rome to add D.O.M. (*Deo Optimo Maximo*), the Benedictine monks' motto, to the bottle's label. For his efforts, Le Grand was given the honorary title of Captain of Firefighters and made a Knight Commander of the Order of St Gregory the Great.

These days, more Benedictine is drunk in the three Lancashire football towns of Burnley, Blackburn and Accrington than in Fécamp itself. On a good night of home wins, more is probably drunk in the three towns than in the casinos of Deauville in a year.

This rather unlikely northern favourite was introduced by members of a battalion of the East Lancashire Regiment, who were camped in the Fécamp area at the beginning of 1919 on their way home from the war. So taken were they with this local concoction that by the time they reached home, local bar and pub owners had little option but to stock up with the stuff. The demobbed soldiers were delighted, and the bar owners added nicely to their profits.

After a busy karaoke night in the Benedictine Lounge of the Burnley Miners' Social Club you can indulge in a Café Benedictine: black coffee with a generous tot of 'Bene' and a scoop of whipped cream on top. And no one in this part of the world is going to sneer if you choose a Bene in preference to a pint.

The cod fishermen of Fécamp needed very clear heads. When France went about colonising parts of North America, they discovered the waters off Newfoundland were teeming with fish. The journey, in small sailing trawlers from France's Channel coast (the Fécamp boats were joined by crews from Saint-Malo, Granville and Jersey), across the North Atlantic was not an easy one. But the quantities of cod in the waters of the Canadian Atlantic made the journey worthwhile. The Treaty of Utrecht of 1713 recognised English sovereignty over Newfoundland but gave France exclusive rights to fish the island's eastern seaboard, and this right was renewed fifty years later at the Treaty of Paris.

A ship would remain in these fishing grounds for up to six months. Six-man open rowing boats called shallops, built of oak planks fastened with wooden pegs, would be lowered over the side of the trawlers to do the work of bringing the cod in. Wise crews would use long lines to tether their small boats to the mother ship for fear of sudden changes in weather. The fish would be beheaded, dried on the rocks and salted. The remains of earth ovens, where the fishermen baked their bread, are still visible beside the coastal paths. Back in Fécamp, there were ready markets for the dried cod at home as well as in Spain and Italy, and the fishing villages of the Channel coast amassed considerable wealth.

However, North Atlantic fishing became steadily less profitable, and by the twentieth century, the fish stocks were running low; so low that in 1992, the Canadian government imposed a moratorium on cod fishing off its Atlantic coast. These days, few fishing boats remain in Fécamp and most of those are refurbished models designed to give tourists a taste of life at sea in earlier times.

Calais and Dover

Calais and Dover have both become towns through which it is more agreeable to pass than to linger. In a sense, this was inevitable once they became popular ferry ports – places merely to depart and arrive en route to somewhere else. There is not much about today's Dover that suggests it is a gate to the Garden of England and little about Calais that suggests the traveller has set foot in *La Belle France*. Yet Calais' history deserves a linger, if only in the mind.

During the early years of the Hundred Years War between the English and the French, Calais was under siege for almost twelve months. The population, who were starving, were eating their horses, dogs and cats. England's King Edward III had himself led an expedition from Portsmouth on 12 July 1346, which had landed at Saint-Vaast-la-Hougue on the Normandy coast. Making his way north, Edward encountered little resistance. He found a ford across the Seine at Blanchetaque, continued through Boulogne and Wissant and arrived under the walls of Calais on 3 September. The place was well defended and gave Edward little choice but to lay siege to the town.

The English king clearly saw Calais as more than a garrison town; rather, he saw it as a valuable and well-positioned commercial base. He was not going to let it go. When the town finally surrendered, eleven months later, Edward was in no mood to treat its people with compassion. They had been an unwanted obstruction – brave and long-suffering perhaps, but a thorn in his side. He decreed that every man, woman and child be killed.

The town's governor, Jean de Vienne, pleaded with Edward's envoys that his townsfolk be spared, but the king stood firm. Sir Walter Manny, one of Edward's two envoys, quietly and diplomatically suggested to his king that one day it might be Englishmen in a similar situation pleading for mercy. It was a brave challenge and one which touched a nerve in Edward. He relented – in part. The town's governor was to bring before him six of the town's principal men with whom he would do as he pleased. When the six burghers were delivered, the rest of the townspeople would be spared.

First to come forward was the town's richest man, Eustache de Saint-Pierre. His example brought, one by one, the remaining five that the English king had demanded. The governor led the six potential martyrs into Edward's chambers.

They shuffled, heads lowered, into the king's presence. Their feet were bare, their heads shaved and hunger had left little flesh on their bones.

It was all too much for Philippa, the king's pregnant wife, who threw herself before her husband, pleading with him to spare them. For what must have seemed hours, the king went into silent deliberation, but eventually his wife's remonstrations left him with little alternative but to concede. The six received clemency, with Philippa charged with deciding what to do with them. She fed and clothed them, and the following day instructed officers to escort them through the British troops so they could choose where next to find a home.

Some question the veracity of the story, suggesting it was all a carefully planned PR exercise to show the king as forgiving and humane. Either way, it prompted the Calais councillors to commission Auguste Rodin to fashion one of his finest sculptures. It recalls the six Calais Principals, with Eustache de Saint-Pierre in the middle, portraying, as only Rodin could, the despair and pain of these brave but wretched men on their way to learn their destiny. The bronze statue of the six burghers stands outside the town hall in Calais, and one of four further casts can be seen in London's Victoria Tower Gardens in the shadow of the Palace of Westminster. A recent plaster cast has been placed in the Rodin Museum in Paris.

Edward now had to set about creating the commercial base he envisioned for the town. First, he built a new town for his army about a mile inland. There were houses, shops and inns fashioned from local wood and topped with thatch. There was a market on Wednesdays and Saturdays. Supplies of bread, meat, fish, wine and beer were shipped from England. Bows and arrows came too.

But Calais itself had been sacked and needed repopulating, so the king offered generous property grants to settlers from England. These were made to merchants from London and from much further afield – Ely, Lancaster, Shrewsbury, Watford and Sandwich in Kent. For them, the promise of a new adventure underpinned with government money proved very attractive.

With the merchants came wool, which was England's major export, and Calais became the port through which it would reach the profitable Flemish markets. The king decreed Calais be the staple – the single licensed trading town – for wool exports to the Continent. This guaranteed that Calais grew to dominate trade in the most important commodity of the day. At the height of the trade, 16,000 sacks of wool a year passed through the town, but when the wool trade began to wane, a new influx of merchants began trading in tin, cloth, feathers and lead.

In further measures to attract new blood to the town, local merchants were exempted from dues and taxes for three years, and pardons were extended to robbers and even murderers who were prepared to settle in this very English town on the French Channel coast. Its Englishness was upheld when Dick (Sir Richard) Whittington became mayor in 1407, while he was Lord Mayor of London. (Whittington was the inspiration for the folk tale 'Dick Whittington and his Cat'.) Archers were paid 6*d* a day to guard the city from behind the castle walls.

By the time Henry VIII had acceded to the English throne, Calais had become the go-to place if you were looking for a professional executioner. And Henry was. His second wife, Anne Boleyn, had failed to give him a male heir (she did give birth to a baby girl, who went on to become Queen Elizabeth I), so he amassed what were probably trumped-up charges of adultery, incest and treason to get rid of her.

Henry had also taken a fancy to one of his first wife's maids of honour: a woman called Jane Seymour. However, Henry wanted Anne's ending to be as painless and as quick as possible. It was common practice in Tudor England to burn female traitors at the stake. But that was neither quick nor painless, so Henry mulled over the alternatives. Beheading with an axe would, in theory, be quicker but if the axe fell just an inch off target, things could get very messy. Beheading with a sword was likely to be the surest way to end his wife's life. But the swordsman had to be a pro. And the best sword-bearing executioners were to be found in France. Calais was still ruled by the English, and it had its share of reliable executioners, so it was to Calais that Henry went looking for his swordsman.

On the morning of 19 May 1536, Anne Boleyn was led out onto London's Tower Green and onto a stage erected for the execution – a raised platform so that all the onlookers might have a clear view of the impending brutal end to a woman's life. The anonymous executioner from Calais used a two-handed sword with a heavy, flat-ended blade. He raised it above Anne's bent body, swung it down hard and fast and removed her head with a single blow. Some onlookers believed they saw the lips quivering as the head rolled across the wooden planks.

The English held on to Calais for 211 years. It made heavy demands on the British taxpayer and was not always the commercial success Edward III had intended. In 1558, in the reign of Queen Mary ('Bloody Mary'), the Duke of Pembroke's army was battling it out with the French at St Quentin. Undetected by British intelligence, the French had amassed a force close to the coast and prepared an attack on Calais. It was a bitterly cold winter and an attempt to rush reinforcements across the Channel was compromised by high rates of influenza among the troops in England. The French force under the Duke of Guise struck swiftly and mercilessly across the frozen marshes. On 7 January, the British surrendered and, by so doing, lost the key to France. The English town in France had become something of an anomaly: to Queen Mary, its loss was an unmitigated disaster.

The Surrender of Calais, a comic opera in three acts recalling the French surrender in 1347 and the beginning of 200 years of British rule over the town, opened at the Little Theatre (later the Theatre Royal) in London's Haymarket on 30 July 1791. Written by George Colman the Younger, with music by Dr Samuel Arnold, the show would have had little trouble in booking the Little Theatre – George Colman the Elder had bought the theatre fourteen years earlier. The playbill records Edward III being played by a Mr Williamson and his queen by a Mrs Goodall, 'with proper Scenery, Machinery, Dresses and Decorations', prompting the artist Thomas Rowlandson to paint the first-night audience in the foyer. The show ran for twenty-eight performances – neither a hit nor a flop in those days – an eighteenth-century reminder of a bitter fourteenth-century siege.

⛵

English lace made its appearance in Calais towards the start of the nineteenth century when a group of Nottingham lace manufacturers moved here. Luddites were smashing their machines at home, and tariffs imposed by the French on lace imports were pricing the Nottingham makers out of business. The immigrant factory owners hired local women and with Paris flourishing as the centre for European fashion, business boomed. The Calais region was soon home to more than 3,000 expats, who had their own church and their own English-language newspaper, the *Calais Messenger*.

But with the Revolution of 1848, the collapse of the economy and a sudden rise in unemployment, foreign workers were seen as taking jobs from the locals. The owners of the lace factories upped spindles and returned to Nottingham, leaving behind a bunch of destitute English lacemakers. Step forward sympathetic machine owner Edward Lander and the British Consul in Calais, Edward Bonham. In the local English church, they met with the heads of lace-making families and hatched a plan to move the lacemakers once more, this time a little further from home. The lacemakers were on their way to Australia.

The British Foreign Secretary Lord Palmerston was petitioned and a fund was created to finance the move (to which Queen Victoria and Prince Albert chipped in with £200) and by 1849, 700 lacemakers were on their way from Calais to the far side of the world.

With the lacemakers went Calais' last remnants of Englishness, although the few remaining expats are quick to point out that the city's Catholic Church of Notre-Dame had been extended in the perpendicular gothic style – an essentially English style of architecture. Still standing 500 years later, the church was host to the wedding of Charles de Gaulle. He married local girl Yvonne Vendroux, the daughter of the local biscuit manufacturer, there in April 1921. He became President of France thirty-eight years later.

4

Cawsanders and Smugglers

Cawsand is not the prettiest of Cornish villages, but it is the keeper of much Channel history, sitting as it does on the western side of Plymouth Sound. It took its place in the history books one icy cold day in December 1577 when Francis Drake sailed on the *Pelican* (he changed his ship's name to *Golden Hind* halfway through the voyage to honour his patron, Sir Christopher Hatton whose crest was a female deer – a golden hind) out of Plymouth Harbour, crossing Cawsand Bay at the start of his three-year navigation of the globe.

Eleven years later Cawsanders would have waved their hats and cheered as the fifty-five Royal Naval ships under the command of Admiral Charles Howard, with Sir Francis Drake at his side, left Plymouth on their way to confront the Spanish Armada further east along the Channel. The Cawsanders would not have known, on that summer's day, that Drake was playing bowls on Plymouth Hoe when the Armada was sighted off the Lizard. Indeed, they might have been aghast to see him finish the game rather than round up his ratings and prepare his ships. But Drake knew his Channel and its tides and had registered the onshore wind blowing in his face. Moving the fleet out of the Sound would simply not have been possible straight away.

The Cawsanders would also have been witness in 1620 to the *Mayflower* setting sail from here with the Pilgrim Fathers on deck, watching their English shore become more distant with every gust of wind – gusts that were infrequent, making the journey slow and miserable. The spectators might have been surprised to learn that in addition to the necessities of exploration, these practically minded pilgrims had with them a printing press with which, having established themselves on the banks of Massachusetts Bay, they printed the Bay Psalm Book to help them spread the word of God. It was the first book to be printed in America.

Two hundred years later, on 26 August 1809, readers of *The Times* in Cawsand would have learnt that a ship that had caught their attention anchored in Plymouth Sound was *Formidable*, just arrived from Gibraltar with a VIP on board. His name was Mirza Abu'l Hassan Khan. Mr Khan, the paper went on to tell its readers, was an ambassador from Persia, 'with dispatches of importance for Government'.

What the paper did not know was that Abu'l Khan was himself an accomplished chronicler who might well have served its pages well. He had, during his diplomatic journey from the court of the Shah to the court of King George, kept a journal, 'In great good spirits the crew brought the ship into Plymouth Harbour, which is one of the ports of England'. He wrote that the anchor was dropped and the ship's captain, Francis Fayerman, struck the sails. And although they had been greeted by very English rain and snow, there was a:

> ... beautiful green field with trees to shame the palm-groves of paradise.
>
> The captain hoisted some flags – a system the English call 'telegraph' – and which, on a fine day, carries the news of a ship's arrival at Plymouth to London in twenty-five minutes. The captain explained that at night messages are sent by means of lights; during the day by flags. On arrival at a port, the ship hoists flags which are seen through telescopes on land; the message is then relayed to London via observation towers which cover the distance at one-mile intervals.

The ambassador went on to tell his journal that the English king had decreed that no one, not even the king himself, could leave a ship within four days of its arrival. Quarantine had to be observed. During the visiting diplomat's time on board, the General of Plymouth Harbour sent city dignitaries to the ship with gifts. Musicians and dancers were sent too, entertaining the passengers throughout the day and evening.

'Later,' wrote Mr Khan:

> I asked the captain why the passengers and crew were not allowed ashore. He explained that there can hardly be a single ship without at least one person on board who is ill or suffering from a fever. As a precaution, for fear lest the sick person should communicate disease to the inhabitants to the town, disembarkation is not permitted until the sick have recovered in the ship's hospital.

This explanation was apparently interrupted when a crowd of women rushed on board from small craft that had appeared at the ship's side. 'With bewitching guile and seductive glances,' wrote Khan:

> They captured the hearts of the seamen. Each one chose a woman and, embracing her, carried her off to his quarters – followed by the hearts of deprived Iranians! Sounds of the seamen's carousing and lovemaking reached even to the ebony wheel of heaven, and the din of their music and singing so excited my comrades that they were overcome by the desire to worship the vine.

Amazed by these goings-on, the Persian Ambassador asked his captain to explain what was happening. It was, Fayerman explained, simply a matter of prudent foresight, 'If these harlots were not allowed to relieve the crew of their money, to empty their pockets as clean as a glutton his plate, the shipowners might be faced with a severe shortage of labour for the next voyage.' It surprised the ambassador that no one who came on board was allowed to disembark for four days. This, he wrote in his journal, is what the English call 'quarantine'.

Six years passed before the Cawsanders were to see HMS *Bellerophon* sail from Cawsand on its way to strengthen the blockade of Rochefort on France's Atlantic coast. It was 28 May 1815, and the Napoleonic Wars were coming to an end. Few bystanders would have taken much notice of the three-masted warship as she made her way out of the bay and turned westward towards the Atlantic Ocean.

When she returned in July, it would be a very different story. The ship would make her way into the Channel and anchor in Plymouth Sound to be surrounded by hundreds of small boats, packed to the gunwales with men and women dressed in their finest as if journeying to a tea party with the lord of the manor. HMS *Bellerophon* would become the object of the greatest curiosity. But that was still a couple of months away.

When *Bellerophon* left Cawsand, the seventy-four-cannon ship, under the command of Captain Frederick Maitland, made good speed down the French coast, arriving off Rochefort on 31 May. She took up her position to monitor and, where necessary, intercept the movements of the French ships. This kept her and her crew busy until the end of June. Intelligence gathered from one of the intercepted French vessels suggested that Napoleon had been defeated by the combined armies of the Prussians and the Duke of Wellington's forces just outside a village in Belgium called Waterloo. The report was confirmed the next day in a letter delivered to Captain Maitland on *Bellerophon*. Further intelligence claimed that Napoleon was on his way to the coast with every intention of escaping across the Atlantic to America. Maitland received orders from the Admiralty that the emperor be prevented at all costs from escaping. *Bellerophon* moved in closer to the French squadron.

On 6 July, Captain Maitland received orders from Rear Admiral Sir Henry Hotham to the effect that Napoleon was indeed planning to make a run for it on one of the frigates based at Rochefort and this should be prevented:

> The Lords Commissioners of the Admiralty having every reason to believe that Napoleon Buonaparte [*sic*] meditates his escape with his family from France to America, you are hereby required and directed in pursuance of orders from their Lordships, signified to me by the Right Honourable Viscount Keith, to keep the most vigilant look-out for the purpose of intercepting him; and to make the strictest search of any vessel you may fall in with; and if you should be so fortunate as to intercept him, you are to transfer him and his family to the ship you command, and there keeping him in careful custody, return to the nearest port in England [going into Torbay in preference to Plymouth] with all possible expedition; and on your arrival you are not to permit any communication whatever with the

shore [...] and you will be held responsible for keeping the whole transaction a profound secret.

As it happened, Maitland and his crew were to be relieved of the need for any risky interception of a French frigate with Napoleon aboard. The emperor had drafted a letter to the Prince Regent (later King George IV) in which he made no reference to his plans to find asylum in America, suggesting instead that the British people might give him hospitality:

Your Royal Highness,
A victim to the factions which distract my country, and to the enmity of the greatest powers of Europe, I have terminated my political career, and I come, like Themistocles, to throw myself upon the hospitality of the British people. I put myself under the protection of their laws; which I claim from your Royal Highness, as the most powerful, the most constant, and the most generous of my enemies.
NAPOLEON.

At daybreak on 15 July, Maitland's Officer of the Watch reported the French brig *L'Épervier* flying a flag of truce and approaching *Bellerophon*. But just at the wrong moment, the tide turned and the wind veered inshore, bringing the French ship to a sail-flapping standstill. Maitland had a strong suspicion that Napoleon was on board the French ship and sent out First Lieutenant Andrew Mott in the ship's barge to meet the becalmed vessel. 'Mott,' wrote Maitland in his journal, 'was the best officer I ever saw in charge of a quarter-deck.' It was not until six in the evening that *Bellerophon*'s barge began its return. At a distance, it was not clear who the passengers were, but as it came closer, one passenger was observed wearing a small, cocked hat with a tricoloured cockade. Maitland focused his binoculars on the hatted figure and caught sight of a sword glinting in the light of the evening sun. He was looking at Napoleon.

Andrew Mott reported that the emperor had been cheered by the company of the *L'Épervier* when he left the ship, with many of the officers and men in tears, and the cheering continued until they were beyond earshot. When the barge arrived, Napoleon's grand marshal and closest confidant, General Henri

Bertrand was first up the gangway and rather grandly announced, 'The emperor is in the boat', which by then was not quite true because the impatient emperor had raced up the gangway and was already on *Bellerophon*'s deck beside Maitland.

'I come,' said Napoleon, removing his hat, 'to throw myself on the protection of your Prince and your laws.'

The emperor was received by Maitland with few of the formalities paid to visitors of high rank. There was a guard of honour but, pointedly, it had not presented arms as Napoleon stepped aboard. 'Buonaparte's dress was an olive-coloured great coat over a green uniform,' wrote Maitland in his journal:

> ... with scarlet cape and cuffs, green lapels turned back and edged with scarlet, skirts hooked back with bugle horns embroidered in gold, plain sugar-loaf buttons and gold epaulettes; being the uniform of the Chasseur a Cheval of the Imperial Guard.
>
> He wore the star or the grand cross of the Legion of Honour, the small cross of the order, and the iron crown. He had on a small, cocked hat, with a tri-coloured cockade; plain gold-hilted sword, military boots, and white waistcoat and breeches. The following day he appeared in shoes, with gold buckles, and silk stockings – the dress he always wore afterwards while with me.

Maitland's observations of his prisoner's clothes are remarkable in their detail, and his description of the man himself confirmed the Englishman's particular interest in the Frenchman's appearance:

> When he came aboard on the 15th of July, Napoleon Buonaparte was within exactly one month of completing his forty-sixth year [...] about five feet seven inches high, his limbs particularly well-formed, with a fine ancle [*sic*] and very small foot, of which he seemed rather vain, as he always wore, while on board the ship, silk stockings and shoes.

Maitland goes on to tell us that the emperor had small hands, light-grey eyes, dark hair without a hint of grey, and a sallow complexion. 'He had good teeth and a highly pleasing smile. Yet he had become corpulent which compromised his physical energy, and he would frequently fall asleep on the sofa in his cabin.'

The cabin that had been allotted to him was Maitland's, which the ship's captain generously vacated in favour of his royal visitor. '*Une belle chambre*,' observed Napoleon, enquiring after the portrait of a woman hanging on the wall. Maitland told him it was his wife, to which the Frenchman replied, '*Ah! Elle est très jeune et très jolie*', the kind of remark Napoleon would make to flatter those in his company.

In time, and after lengthy discussions between Bonaparte, his aides and the most senior British naval officers, *Bellerophon* received orders to set sail for England with the prisoner on board and, in the first instance, to make for Torbay. Napoleon had negotiated that he be accompanied not only by General Bertrand but by Bertrand's English-born wife, Fanny, and their three children. With them came General Montholon, his wife and son, sundry aides, valets, chefs and the Comte de Las Cases, a historian and long-time admirer of the emperor.

On the journey to England, Maitland and his crew treated Napoleon like a king. Dinner, prepared by the prisoner's cook Le Page, was at 5 p.m. and it was Napoleon himself who led the captain's party into the dining room. He sat himself in the centre of the long table, suggesting Rear Admiral Sir Henry Hotham be seated on his right. In good humour, both Hotham and Maitland acquiesced to the ex-emperor's regal behaviour. Eating turbot one night, Napoleon suggested that the fishermen who worked the Channel were smuggling at the same time, 'At one time a great many of them were in my pay for the purpose of obtaining intelligence'.

Much claret was consumed by his entourage, but Napoleon rationed himself to half a pint during each meal, at ten in the morning and five in the evening. He feasted on fish, beef, duck and various fruits, rounding off his meals with strong coffee and cordials. When they all played *vingt-et-un*, the popular French card game, Maitland excused himself, saying he had left all his money with his wife, at which Napoleon offered to lend him enough for the game.

Few defeated generals could have been better looked after by their captors. Napoleon would retire early to bed, leaving the French ladies and officers in the wardroom drinking 'bishops' – a cocktail of port, madeira and nutmeg.

According to Maitland, Napoleon showed no signs of depression on the journey to England. 'He never in my hearing threatened to commit suicide,' wrote Maitland, although he concedes that his prisoner's blunt refusal to go to St Helena could have been construed as a suicide threat.

When playing cards was no longer a distraction, the ship's crew put on a couple of shipboard theatricals. George Colman's 'The Poor Gentleman' cast the ship's assistant surgeon, Ephraim Graebke, as the mischievous Corporal Foss, with the smooth-chinned ratings taking the female roles. Countess Bertrand, at Napoleon's side, relished her role as interpreter. By all accounts, the show was a great success, causing the emperor to laugh heartily and surprise everyone by remaining in his place for all three acts, in stark contrast to his visits to the Paris opera, where he would invariably leave after Act One.

When *Bellerophon* reached Ushant at dawn on 23 July and turned into the Channel, Napoleon appeared on deck, much to the surprise of midshipman George Home, who had just taken his place on the morning watch as the cleaning of the decks had begun. Home saw their prisoner emerge a little bleary-eyed and make for the poop deck, and apparently became concerned that the emperor might lose his footing on the slippery deck. 'What must have been his feeling who had lost the fairest empire on the face of the globe?' mused Home. 'Nay who had lost a world?' As those thoughts came to the midshipman, he rushed to offer his arm to their prisoner. Bonaparte took it immediately, smiling his 'thank you' and ascended the ladder to the poop deck. There, he took out his pocket-glass, peered at the French coast and remained there with his retinue coming one by one to stand respectfully at a distance behind him until France was no longer in view.

Napoleon may not have been suicidal. There was, however, someone on board who was. By the time *Bellerophon* anchored in Plymouth Sound, rumours had reached the media that Napoleon was to be exiled to St Helena, a volcanic island in the South Atlantic, off the coast of Angola. Fanny Bertrand, the wife of Napoleon's trusted grand marshal, assumed her husband would have to go too. There were histrionics, and she remonstrated furiously with Maitland and anyone else who would listen. Finally defeated, she charged into her cabin, slamming the door behind her.

Concerned at this extreme behaviour, Count de Montholon, one of Napoleon's senior aides, gave chase to find the duchess dangling halfway out of the porthole. Much shrieking brought others into the cabin, who found the woman's legs in Montholon's grasp but the rest of her body still hanging out over the water. Dragging her back in was made difficult by the bar across the porthole designed to prevent people falling out whenever the ship listed heavily. However, she was eventually retrieved, telling Maitland she was driven to despair while abusing Napoleon in breathless English.

Bellerophon lay at anchor off Plymouth while the British government decided what to do with Bonaparte. As news of the Frenchman's presence began to leak, curious bystanders started to circle the ship in all manner of small craft. Napoleon had become a celebrity. Ladies had put on their finery, which amused and flattered Napoleon, who doffed his hat to his bobbing admirers at every opportunity. (The emperor, being a good flatterer himself, at one stage told a bemused Maitland that it was time he, Maitland, was promoted to the rank of rear admiral.)

The more adventurous admirers tried to climb aboard, and Maitland, fearful that his prisoner might effect an escape, had to launch as many tenders as he could to provide a protective ring around his ship. The thought of the French emperor escaping into the English Channel became a nightmare for Maitland.

Bonaparte was in no mood to contemplate spending the rest of his life on an island miles from anywhere. 'The idea is perfect horror to me,' he told Maitland. 'Had they confined me in the Tower of London or one of the fortresses in England, I should not have so much cause of complaint.' Bonaparte wrote again to the Prince Regent but to no avail. On 31 July, orders arrived to prepare for Bonaparte's departure, together with four of his officers, their families, his bodyguard and a small band of cooks and valets. *Bellerophon* was considered unsuitable for the long journey to the South Atlantic (much to the chagrin of Maitland and his crew) and arrangements were made to transfer the French party to HMS *Northumberland*. Napoleon finally left British waters on 7 August for the ten-week voyage. He spent six years on St Helena and died there at the age of 51.

※

Napoleon had been dead ten years when Charles Darwin, aboard *Beagle*, under the command of Robert Fitzroy, set out from Barn Pool on the western side of

Plymouth Sound for his epic voyage around the world, during which he would form his theory of evolution. Darwin was just 22 years of age, the second son of Robert Darwin, a well-to-do physician, and Susannah Wedgwood, daughter of the potter and industrialist Josiah Wedgwood. Fitzroy, who had taken over from *Beagle*'s former captain Pringle Stokes, who had shot himself on the previous voyage, was only 27. (Nearly a century and a half later, in 1967, Francis Chichester sailed *Gypsy Moth IV* triumphantly into the Sound after his nine months and a day solo circumnavigation of the globe.)

Beagle had left Devonport on 23 November 1831 but dropped anchor at Barn Pool under Mount Edgcumbe, where Fitzroy waited for a favourable north-easterly wind to take him into the Channel and onwards into the Atlantic. Darwin went for long walks while waiting to join the ship. Visiting Cawsand, he found it in 'a very pretty little bay, which shelters numerous fishing and smuggling boats'. He walked often into Plymouth, where he spent time with Charles Hamilton Smith, a local botanist. Spending more time on *Beagle*, Darwin had trouble finding his sea legs and wasn't used to sleeping in a hammock. 'I experienced a most ludicrous difficulty in getting into it; my great fault of jockeyship was in trying to put my legs in first' (Darwin's Journal: 4 December 1831). Over Christmas, *Beagle*'s crew went on a drinking spree, and Darwin went to church. On Boxing Day, the weather was perfect, but the crew would have even greater difficulty in getting into a hammock than the sober Darwin. However, they had sobered up by the morning of the 27th and *Beagle* urgently set sail, leaving Darwin having to catch up on a friend's yacht.

When *Beagle* finally left Barn Pool on that day just after Christmas, folk in Cawsand would have been more interested in that night's prospects for the smuggling boats Darwin had clocked on his walks than in the botanist himself. Smuggling had become big business for the Channel ports after the government had imposed heavy import taxes on all manner of goods from the Continent. The smugglers were ambitious and fearless. On one night alone, the Cawsand smugglers collected 300 ankers (ten-gallon casks) of contraband brandy from Jersey. They were back again just weeks later for another load.

However, not all their criminal escapades were without incident. The following year, smugglers returning under cover of darkness from France in their cutter *Two Brothers* were chased by customs officers onto a breakwater which smashed their ship and threw them into the water. At daybreak, there were still members of the crew clinging to the wreck. There was spilled brandy everywhere, but 100 casks were still intact, providing a bonanza for local folk who lost no time in taking advantage of the illicit cargo conveniently resting on the beach.

Dozens of vessels based at Cawsand were involved in the smuggling business. The crews were made up of experienced, canny and enterprising sailors who enjoyed the backing of local people, who were being taxed to the hilt on essential imports like tea, sugar and salt. Coffee, spirits and chocolate attracted even higher taxes. Gamekeepers often turned to poaching as disaffected customs officers changed sides and joined in the fun.

The eighteenth century became the golden age of smuggling, triggered by the duties imposed by the British government to finance the wars with France and the United States. Smuggling provided jobs too, and the people of the Channel coasts of both England and France were profiting handsomely. Tea, wine, spirits, tobacco and lace were brought to England by highly organised gangs, well managed and loyal, stretching well beyond Cawsand to the east and west.

A 'spotsman' was responsible for directing the smugglers' ships to selected hidden coves; the 'lander' organised the unloading; a 'tubsman' did the heavy carrying while a 'batsman' watched his back. On shore, the local villagers turned a blind eye to these illegal goings-on, nor was it unusual to find onetime revenue officers or even coastguard officers lending a helping hand in return for cheap goods.

Hundreds of barrels of liquor would be carried in a single Channel crossing. Wine and tobacco were stored discreetly in parts of ships that appeared solid. Tobacco was stuffed into the ships' fenders, and jewellery was hidden in fishermen's boots. When the crew's womenfolk came aboard, silk stockings were sewn into the linings of their petticoats and under the petticoats, pigs' bladders were filled with spirits. The tops of tea chests were made to resemble the ships' decks. Hollow keels held gin and brandy.

The Hawkhurst smugglers, taking their name from the Sussex village some 14 miles from the Channel coast, were among the most brutish and successful of the gangs, ruthlessly controlling the smuggling business on the Sussex and Kent coasts. The landlords of two pubs in Hawkhurst turned a blind eye to smugglers who would gather nonchalantly to plan their operations with their firearms casually placed on the tables in front of them. When a seizure of contraband tea was made on the Dorset coast in 1747, it was the Hawkhurst gang who rode the 140 miles along the coast with cloaks flying and pistols stuffed under their belts, brazenly attacking the Customs House at Poole to recover it.

The acquiescence of local church wardens and country landowners who would find smuggled goods stored in their churches and barns made the handling of contraband simple and beyond suspicion. Corruption around smuggling ran deep and encompassed even the City of London, where financiers who bankrolled the gangs made handsome profits. Dragoons who managed to successfully relieve smugglers of their bounty might reasonably have been expected to turn the goods over to the customs men. They did – but they made the customs men pay for it.

While the Hawkhurst gang was working the eastern reaches of the Channel and the Cawsand men ruled the western reaches, John Rattenbury was single-handedly giving customs officers along the Devon coast a run for their money. Mr Rattenbury hailed from Beer, a village in Lyme Bay on the Jurassic Coast, where he was born in 1778, the son of a shoemaker. He fell in love with ships and learned early in life that smuggling was the most profitable way to enjoy a life at sea. 'I engaged ostensibly in the trade of fishing,' he wrote at the time, 'but in reality, I was principally employed in that of smuggling.' With the help of a local priest he wrote his memoir, recalling how he would bring kegs of brandy from Cherbourg, run the gauntlet of customs officers and have to drop the kegs over the side with tiny marker buoys attached if the customs clippers got too close.

The Channel Island of Alderney proved a happy hunting ground for Rattenbury, being a halfway house for contraband from France destined for dealers on the English coast. He was constantly pursued by the customs men at sea and on land, and on one occasion, in the small hours of the morning in Weymouth, spent an hour up the chimney of a public house while the building was searched. In his *Memoirs of a Smuggler*, published in 1837, Rattenbury recalls, with a degree of mischievous pride, a close run-in with a dead goose:

> On one occasion I had a goose on board which the master who overhauled the vessel was very desirous of buying; but I was too well aware of the value of the stuffing to part with it, for instead of onions and sage it consisted of fine lace.

He was less sure that a tin box would go unopened:

> I had stowed some valuable French silks in a tin box which being soldered to prevent water getting in. While an officer was searching another part of the vessel, I contrived to throw it overboard, having previously attached it to a stone and a buoy, by which means I recovered the silks perfect and uninjured.

When he was finally caught, he was treated leniently by magistrates, who enjoyed easy access to his French brandy.

Rattenbury also had a nice line in people smuggling. In 1808, he was involved with smuggling French officers, prisoners of war, back to France. And he came back with French aristocrats escaping the Reign of Terror during the French Revolution. (*The Scarlet Pimpernel*, Baroness Orczy's fictional hero, followed Rattenbury's example of smuggling French aristocrats out of France.)

During his law-abiding days, Rattenbury was party to plans to improve the harbour at Beer and supervised the building of a canal from there to Thorverton, 42 miles away. At a hearing into the scheme in Westminster Hall, he was questioned about his trade. 'I told them sometimes fishing, sometimes piloting and sometimes smuggling.' Rather than outrage, the assembled dignitaries received this information as a matter of fact and Rattenbury went on unhindered to give evidence about the suitability of the harbour scheme.

The Beer Harbour scheme hearings brought Rattenbury into contact with Lord Rolle, a prominent Devon landowner with a country seat close to Beer. Rolle had been the local MP before his elevation to the peerage, during which time he had occasion to write to the prime minister, William Pitt the Younger, about local smuggling, underlining the degree of freedom smuggling in general enjoyed, 'Upwards of 56 horses loaded with brandy and tobacco passed my house yesterday. They are too powerful for the Revenue Officers to contend with.'

> Five and twenty ponies,
> Trotting through the dark –
> Brandy for the Parson, 'Baccy for the Clerk.
> Them that asks no questions isn't told a lie –
> Watch the wall my darling while the Gentlemen go by!

So wrote Rudyard Kipling in *A Smuggler's Song*. Kipling lived in Burwash, halfway between Tunbridge Wells and the Channel coast on the smugglers' route to London. A bronze statue of Kipling seated on a bench pays tribute to the village writer.

If John Rattenbury was brazen about his smuggling exploits, then a certain Madame Durfort was recklessly arrogant. According to the Cambridge historian Renaud Morieux, it was Madame Durfort, the wife of the Comte Louis de Durfort, who claimed she had been told of the sudden death in England of her father, the British politician Henry Seymour. She duly bought some black silk with which to have a dress and cape made which was suitable for mourning. The bereaved woman booked her passage on the Calais–Dover packet, and on her arrival was delayed by customs officers who were curious about the contents of her trunk. Whereas Madame Durfort volunteered information about her 'five black silk gauze coats, four pair [sic] of lace sleeves and four lace handkerchiefs', all of which, she claimed, were old and worn, she was less forthcoming about the contents of the trunk. The customs officers had their suspicions about Madame Durfort confirmed when they discovered the trunk had a false bottom and false sides, which were all full of dutiable goods the good lady had failed to declare.

The articles seized by the customs men included:

67 pairs of lace and needle work sleeves
35 habit shirts

10 handkerchiefs
4 caps and 25 tops for dresses
62 yards of black silk lace
54 yards of velvet trimming
168 fans
37 necklaces
35 pairs of ear drops
and 6 embroidered girdles.

The outcome of Madame Durfort's deception is not recorded. But corruption among customs officials being rife, her freedom was likely secured in exchange for a few pieces of silk, a necklace or two and even perhaps an embroidered girdle for a dutiful wife.

Daniel Defoe, the eighteenth-century writer who created Robinson Crusoe, described Lymington in his *A Tour Thro' the Whole Island of Great Britain*: 'I do not find they have any foreign commerce except it be what we call smuggling and rogueing which I may say is the reigning commerce of all this part of the English coast from the mouth of the Thames to the Land's End in Cornwall.'

Wrecking was another part of the Cornish smuggling trade, as goods that were washed ashore from a wrecked ship were regarded as common property. The sight of a ship foundering would bring the nearby population to the beach and before long, using pickaxes and hatchets, they would dismember the ship and make off with any goods they found.

The law deemed it illegal to claim salvage from a wrecked ship if anyone was alive on it, which meant the law virtually condemned survivors to death. There are legends that lights would be tied to horses' tails to lure the ships onto the rocks, but beacons lit strategically on shore, falsely suggesting safe passage, were more common means of luring the ships onto the rocks. It was a merciless but profitable business.

Gabriel Tomkins was a bricklayer. But he found bricklaying in Tunbridge Wells dull and unrewarding; Tunbridge Wells was inland, yet the townspeople were accustomed to the sight of horses laden with smuggled goods making their way to London. And like other towns on the route from the south into London, the otherwise genteel country town was also witness to the process of mixing contraband goods with those on which duty had been paid, making it impossible for customs officers to detect the one from the other. This was eighteenth-century organised crime.

The law enforcement agencies were outnumbered and outwitted. For Tomkins, the switch from bricks to brandy was irresistible. He joined the Mayfield Gang, which took its name from its base in the village of Mayfield, some 8 miles south of Tunbridge Wells. The gang, less brutal than some of the others, tying up customs men rather than thrusting pikes through them, set the trend for wool smuggling on a grand scale. And Tomkins was soon running the operation.

It was the tax on wool exports, imposed by Edward I in 1275 to support a hard-pressed crown, that proved the trigger for this wool smuggling across the Channel, where merchants in France, Flanders and Italy were eager to get their hands on the top-quality English product. They called these wool smugglers 'Owlers', although quite why is not clear unless they communicated with each other through the darkness of night with owl-like hoots. What made life even more attractive for 'Owlers' was the ideal sheep grazing afforded by the marshes at Romney, in the south-eastern corner of Kent. (The sheep from here did some legal travelling of their own: the Falkland Islands halfpenny stamp printed in 1933 bears a picture of a Romney Marsh ram.) The marshes presented a long stretch of coast that was difficult for the customs men to patrol, and it was a relatively short hop across the Channel. The sheep farmers were happy to accommodate the smugglers, who were getting them much higher prices for their wool abroad than the textile industry at home was prepared to pay.

Business for the 'Owlers', often men of Huguenot families who had crossed the Channel to escape religious persecution at home in France, fluctuated with the changes in the domestic price for wool. However, sheep farming took hold across the country, and during the plague of 1349, when more than 2 million people died, it gained a further advantage as shepherding and the husbandry of sheep could manage with fewer farm hands than the other agricultural pursuits.

By the seventeenth century, wool exports were the country's most important foreign currency earner, putting a strain on supplies for the domestic textile industry. By the end of the century, Parliament, recognising the difficulties for the textile business at home, had passed an act preventing the export of wool, with the death penalty for smugglers caught in the act. But prices on the Continent remained high and the smuggling gangs, seemingly undeterred by the death penalty, thought nothing of killing poorly armed customs officers who got in their way.

Churches were often the hiding places for sacks of wool avoiding duty on their way out of the country, and for tea, tobacco and brandy coming in. At Ivychurch, the vault beneath the church's north aisle proved an ideal repository for contraband in transit.

From Dymchurch, the Aldington gang worked the coast from Camber to Deal, turning Deal into a town that thrived on smuggling, becoming lawless and morally corrupt in the process. The town's boat builders specialised in long, slender galleys with oars for twenty men or more – the boats of choice for the smuggling of gold guineas to France during the Napoleonic Wars. The French currency had collapsed, and a guinea sold at a 50 per cent premium across the Channel. Passing through Deal on his journey round Britain in 1704, Daniel Defoe wrote of his unfettered dislike for the place, 'The barbarous hated name of Deal shou'd die or be a term of infamy. And till that's done, the town will stand a just reproach to all the land.'

Admiral Lord Nelson would have tempered any criticism of Deal because he used its smugglers as pilots to guide his ships through the more hazardous waters of the Channel. The admiral set up his shore headquarters in Deal's Three Kings Hotel on the seafront (now the Royal). Once settled in, he invited his brother William and his family, along with the family of his mistress Emma Hamilton, to come and stay at the hotel. The hotel's food was not up to much, prompting Nelson and his guests to eat out, greatly disappointing the hotel's chef.

The houses along the town's Middle Street and on both sides of Gold Street and Silver Street had cellars with street-level openings protected by retractable grids through which barrels of smuggled brandy and French wines slipped to safe and hidden storage. The more law-abiding seafarers and fishermen often fought with the smugglers, so Deal had an ambivalent reputation of which Nelson would have been aware.

After the Battle of Copenhagen on 2 April 1801, Nelson suffered a serious inflammation of the lungs, and in Deal was feeling run down and depressed. Deal, he told friends, 'was the coldest place in England'.

He had to wait four years before he could lead the British Navy into a stunning victory against the combined fleets of the French and Spanish off Cape Trafalgar, but for Nelson, victory came with a heavy price. Closing with enemy ships at Trafalgar, he was hit by a musket ball which severed an artery. As his life was slipping away, he repeatedly thanked God that he had been able to do his duty.

Meanwhile, Gabriel Tomkins had become so successful at shipping untaxed wool out of the country and bringing liquor, tobacco and silks back in (he boasted that he smuggled in 11 tons of tea and coffee in a single year) that the authorities had little choice but to offer a reward for his capture. In 1788, the *London Gazette* announced, 'A reward of £100 is offered for Gabriel Jarvis alias Tompkins, notorious smuggler & Owler transported in the 8th year of the late King for transporting wool and resisting Customs Officers. He has returned from transportation and is again smuggling in Kent & Sussex.' Tomkins was described as 'something pitted with smallpox, has a very large dark eyebrow and usually wears a light wig and fustian frock, is now supposed to wear his own dark brown hair. He is a tall well-made man shot through the left arm with a brace of bullets.'

With an eye on the reward, a small group of vigilantes chased and caught Tomkins' half-brother, Edward, hoping he would lead them to their reward. The leader of the vigilantes, one John Rogers, a hop grower who moonlighted for the revenue service, took Edward before the magistrates, who promptly threw the bewildered Rogers in jail for executing an illegal arrest. Edward Tomkins was handed over to the local constable who, being short of a bob or two, 'discovered' Mr Tomkins was no longer in custody.

Mr Rogers was released and continued his hunt, accompanied by a party of Grenadiers under a Lieutenant Jeckyl. And this time they had better luck.

They found Gabriel Tomkins with a bevy of smugglers at Burwash, later the East Sussex home of Rudyard Kipling. An 18-mile breathless chase on horseback ensued with Tomkins and his cohorts finally being surrounded at Nutley in Ashdown Forest. Once again, Rogers took his captives to the magistrates, this time sitting in Horsham. With a bunch of known villains in front of them, the magistrates had little choice but to lock them all up.

Tomkins again had little difficulty in bribing his gaoler, who 'accidentally' left the key in the lock of his cell door. The arch-smuggler and gang leader was eventually recaptured, tried and hanged in London. Yet his demise did little to interrupt wool smuggling along the Romney Marshes, and in 1788 the front page of the *Illustrated London News* proclaimed, 'Smugglers Rule in Romney Marsh', with a cartoon showing a smuggler lauding it over a hapless customs officer.

However, the beleaguered customs men were not entirely without their successes. In the thirteen-year period from 1723, records show that 200,000 gallons of smuggled brandy were seized and 2,000 smugglers and their accomplices prosecuted. Even so, the eighteenth century had provided rich takings for the thousands involved in smuggling on both sides of the Channel.

※

It would, however, be a mistake to assume that all the alcohol that reached England's southern shores did so illegally. As far back as the 1300s the equivalent of 4.5 million bottles of red wine were shipped from Bordeaux to Winchelsea. The small Sussex coastal town still has a fine network of cellars which bear a striking resemblance to the architecture of Monpazier, the picturesque *bastide* in the Dordogne. These cellars were probably used to store the wine at a cool, constant temperature before being distributed across the south of England. Hauling the barrels up the hill from the harbour would have needed the efforts of the town's fittest and strongest, but in the locked cellars the wine would be safer than in the warehouses at the docks.

In the Middle Ages, the wine was a spiced rosé, light in colour. (Today's claret, from the French *clairet*, meaning pale, began its transformation into the wine we know now during the eighteenth century.) The early light wine did not keep and was drunk young, directly from the barrel.

Winchelsea itself, one of the later Cinque Ports, came under attack from the French at the start of the Hundred Years War. Today, the twenty-first-century Sussex town is home to a population of around 600, who will tell you proudly that their town was originally the creation of King Edward I who modelled it on the French *bastides* with which, as Duke of Aquitaine at the end of the thirteenth century, he had become well acquainted.

It was Prime Minister William Pitt the Younger who lowered import duties at the beginning of the nineteenth century and hastened the end of the swashbuckling smuggling gangs. It was also the end of cheap luxuries for all those who smoothed the smugglers' way. A century later, it was the turn of the Channel drug smugglers and people smugglers, who think nothing of sending hapless asylum seekers to their deaths in overcrowded small boats unsuitable for the Channel crossing.

5

The Channel's Islands

The Channel Islands

'Fragments of France which fell into the sea and were gathered up by England' is how Victor Hugo described the Channel Islands. That prince of French literature, whose political views were out of step with Louis Napoleon's dictatorship of the 1850s, found himself seeking refuge in Jersey. But the islanders were uncomfortable with a revolutionary in their midst so, from there, he moved to Guernsey. He bought Hauteville House, a substantial four-storey property overlooking St Peter Port, and put his mistress, Juliette Drouet, in a smaller house just down the road and close enough to be on the route of his daily walk. Here at Hauteville, upstairs in a room with a view, he wrote, among much else, *Les Misérables*, his stunning blockbuster portrait of social strife in post-Napoleonic France.

GUERNSEY

There are about 30 square miles of Guernsey. It is a self-governing British dependency in the Gulf of Saint-Malo, having come under the Crown in 1066. Before that, it was part of the Duchy of Normandy, and the sovereign, King Charles III, is, as far as the island is concerned, still Duke of Normandy: the loyal toast on the island is 'The King, our Duke'. Within the bailiwick (jurisdiction) of Guernsey come Alderney, Herm, Brecqhou, Jethou and Lihou (Sark and Jersey have their own administrations).

The Channel Islands had not much concerned the English Crown before Richard II became King of England at the age of 10. He did not know it then, but he was later to set Guernsey on its path to becoming a prosperous tax haven. During young Richard's first years on the throne, Regency councils did most of the sovereign's work. Yet, by the time he reached his teens, he was playing a not insignificant role in the affairs of state.

The Guernsey islanders were loyal subjects and in return, Richard, with the maturity that came with reaching his twenties, granted the island a new charter exempting it from paying English tolls and customs duties. Encouraged by their new tax arrangements, the islanders became accomplished shipbuilders, finding customers at home as well as in France and England. They built boats to withstand the seas of the North Atlantic, and their fishermen joined their French and English counterparts who went fishing off the coast of Newfoundland. The salted, dried cod they brought back found markets in France, Spain, Italy and the Netherlands.

The trade remained commercially important until the beginning of the eighteenth century when the fishermen discovered smuggling, which they found to be markedly more profitable than fishing. This did nothing to harm the island's boat-building industry – smugglers needed boats as much as anyone. With the added need for fighting ships as wars continued against France and Spain, the Guernsey shipbuilders found their order books full to the gunwales.

Experienced sea captains took advantage of the wars by applying for 'letters of marque' to become privateers licensed to attack vessels of countries with whom England was at war and rob them of their cargoes. Francis Drake set the standard for privateering by providing his queen, Elizabeth, with all manner of treasures – a service for which he received a knighthood, an honour which ostensibly recognised his skills in navigating the globe. Kneeling before his queen on the *Golden Hind*'s deck, he withheld a wry smile as the royal sword was passed to the ceremony's honoured guest, the French Ambassador. Elizabeth persuaded the ambassador to do the honours on her behalf lest she be seen to condone her sailor's privateering in the eyes of the Spanish. Whether the ambassador, the Marquis de Marchaumont, or the queen herself commanded *Sir* Francis to arise remains a sensitive detail lost in the smoke and mirrors of diplomatic history.

⚓

While Richard II had laid the foundations for the island's tax-free economy, Elizabeth's half-sister Mary, from whom she inherited the crown, found in Guernsey three martyrs whose deaths were to leave a dark and indelible stain on the island's history. With the restoration of the Catholic Church, Queen Mary made sure prominent Protestants lived in fear. Or did not live at all.

Mary had already ordered the bishops Ridley and Latimer to be burned at the stake in the street outside Balliol College in Oxford. A year later came one of the ugliest incidents of 'Bloody Mary's' reign. It was 1556, the year that Archbishop Cranmer was finally martyred. On the island of Guernsey, three women were taken to court for receiving a stolen goblet. On that charge, the three Protestants were found not guilty but were subsequently found guilty of heresy by an ecclesiastical court. The three heretics, Katherine Cawches and her two daughters, Guillemine Gilbert and Perotine Massey, were brought from prison to the Tower Hill steps in St Peter Port. The execution was carried out on or around 18 July.

All three were burnt on the same fire. They ought to have been strangled beforehand, but the rope broke before they died, and they were thrown into the fire alive. The sixteenth-century historian John Foxe recorded that Perotine was 'great with child' and that 'the belly of the woman burst asunder by the vehemence of the flame, the infant, being a fair man-child, fell into the fire'. The baby was grabbed from the edge of the flames by one William House and laid on the grass. But the screaming infant was immediately seized by the Provost, who carried the tiny piece of defenceless humanity to the Bailiff, who ordered that 'it should be carried back again, and cast into the fire'. It was not Guernsey's finest hour.

⚓

Three hundred years later, shame and disgust had been diluted by the passing of time and there was evident pride among the islanders when they welcomed the French literary giant Victor Hugo at the end of October in 1855. He had been thrown out of Jersey for supporting the local newspaper in its decision to run a letter from a known Republican who was libelling Queen Victoria. But, for the islanders of Guernsey, he was a star not a political troublemaker and they quietly lined the streets from the dockside to the hotel, doffing their hats as he passed. The following day, he wrote to his wife who had remained temporarily on Jersey:

> Dearest, we have made it and are landed, not without a bump or two. Huge swell, wild wind, cold rain, black fog. Jersey is no longer even a cloud, Jersey is nothing; the horizon is empty. I feel as though I am in suspended animation; when you are all here with me life will begin again.
>
> We were well received. There was a crowd on the quayside; silent, but sympathetic, at least so it appeared; everyone had taken off their hats as I passed.
>
> As I write to you, I am looking out at a superb view. Even in the rain and fog, the entrance to Guernsey is magnificent. [François]-Victor is in raptures. It is a real old Norman port, hardly any English influence. The consul, wearing a white tie [...] was present as I disembarked.
>
> Someone told me he too had acknowledged me as I went past. It would seem that the local authorities have said that we will be left in peace here, as long as we don't cause any trouble. They treat us as though we were common criminals. But bucketsful of water won't put out volcanoes.

Comfortably installed at Hauteville House, he planted an oak in the back garden in the fervent hope that the continent of Europe would be in unison by the time the tree was fully grown. Inside, he decorated the house in ways that prompted his son Charles to suggest that the interior became 'a veritable three-storey autograph, a poem in several rooms'.

It was in his room on the third floor of this 'autograph' that Hugo wrote *Les Misérables*. He had made a start before he arrived, but the bulk of his magnum opus was written and published in 1862, during his exile on the island. He painted here, too, paintings that dug into his subconscious. They were personal and private paintings put on paper with ink using the nib and feather of a quill – the same tool he used to write with.

Pen and ink were conventional enough, but Hugo's art was experimental, as were some of his materials. In addition to conventional pens and brushes, he used leaves, lace and his fingertips, and, according to son Charles, was not averse to adding a 'light shower of black coffee' to finish off a painting. On seeing a selection of Hugo's work, the leading romanticist Eugène Delacroix wrote to him in the most complimentary terms suggesting he could have been one of the leading artists of the century.

Hugo returned to France in 1870, the year Guernsey's first gentlemen's club was launched. Might he have been invited to be its first president? He was, after all, held in great respect in terms of his writing, but his strident political views might have troubled the more conservative founding members of the club – members who provided a reflection of Guernsey's growing prosperity during the nineteenth century.

Victor Hugo, who might have become an artist rather than a writer, was followed to Guernsey by an artist who might have become a chorister. It was 1883 when 42-year-old Pierre-Auguste Renoir took lodgings in the island's capital, St Peter Port. From there, it was a comfortable walk to the bay at Moulin Huet, where, over a period of five weeks, Renoir completed some fifteen canvases. They were not his finest paintings and not representative of the work for which he became best known. And why he chose to travel to Guernsey when the coast of Normandy provided ample opportunity for landscapes and seascapes remains a mystery. Yet, his study of a nude bather seated on a rock at Moulin Huet did presage his later prodigious output of sensuous female nudes.

⚓

This second-largest Channel Island, just 30 miles from the Normandy coast, found its place in the fast-growing global maritime trade with its own William Le Lacheur trading in coffee between Europe and Costa Rica. Trade in luxury goods – wines, spirits, silks, lace and tobacco – was conducted openly, unhindered by the tariffs that operated at English ports. Nurserymen covered the island with glasshouses to grow tomatoes while quarrymen exposed and chipped away at its top-quality granite, much of which found its way to London. The steps of

St Paul's Cathedral were built with Guernsey stone, as were the road surfaces of London Bridge, the Strand and the Thames Embankment.

But what gave a massive economic boost to both Guernsey and Jersey was smuggling. The islands became ideal staging posts between England and France. There were plenty of experienced seafarers, an established boat-building industry and ready markets for luxury goods on both sides of the Channel. For the islanders, smuggling became an open and accepted way of life.

Jersey

It was not until late in the seventeenth century that the English authorities became aware that Jersey merchants were buying far more tobacco than their islanders could possibly smoke. The trick was to buy bales of tobacco in England duty free as it was bound for Jersey. Then, having sailed far enough into the Channel to allay any suspicions, turn around under cover of darkness and deliver the cut-price goods to land-based smuggling gangs waiting in sheltered coves along England's south coast.

Customs authorities in Southampton made the strongest representations to Jersey's Lieutenant Governor Edward Harris, but to no avail. Mr Harris had too much to lose by standing in the way of the smugglers. Such was the support in the community for the smugglers that lesser mortals than the governor risked having their ears cut off if they got in the way of the law breakers. The island's coopers made small kegs specially designed for smugglers to carry single-handed, while warehouses were set aside to store the contraband goods.

With England at war with France, William of Orange (William III) banned the import of French goods, which were benefitting the enemy's economy. This was good news for smugglers, as was war with Napoleon a century later, when British smugglers were subverting national borders and national identities by trading with the enemy.

Estimates suggest 20,000 British nationals were involved in smuggling during the eighteenth century. The most lucrative goods were those that attracted the highest duties. Brandy, wine, tobacco, lace, silks and gin gained the best returns, and the smugglers used the Channel Islands as a staging post. During 1812, 604 smuggling ships loaded with brandy, wine, textiles and leather goods left the ports of Jersey. In one nine-year period during the latter half of the eighteenth century, 7.5 million pounds of tea, enough for more than a trillion cuppas, were smuggled into England, much of it through the Channel Islands. This was organised crime operating unchecked under the eye of the authorities.

However, when William Pitt the Younger, the British prime minister, slashed the tea import duty in 1784, its value for smugglers fell dramatically. But that did not mean the smugglers were suddenly out of business. Not a bit of it. There was much else to smuggle.

With the Royal Navy occupied fighting Napoleon's navy, the customs cutters of the Preventive Water Guard were on their own and no match for the canny smugglers. In 1811, Napoleon lent the smugglers a hand by opening Gravelines, the port between Calais and Dunkirk, to their ships. The English smugglers delivered 1.6 million gold guineas to Napoleon through Gravelines, a haul then worth more than 42 million francs. And the smugglers boosted their takings by hiding escaped French prisoners of war among the contraband and charging the soldiers 300 guineas a trip. Locals rejoiced in hailing the port the *ville des smoglers*.

⚓

Smuggling cows was never going to be a practicable exercise, even among the most resourceful of criminals, so the cows that found their way from Normandy to Jersey will have done so legally. The cows that arrived on the island were identified as a breed during the eighteenth century, and from the beginning of the nineteenth were protected from any crossbreeding by a ban on cattle imports.

The pure-bred Jersey cow was comparatively small, generally fawn in colour and docile by nature. It also proved to be extraordinarily efficient. A healthy Jersey cow produces some 8,000 litres of milk a year, feeding on relatively small areas of pasture during the spring and summer. Its diet is supplemented later in the year with silage and nuts. The milk, drawn from capacious udders which have particularly well-developed milk veins, is highly nutritious, containing more protein, calcium and butterfat than the milk of other breeds.

Exports of milk plus the export of the cows and the bulls' semen has turned the animal into a top earner for the island's dairy farmers.

Jersey cows are now farmed across continents, becoming the world's second-largest dairy breed after the Holstein Friesian from Holland. In the United States, the Channel Island cows grab the red rosettes with almost indecent regularity. 'Brown Bessie' started the run of successes at the Chicago World's Fair in 1893 when she produced 17.5 litres of milk every day for five months. Jerseys look good and their milk tastes good. Today, there are still thirty herds on Jersey, totalling 3,000 animals.

⚓

On a very small island in the English Channel, cows, potatoes and financial services make unlikely economic bed fellows. Hugh de la Haye was a potato farmer. And he liked a party.

In 1879, with the ploughing over for another year, Mr de la Haye organised a dinner for his friends and helpers, an annual event they called *La Grande Charrue* (the big plough). With the hard work behind them, their spirits were high as they chomped through platefuls of all the good things this farming island in the Channel could produce, washing it down with goblets of wine from their French neighbours across the water.

At some point during the festivities, the host rose to his feet and, when the room fell into silent expectation, produced two very large potatoes he had been given as curiosities: one had no fewer than fifteen 'eyes', each with their little stems. It was his intention, he told his guests, to cut the potato into fifteen parts and plant each separately in the most fertile part of his land above the valley at Bellozanne, north-west of St Helier. This he proceeded to do.

The winter passed uneventfully and, come the spring, his fifteen experimental plants not only appeared early but one of them had produced a small kidney-shaped, thin-skinned potato, the like of which no one had seen before. What's more, it had a flavour that set it apart from other potatoes.

It was not immediately apparent, but Mr de la Haye had a farming phenomenon on his hands. The new potato got the name 'Jersey Royal Fluke', but marketing-savvy members of the island's farming community quickly dropped the 'Fluke' and began planting their new discovery on *cotils* – steep, south-facing slopes which drained well and are bathed in the warmth of the morning sun.

Generations of Jersey families have continued to grow Jersey Royals, planting and harvesting by hand, fertilising them with seaweed and profiting handsomely. Figures from the Jersey government suggest that around twenty-five families produce some 30,500 tons of potatoes in a good year, contributing £32 million to the island's economy.

⚓

When Hugh de la Haye was growing his first Jersey Royals, many of his farming colleagues were dividing their time between working on the farm and going to sea to fish for cod. George de la Haye, born in 1847 (a distant relative and younger than Hugh), acted as chief officer on the two-masted schooner *The Comet of Jersey*. He would join French and British fishermen in the cod-rich waters off the coast of Newfoundland, leaving Jersey in the spring and returning, with dried and salted cod for markets across Europe, in time for the autumn ploughing.

The tongues and the swim bladders (the cod's internal oxygen-filled organ that helps the fish control its chosen depth) were considered a luxury and recipes abound for both. The bladders are apparently equally good fried, stewed or curried. One old-time recipe suggests, 'Rub the bladder with salt, stew in a seasoned white gravy. Add cream, butter and flour and bring to the boil. Finally add lemon peel, nutmeg and mace from the coating of the nutmeg seed.' The tongues could be rolled in a mixture of flour, eggs, cream, salt and pepper and then fried in olive oil and butter.

Cod fishing proved considerably more profitable than potato growing, enabling the most successful fishermen to build grand houses on all sides of this small island – just 9 miles long and 5 wide – showpiece houses that became known as 'cod houses'. The bannisters in an old Jersey house reveal much about the money used to build it. A newel-post at the bottom of the main staircase with an inlaid button of ivory or mother-of-pearl on top signified the house had been paid for without a mortgage. And rafters fashioned from offcuts of wood from the shipyards provided further clues to the provenance of the house.

In the twenty-first century, these grand 'cod houses' are no longer occupied by fishermen but by bankers and financiers: Jersey's new rich, who trade on the island's tax system. Jersey is a British dependency with its own parliament and its own courts. It owes its independence to a persuasive King John in 1204 who, having been defeated in battle by the French King Philippe-Auguste, seduced the islanders, who might otherwise have fallen in with the French, with the right to self-governance. Britain would still act in the island's defence, so Jersey duly signed up to life under the British Crown, with subsequent royal charters confirming the island's constitutional status.

This has left the island in control of its fiscal policy. There is no capital gains tax or inheritance tax, with income tax pegged at 20 per cent. GST (the island's equivalent of VAT) is set at just 5 per cent. The financial services industry pays just 10 per cent corporation tax.

These tax advantages date back to 1349 when goods imported into Jersey were exempted from duty by King Edward III. This tax break was followed later by the additional lifting of duties on goods exported to the colonies from the island. The downside to these tax advantages – and they are considerable – is that the islanders do not enjoy the protection of a welfare state. Its health services are not part of the NHS.

The Channel Islands at War

The Channel Islands got off lightly during the First World War, but its soldiers paid a heavy price on the battlefields of northern France. The Royal Guernsey Light Infantry fought near Ypres and then at Passchendaele. Just fifty-five of the 1,200-strong force survived. Jersey lost fewer men but had to contend with a prisoner-of-war camp. It was so well appointed with electric light, heating, laundry facilities and kitchens that the islanders began to envy their German guests. They nevertheless treated them with respect, and when one young soldier died in the camp after an epileptic fit, his colleagues were allowed to draw his funeral carriage from the prison through the town for burial at the local church, with islanders providing a guard of honour.

The Second World War found the Channel Islands in a vastly different position. Jersey, Guernsey, Alderney and Sark were occupied by Hitler's army. And things started badly. The Wehrmacht had not been told that the islands had been demilitarised and began with bombing raids over both Jersey and Guernsey.

At St Peter Port, the Luftwaffe mistook tomato lorries for troop convoys. Forty-four islanders died, their blood mixing with the juice of tomatoes covering the granite of the harbour streets with puddles of deathly crimson.

Just days previously, General de Gaulle had touched down for coffee in Jersey on his way to London and his BBC broadcast to the French people. The British government decided the islands could not be defended, and the invasion which followed the bombing raids was swift and decisive. More than 22,000 islanders were evacuated, church bells summoning people to the dockside and to ships that would take them to Weymouth. The islanders made little objection to the evacuation.

The Germans insisted that local time be changed in line with continental time, and motorists had to switch to driving on the right. Sundry new laws were introduced, including one that offered rewards to anyone informing on fellow islanders who daubed walls with 'V' for 'Victory' signs in favour of the Allies. Sheltering escaped slave labourers, spreading news broadcasts by the BBC and helping German soldiers to desert were crimes that resulted in sentences in German prisons or concentration camps. A woman who slapped a German officer in the face when he made improper advances was arrested and sent off to Germany. Another spirited woman, Winifred Green, working as a waitress in Guernsey's Royal Hotel, greeted diners with '*Heil Churchill*', which cost her six months at the German prison in Caen, where she joined sixty-six others from the Channel Islands.

Twenty-two Jersey islanders died in German prisons. The island's cinema projectionist ended up in Buchenwald, and a young school teacher, Harold Le Druillenec, was sent to Belsen for radio offences. He was found alive when the camp was entered, but his sister, Louisa Gould, who had been arrested with him for sheltering two escaped Russians, died in the gas chamber at Ravensbrück.

Frank Falla, a Guernsey man, survived his incarceration. Frank was the driving force behind GUNS – the Guernsey Underground News Service. He and four others who ran the service, disseminating news from the BBC, were betrayed by an informer. Two of the five died in prison. Frank survived the sixteen months of beatings, starvation and sickness in the prison at Naumburg and made it his life's work to win compensation for the families of the islanders who never returned.

The Germans confiscated weapons, boats, radios, motor vehicles and cameras. Local people were forced into helping build a mass of fortifications, even though such coercion was in violation of the Hague Convention. Algerians, Moroccans and Jews from France were shipped in from Saint-Malo to help and were joined by 2,000 Spaniards who were handed over to the Germans by the Vichy government. Meetings of more than three people were prohibited, drinking was restricted and access to beaches monitored. Medicines became less freely available.

Camps were built for the German troops and their civilian labourers. Light railways connected the coastal fortifications. And in time, German officers were connecting with the islands' women – married and single – and, according to a Ministry of Defence report, from 'all classes and families'. The fraternising led to several hundred illegitimate births to German fathers, countless abortions and some bitter recriminations.

But there were soon to be fewer women to fraternise with and fewer island men to do the heavy work of building the islands' defences. In the autumn of 1941, hundreds of Germans working in Iran had been arrested by the British authorities after an Anglo-Soviet advance, designed to secure the country's oil supplies and protect Allied supply routes through the country. Hitler retaliated by ordering that for every German taken in Iran, ten Channel Islanders born in Britain were to be rounded up and shipped to Germany.

Nothing happened until more than a year later, when some 1,200 Jersey islanders were deported, along with 850 men, women and children from Guernsey and nine from Sark. Eleven Sarkees were on the original deportation list but a certain Major Shelton and his wife cut their wrists to avoid being taken away. Mrs Shelton survived but her husband died. The deportees were taken first to Stalag VI-J in Dorsten on the Lippe River in north Rhine-Westphalia and then distributed to rather more pleasant camps on the borders with Austria and Switzerland.

For those left behind, the shortage of food and other essentials began to weigh heavily. Potatoes and swedes were the staple. Bread was scarce and meat rationed to 4oz per person per fortnight. Tea was made by recycling dried leaves or with bramble leaves. With no sugar, islanders made syrup from sugar beet. Rabbits were bred and sparrows snatched from the hedgerows. There were mussels and crabs in the shallows, but the beaches were mined. And when there *was* food, there was little or no fuel to cook it with.

Toothpaste was made by mixing soot with chalk dust. There were no nappies. Crisis levels were reached after D-Day at the beginning of June in 1944. Supplies from occupied France dried up completely, and with the occupying forces still in place – and they stayed for another year – the shortage of food became acute.

German soldiers rounded up stray cats, skinned and ate them. When there were no strays left, they stole the islanders' pets and ate those. Red Cross parcels from Portugal did eventually arrive but much of the food was too rich for people whose stomachs were accustomed to little other than meagre root vegetables and an occasional piece of sparrow.

Alderney, the most northerly of the Channel Islands, was evacuated prior to the German occupation with only a handful of farmers staying behind. What the Germans did here is disputed. To find out, a 23-year-old Army Intelligence Officer, Captain Theodore Pantcheff, was sent to the island in the summer of 1945 to talk to prisoners and interrogate their German guards. He spoke German and had spent holidays on Alderney. He had also interrogated Germans held in London.

On the island, he found labour and concentration camps built by prisoners shipped to the island – Russians, Poles, Ukrainians and French Jews, forced into slave labour by the Nazi military engineers. The Germans had established their island headquarters in the Lloyds Bank building, from where they perpetrated a regime of appalling brutality against their prisoners, mostly Russians and Ukrainians, including those who had built the camps.

Pantcheff was not entirely forthright in his findings, 'It has been established, I think, that crimes of a systematically callous and brutal nature were committed – on British Soil – in the last three years [1942–45]'. Prisoners told of starvation diets that constituted half a litre of black coffee for breakfast, thin cabbage soup for lunch and a portion of soup with a kilo of bread between six of them for dinner. And this was a diet on which the prisoners were expected to build the defences Hitler demanded be built on the island.

Pantcheff reported that these slave labourers were beaten for the most minor of offences, such as trying to extract food from the garbage pail. Others who collapsed through exhaustion were shot. One prisoner told the young intelligence officer that he witnessed a German shoot a Russian who was simply collecting potatoes.

Nazi officers would tempt prisoners to break rules. Pantcheff reported that the SS 'competed in getting leave by shooting prisoners for the smallest offences, e.g. they threw away cigarette ends and as soon as an inmate bent down to pick them up they shot them'. The head of the SS gave 'a bonus of 14 days' leave, extra food and drink to guards for every five dead prisoners'.

Pantcheff found burial grounds on which he based the number of prisoners who died at the hands of the Nazis on Alderney. But it seems likely that the Germans dumped bodies in the sea, masking the true number of deaths. Historians have more recently revised Pantcheff's estimate of 372 by a factor of three. But they acknowledge that the true number may never be known.

In 1947, the British Army was asked by their French counterparts whether the British government intended to try the Nazi officers responsible for the war crimes. The response from the office of the Judge Advocate General pointed out that most of the prisoners had been Russian, so reports on the operation of the Alderney camps had been handed to them:

> The only information we can give on this matter is the general statement that the Russians were treated with great cruelty and that not only were many tortures inflicted on them, but many were allowed to die as a result of starvation.

Pantcheff went on to enjoy a distinguished career in the Secret Intelligence Service, MI6. In 2023, Lord (Eric) Pickles, the Conservative peer and the UK's Special Envoy on Post-Holocaust issues, ordered an inquiry into the wartime atrocities on Alderney.

Sark

Life for the post-war Channel Islanders took some getting used to. They had been spared bombing raids but many harboured guilt at acquiescing with the occupying forces. Most of the islanders had faced months of hunger, and many were forced to have German officers living with them. Hundreds of children were born illegitimately to German fathers. Yet, life did return to normal, with tax breaks attracting the financial services industry to the islands.

Tax breaks and the pre-war prosperity delivered first by the cod fishermen and then by the oyster fishermen, boat building, mahogany, potatoes, cider making and entrepreneurial French Protestants escaping religious persecution at home all contributed to restoring the islands' economies. On top of this, the semi-detached nature of the islands' constitution gave their people independence and with it a sense of solidarity. And, with England keeping a post-war eye on its defence needs, a sense of security.

In some parts of the world, the first quarter of the twenty-first century saw democracy come under pressure. In India, Narendra Modi's ruling BJP drew several state organisations under its wing; Poland and Hungary, two stars of post-Soviet eastern Europe, showed an increasing contempt for democratic norms; Hong Kong felt the heavy hand of Beijing; the Turkish president threw people who questioned his policies into prison; President Trump issued executive orders faster than he could down a can of Coke.

But Sark, one of the smaller Channel Islands, bucked the trend. It embraced democracy for the first time, giving the 500 islanders a voice in 2008 as the island held its first general election. Never before had the people of this car-free and almost tax-free island been given the opportunity to choose the twenty-eight members (slimmed down now to eighteen) of the Chief Pleas, the island's government.

Sark's administration had been in the hands of the island's forty land-owning farming families who had presided over Europe's last remaining feudal community. The island's two volunteer police officers made sure everything was above board on polling day. (The officers also issue bicycle licences for £14 and licences to drive a horse and carriage for £12.)

The election split the islanders into two factions: those for and those against the designs of the billionaire Barclay brothers, Sir David and Sir Frederick. The property magnates and then owners of the *Spectator* magazine, the *Telegraph* newspaper group and previously owners of the *Scotsman* and London's Ritz hotel, had plans for the island. The twins (David arrived a minute or two before Frederick) had long been visitors to Sark. In 1993, they bought the neighbouring islet of Brecqhou for about £3 million, a modest sum for the brothers who, in the *Sunday Times* Rich List of 2020, were estimated to be worth £7 billion.

They knocked down the islet's few existing buildings and built a ninety-room Gothic palace, in part to accommodate their desire for privacy. They invested heavily in neighbouring Sark but became frustrated at the island's feudal way of running things and began to lobby for a more democratic means of managing the island's affairs.

The brothers (they disliked being referred to as 'the twins') found themselves up against formidable opposition in the person of Michael Beaumont. Beaumont was the grandson of the Dame of Sark, Sibyl Hathaway, who had been the island's redoubtable *Seigneur* (feudal head) for forty-seven years. When an election for a twenty-eight-seat parliament was finally agreed to in 2008, the brothers weighed in heavily, using the local paper, which they owned, to support their preferred candidates and castigate the rest.

The belief among many Channel Islanders was that the Barclays had plans to turn the island into their own version of Monaco, using it as their personal tax haven. But at the same time, there was the worry that if the Barclays' favoured candidates failed to get elected, the brothers would pull out of their commercial interests and cease any further investment in the island.

This is more or less what happened. Their preferred candidates did lose and in a fit of pique, the unwelcome pair even ripped up the small vineyard they had established.

The brothers were 86 when Sir David died in 2021 after a short illness. A year later, Sir Frederick was facing a multimillion-pound legal battle over the divorce settlement involving his wife of thirty-four years; and in 2023, the family was facing losing control of the *Telegraph* newspaper group.

Sark, meanwhile, remains much as it has been for centuries: otherworldly, with a beauty sometimes gentle and sometimes stark, undisturbed by the sights and sounds of the twenty-first century. There is a school, a medical centre with the island's sole GP (but no dentist), three pubs and a sub-post office.

Taking on debt was prohibited until 2021 when the island's government sanctioned the use of mortgages to buy property. And divorce was permitted from 2003. Tractors must not exceed the 10mph speed limit on public roads unless pulling the emergency ambulance or either of the two fire trucks. Dangerous cycling is a punishable offence, as is riding a bicycle without a bell.

Yet all of this was not enough to dissuade one newspaper columnist from describing Sark as the second-most boring place in Britain. Seemingly, boring can have its advantages. Sark was one of the very few places in the world to escape the coronavirus. It went into lockdown in March 2020 on orders from its Pandemic Emergency Committee, a hastily assembled body made up of the vicar, the island's safeguarding officer and a bunch of local parliamentarians. During the pandemic, home delivery for food was made twice a week by tractor, and the annual sheep-racing event was cancelled. The island resurfaced Covid-free, which must have been a blessing for the island's one doctor. The 400 islanders pay no income tax, no VAT and no stamp duty, but that leaves them with no welfare services and no access to the NHS. A ride on the Sark Shipping Company's *Sark Belle* is the route to Guernsey's hospitals.

Herm

In the same election year that brought democracy to Sark, the forty-year lease for Herm, another of the Channel Islands, became available. It was bought by a charitable trust managed by Guernsey resident John Singer for £15 million. John and his wife, Julia, manage the 1.5-mile-long island, which they must open to the public during daylight hours. It has a population of about sixty, a one-room junior school, one shop, a pub and a hotel.

Compton Mackenzie, the Scottish writer of comic novels – *Whisky Galore* and *Monarch of the Glen* are his best loved – held the lease in the 1920s. He and his first wife, Faith, lived in a bungalow surrounded by flowers and fruit trees. When he was not writing, he knocked croquet balls through their hoops, indulged his Siamese cats and made vocal his support for Edward VIII, later the Duke of Windsor.

Jethou

Neighbouring Jethou – all 44 acres of it – is more private. Having once been home to the abbots of Saint-Michel, it was leased by the Crown in the early

1990s to Sir Peter Ogden, the co-founder of Computacenter, one of Britain's largest-quoted IT companies.

The Isle of Wight

If the Channel Islands are, in Victor Hugo's words, a fragment of France in the Gulf of Saint-Malo, then the Isle of Wight is a fragment of Hampshire in the waters of the Solent. It too has faced occupation – not by the Germans but many centuries earlier, by the Romans. And then towards the end of the nineteenth century, it felt to many islanders as though it had been occupied by the court of Queen Victoria.

Brighton had gained in popularity with Victoria's subjects, but for the queen herself, its attraction as a respite from the demands of London and Windsor waned. And she felt uncomfortable about its association with her profligate uncle. So, she upped sticks and, rather than leaving the Pavilion to the people of Brighton, made them pay through the nose for it – after first stripping it of its fittings and furnishings.

But her affinity with the English Channel held fast, and her prime minister Robert Peel found for her an estate on the Isle of Wight. The Osborne Estate, to the east of Cowes and with its own private beach, already had a fine, three-storey mansion at its heart. But try as he might, Thomas Cubitt, the celebrity master builder responsible for much of the grandeur that is today's Belgravia, could not find a way of satisfying Victoria's aspirations without pulling the house down and starting from scratch.

Prince Albert supervised the layout of the gardens, creating terraces in the Italianate style and filling them with myrtle, with all its symbolic promise, magnolia, lime-green ipomoea, dark-leaved perilla and a host of evergreens. Together, Albert and Thomas Cubitt created a spectacular palace surrounded by lush gardens that Victoria was to use for fifty years. For her many children, she had a prefabricated Swiss chalet shipped in and erected as a playhouse. She extended one of the staff dormitories to house her Indian servants and commissioned a private chapel.

She used her bathing machine for summer dips in the chilly waters of the Solent but rarely ventured into waters deep enough to swim. For Victoria, a tentative cooling off in the shallows was all the amusement she needed from the waters of the English Channel.

When she wasn't dipping herself, she would enjoy watching her children paddling and making sandcastles, or she would sit at the top of her little beach content with her easel and watercolours. She loved her time at Osborne and, while Albert was alive, spent some of her happiest days here.

Albert twice visited the poet Alfred Lord Tennyson, who had a house at Freshwater on the west of the island, and it seems more than likely that Albert was largely responsible for Tennyson becoming Victoria's Poet Laureate, succeeding William Wordsworth. His most admired work was 'In Memoriam A.H.H.', the poet's elegy for his young Cambridge friend, Arthur Henry Hallam. Although Victoria herself never visited Tennyson, she did admire his work, and when Albert died, she copied out extracts from 'In Memoriam' into her journal which, she said, gave her great comfort.

Albert's visit to Tennyson in 1856 was preceded by one made by Giuseppe Garibaldi. It was two years previously that the Italian general and unifier had visited Tennyson while staying on the island with the industrialist and Liberal MP Charles Seely. Charles and his wife Mary had long been admirers of the Italian general and had written to him often in warm and encouraging tones. Believing Garibaldi would be continually mobbed if he spent his visit to England in London, Charles and Mary invited him to stay with them on the Isle of Wight at Brook House.

The Italian hero arrived at Cowes on Monday, 4 April 1864, to be greeted by cheering crowds waving banners and flags. J.S. White, the shipbuilders, gave their staff the afternoon off and many of them were among the crowds that escorted Garibaldi on horseback on the 15-mile journey to Brook House. *The Isle of Wight Observer* gushed, 'Cowes claims the honour of being the first spot in the Isle of Wight trod by the greatest man who ever set foot on our soil.' Which, while possibly true, was a little tactless, as Victoria's consort, Prince Albert, had already set foot on the island to work on Osborne House.

From the Seelys' house at Brook, Garibaldi travelled on horseback for the short ride to the Tennysons' house at Farringford, where crowds had gathered at the gate to catch a glimpse of the Italian general. Emily Tennyson was alerted to his arrival by the welcoming cries of the people in the village, but before she had a chance to welcome their guest properly, her husband had led him off to his study, where he advised Garibaldi not to talk politics while he was in England.

Neither spoke the language of the other and the meeting had its awkward moments until they started reciting poetry to each other. Garibaldi was still speaking Italian, but the poetry did not demand a verbal response: the fact that Tennyson did not understand a word seemed not to matter.

There was further awkwardness as Garibaldi was taking his leave. A woman, appearing wretchedly clothed and unwashed, went down on her knees in front of him at the front door, holding out her stained hands. The general took her to be a beggar. She was nothing of the kind. Julia Margaret Cameron was one of *the* portrait photographers of the Victorian era and happened to be a neighbour of the Tennysons. With their encouragement, she was beseeching Garibaldi to sit for her at her studio nearby. But mistaking her for a local tramp, albeit a well-spoken one, he brushed her aside.

Before leaving the Italian general planted a wellingtonia in the garden (he had planted an oak in the Seelys' garden at Brook House) and left for London in the company of Charles Seely. Queen Victoria, not best pleased to have her British workmen and their trades unions fired up by a revolutionary, wrote to Seely asking him to ensure Garibaldi left the country as soon as possible. Victoria would not have been amused to learn that a biscuit manufacturer in Bermondsey, managed by James Peek and George Frean, was creating a Garibaldi biscuit based on the kind of raisin bread the generalissimo had fed to his troops.

⚓

Looking across the water from Osborne House towards nineteenth-century Portsmouth, the view might have brought to Victoria's mind the plight in 1545 of a British fleet of sixty ships, caught between her vantage point and the mainland, confronted by some 225 French warships and 30,000 men. Henry VIII was watching the battle from dry land at Southsea as the French were driven back by the English in their man-powered galleys. Unwilling to accept defeat, the French ran riot through the north-east coast of the Isle of Wight, burning villages as they went.

They were mistaken in their belief that they had sunk one of England's finest ships. The *Mary Rose* had, in fact, capsized while effecting a change of direction and catching a strong gust of wind. She sank, taking more than 500 soldiers and sailors with her. When she was finally brought to the surface in 1982, seventy-eight skeletons were found in the wreckage, along with all manner of artefacts which reflected the way of life on board the sixteenth-century queen of the seas.

⚓

Victoria died at Osborne in 1901. Not only was it the end of an era, but also the temporary end of the Isle of Wight as the place to be seen during the summer season. The estate was left to the nation. Part of it became a convalescent home for officers and the rest was turned into a naval college. Neither exists today at Osborne. The main house is now in the care of English Heritage and open to the public.

The presence of the queen and her consort during the summer months of the mid-1850s drew ever more visitors to the Isle of Wight – not least those with yachts. West Cowes, which had been a quiet village and home to a handful of fair-weather sailors, was soon a centre for serious yachtsmen encouraged by the presence of royalty and their aristocratic, well-heeled friends with their well-trimmed yachts.

So it was that during the summer of 1815, in a London pub – the Thatched House Tavern behind a row of shops in St James's – a yacht club for gentlemen was launched. Forty-two yachtsmen who owned yachts of 10 tons or more agreed to meet for dinner twice a year: once in London and once in Cowes. They called their nascent group the Yacht Club and set it on a course to become the world's most exclusive club for sailors. When King George IV became a member it became the Royal Yacht Club, and his successor William IV, who had joined the navy at the age of 13 but was never thought capable of commanding a ship, elevated the club to the Royal Yacht Squadron.

Cowes and the Royal Yacht Squadron brought to the English Channel summers of wind-filled sails that perfectly complemented cloud-dusted blue skies and a sea ruffled by gentle, white-topped waves. The yachts were magnificent. They were gaff rigged in the early days with a host of sails catching every small breeze, two-masted with lengthy bowsprits to accommodate the several foresails, and their hulls, crafted in oak and larch, cutting so stylishly through the water. Blue shirts and white flannels were de rigueur for the crews, and for members of the Royal Yacht Squadron there was the privilege of flying the white ensign – the cross of St George with the union flag in the canton – now the ensign of the Royal Navy.

Yet the Isle of Wight, and Cowes in particular, were to enjoy still further attention from the best of the world's yachtsmen. The Prince Albert-inspired Great Exhibition of 1851 drew visitors to London's Hyde Park from far and wide. Among them was John Cox Stevens from New York. Stevens was the sports enthusiast of a prominent east-coast family of industrialists. He had served as the President of the Jockey Club of America and had introduced cricket to his sport-loving countrymen. He also built yachts, and while in London for the Great Exhibition challenged the Royal Yacht Squadron to a race in the English Channel around the Isle of Wight. The Squadron's committee readily agreed, and Stevens, by then the commodore of his own yacht club in New York, put together a crew under an experienced deep-sea pilot called William Brown.

Stevens waved his schooner farewell from the New Jersey shore of the Hudson River in the summer of 1851. *America* and her thirteen-man crew sped across the Atlantic, at times clocking up more than 200 miles in a day. They arrived at Le Havre in high spirits after the twenty-day crossing, repainted *America*'s black 100ft hull, restepped her two masts and prepared her racing sails. And with confident anticipation, muted a little by a clinging fog that had rolled up the Channel, crossed to the Isle of Wight.

When *America* arrived off Cowes, Britain's finest yachtsmen were in awe. One hundred feet in length and with over 5,000 square feet of machine-woven cotton sails, this schooner from New York was something special. For days, the competitive courage of the host yachtsmen dissolved into feeble excuses to cancel the idea of a race. Until *The Times* thundered that it was unimaginable that the English 'would allow an illustrious stranger [Stevens] to boast he has flung down the gauntlet to England and had been unable to find a taker'.

The paper went on to compare English yachtsmen with a flock of pigeons paralysed by fear at the sight of a hawk. This was all too much for Robert Stephenson, the son of George, the railways man, who offered to take up the challenge in his yacht *Titania*, only to be upstaged by the Royal Yacht Squadron (of which Robert was himself a member). The Squadron decided that eight cutters and seven schooners would accept the challenge and race *America* around the Isle of Wight, covering the 53 nautical miles in a clockwise direction.

The wind was from the south-west on 22 August, and with the help of a strong tidal stream the yachts got off to a good start from Cowes. *America*, with Richard Brown taking the tiller under his arm in the cockpit, inched into the lead at about eleven o'clock as the boats passed Ryde. They headed south, with *America* taking a passage between the Nab Lightship and Bembridge. This raised eyebrows as some of her challengers had taken the longer route around the lightship, although no instruction to do so had apparently reached *America*'s crew. At Sandown, *Wildfire* drew level but was disqualified for using moveable ballast to enhance her performance. *America* kept her nerve tacking down the coast. Short of Ventnor, the cutter *Alarm* went aground and *Arrow* went to her aid, effectively taking them both out of the race.

More red faces were evident when the cutter *Freak* misjudged her tack, narrowly avoiding ramming *Volante* but managing to snap off her bowsprit to which her foresail was attached. Off St Catherine's Point, at the southern tip of the island, *America* was well in the lead. But as the yachts turned north up the west coast of the island the wind dropped.

Aurora, the small cutter owned by Thomas Le Marchant, a naval officer from Guernsey, was now the Squadron's last hope. It was a mile behind the American yacht, whose crew promptly goose-winged her two main sails to catch every breath of the south-westerly breeze. Yet *Aurora* managed to make up some lost water.

But it was not enough, and at 8.37 p.m., with Queen Victoria and Prince Albert watching from the Royal Yacht, and the last vestiges of summer light illuminating the Cowes waterfront, *America* crossed the winning line in front of Cowes Castle a full eight minutes ahead of *Aurora* to win the first America's

Cup. Commodore Stevens proudly accepted the 12oz silver ewer, which he took home with him to New York, where he left it in safekeeping with the city's yacht club.

Having begun in 1851, the America's Cup is still yachting's premier competition, although today's yachts have come a very long way since the cutters and schooners of the 1850s. A British yacht has yet to win an America's Cup, and the race has never returned to the Solent. In the spring of 1942, snowstorms lashed the eastern seaboard of the United States. On the waterfront at Annapolis the weight of snow brought the roof of a nondescript boathouse crashing down on the hull of the yacht inside. That hull belonged to *America*, once invincible, now a beaten wreck.

Although a British yacht has yet to win the America's Cup, a British yachtsman has captained three winning yachts – all American-owned. Charlie Barr was a diminutive Scot, born in 1864 just outside Glasgow into a family of successful and respected yachtsmen. His mother, fearing she would lose another son to a life at sea, found him a job as a clerk at the local greengrocers. But the sea was in his blood and young Charlie, all 5ft of him, ran away, finding a berth on a coaster working out of Greenock. He may have been small, but Charlie Barr was tough, agile and fearless.

His elder brother knew his worth and took him with him to deliver a new yacht to customers in America, where Charlie honed his yacht-racing skills – which did not escape the notice of the wealthy east coast yachting fraternity. With Charlie at the helm, American boats won the America's Cup for three consecutive years from 1899.

⚓

Medicine has come a very long way since the end of the nineteenth century when Alfred Beken ran a small pharmacy in Canterbury, in the shadow of the city's great Anglican cathedral. As much as he enjoyed serving the people of Canterbury, Alfred Beken was getting restless and decided to move further south, eventually finding himself on the Isle of Wight, where he bought an existing chemist's shop in Cowes, supplying, among others, the royal household at Osborne. There were rooms above the shop for the family, and Alfred's son Frank had a bedroom with a view out over the Solent.

Frank was captivated by the never-ending sight of ships, yachts and small boats framed in his bedroom window. Photography was still in its infancy and the cameras of the day were not suitable for marine photography. But young Frank decided this yachting spectacle had to be recorded on film – and he was the man to do it.

Frank designed and built his own camera. What evolved was a heavy mahogany box – or two boxes, one on top of the other. The one on the top housed the viewfinder and the larger one underneath housed the lens, the shutter and 8in x 6in glass plates. Connected to the shutter was a cable with, on the other end, a small rubber balloon which Frank kept in his mouth. This left his hands free to steady the weighty camera. When he had composed his picture, he would bite the balloon, which sent air down the cable to fire the shutter and make the exposure.

Marine photography became Frank's life, and he soon won the respect of the sailing fraternity, taking to sea attired in a sailor's cap and bow tie. Keeping himself steady in his small launch was a challenge he largely overcame by training his body to behave like a gimbal – staying upright when the boat rocked from one side to the other. The result of all this was beautiful black and white pictures of yachts at their very best, their sails full and their hulls creaming through the water with their topsides dipping close to the waves.

Frank printed his own pictures and sold them through the pharmacy. The shop conveniently had two front windows: the one on the left of the door was filled with pills, perfumes and potions while the one on the right showcased prints and photographic materials. 'Beken and Son', it said above the door, 'Chemists and Photography Depot'. And above the family name was a large wooden version of Queen Victoria's royal warrant announcing the Bekens were, by appointment, chemists to Her Majesty.

The son referred to on the shop sign was Frank's son, Keith, born at the beginning of the Great War, who had taken himself off to the mainland to study pharmacy in London and returned, qualified, to help his father in the shop. But his father enthusiastically passed on his photographic skills and son Keith was soon spending more time taking pictures out on the water than standing behind the pharmacy counter. His pictures of *Endeavour*, *Britannia* and *Shamrock V* – yachts owned respectively by Sir Thomas Sopwith, father of the racing driver, King George V and Sir Thomas Lipton, the tea magnate – became standards for classic yacht photography. And having photographed all three of the late Duke of Edinburgh's yachts and towed the duke and boat designer Uffa Fox to safety after their racing dinghy capsized throwing both men in the water, Prince Philip awarded Beken his royal warrant.

Keith's son, Kenneth, joined the photography side of the business and brought new technology and colour with him. His launch, with a watertight box for the cameras, used to weave at speed around the yachts, always careful not to impede their progress. But in his seventies, and with camera phones democratising photography, Kenneth decided to give up the chase to work instead on the family's photographic archive of glorious black and white pictures that imprinted the Beken name in the annals of world yachting.

⚓

While the Isle of Wight drew yachtsmen from clubs around the world into the English Channel, the island had also long been a meeting place for villains. Robben Island, in Table Bay off the coast of South Africa, Alcatraz, in San Francisco Bay, and Rikers, in New York's East River between Queens and the Bronx, all had water beyond their prison walls to deter potential escapees. Parkhurst Prison, just outside Newport on the Isle of Wight, enjoyed the same safety moat in the form of the English Channel.

Parkhurst was first used as a prison in 1838. Its inmates were all young boys. Hundreds of these young offenders went from here to penal colonies in Australia and New Zealand. In 1845, Queen Victoria paid the prison a visit during her summer sojourn at Osborne House. She recorded the occasion in her journal:

> We visited the dining hall in which all the poor boys were ranged in rows, who sang 'God Save the Queen' [...] We saw the cells where the boys are kept in solitary confinement – very lonely without any lookout [...] We afterwards saw them at school [...] They receive a most admirable education and

we saw them at work tailoring. They make all their own clothes. We were all struck by their being the plainest set of boys we had ever seen – really frightful and we were told that this is the case with all children of a criminal class [...] I asked that the most deserving boy in each ward should be pardoned. It was a most interesting experience.

Two hundred years earlier, Charles I experienced prison on the Isle of Wight but, by all accounts, found it anything but interesting. He spent 1648 incarcerated in Carisbrooke Castle, the home of the island's governor, Colonel Robert Hammond.

Charles had been under house arrest at Hampton Court having lost a long and bitter civil war against Oliver Cromwell and his Parliamentarians. When news reached him that he was to be tried for treason and, if found guilty, executed, he decided to make a run for it. Under cover of darkness, and with the aid of loyal and trusted friends, the king evaded his guards, slipped down Hampton Court's back stairs and out into the courtyard.

Horses and escorts were waiting for him, and with little more than the clothes on his back, Charles rode off into the night, heading, he believed, for the south coast. But the denounced monarch had not only lost the support of his countrymen, he had also lost his sense of direction. He galloped through the darkness, leading his hapless band of brethren round in circles, driving the horses until they foamed, exhausted, at the mouth.

Things improved with the coming of daylight, and with fresh horses, a more assured course was set for Southampton and the Solent.

Messengers were sent to the governor of the Isle of Wight asking that the king be allowed to reside on the island while he negotiated with Parliament. Governor Hammond agreed and crossed the Solent to greet him. At first, the governor was friendly enough, providing Charles with comfortable quarters in Carisbrooke Castle. But when the king rejected the proposals for power-sharing from Parliament, he lost the support of Colonel Hammond and found himself once again under house arrest with guards sleeping outside the doors to his room.

However, in the young Henry Firebrace, Charles still had one loyal and determined servant intent on securing the king's liberty. Or so he thought. But

Firebrace was a double agent. He was at heart a monarchist and as page of the royal bedchamber became close to the king. Yet, close as he was, he was passing intelligence to the Parliamentarians.

On this occasion, however, he was bent on setting Charles free and hatched a plan that involved the king shinnying down the castle's wall on a rope. But first Charles had to squeeze through the window of his apartment. It was either by virtue of blind desperation or by simple miscalculation that halfway through the window His Royal Majesty got stuck. He just managed to wriggle himself back into the room, and after considering plan B, which involved laboriously filing through the window's bars, he decided to give up on his escape attempt.

Charles was eventually transferred to London and executed for treason there in front of the Banqueting House in Whitehall on a cold January Tuesday in 1649.

⚓

Towards the end of the twentieth century, Parkhurst, less than 3 miles from Carisbrooke, was home to many of the most dangerous criminals in England. Moors murderer Ian Brady, the Yorkshire Ripper Peter Sutcliffe, the south London gangsters Charlie and Eddie Richardson, and the notorious Kray twins all served time here. However, the prison was downgraded after a daring and spectacular escape, which was foiled eventually by the waters of the Channel.

It was an escape long in the planning. Matthew Williams was 19 when he was arrested. He had been robbed when he was a student and decided to exact revenge by placing a homemade bomb under the park bench where he was mugged. The bomb was found and defused, and police found the bomb-making equipment in Williams' digs. Convicted of the bomb making and other offences, he very nearly escaped from the prison van by stabbing himself with a hypodermic needle he had concealed in his sock and lying to the terrified guards that he had AIDS. The prison van raced to the nearest police station, where Williams was overcome, patched up, put back in the van and taken to Parkhurst.

He wasted no time in looking for ways to get out. He met up with a bearded Keith Rose, who had been convicted of the kidnap and murder of the wife of a supermarket owner by shooting her in the back. Together, they plotted their escape. Williams had noticed the locks on the doors and gates were colour-coded, some green but the majority red, and assumed correctly that one key fitted all the red locks and a second fitted the green.

He spent hours looking at the red keyhole in the door to the music room, and when a prison officer left a key exposed while dozing, he impressed on his mind a picture of the key's structure. He took these mind pictures into the metalworking shop and, under the negligent eye of the shop's single prison guard, began to fashion a key for the red locks. The key worked on the music-room door but stuck as Williams tried to extract it. He struggled, terrified of being caught, and finally pulled it clear.

After more refining touches in the workshop and in his cell, he tried again, and this time it worked perfectly. And in more than the one red-coded lock. His early theories about all the red locks using the same key were proving founded.

But there were still cameras, guards, a 20ft steel fence topped with razor wire and a 24ft concrete wall capped with a grapple-proof hood. And then there was the Channel.

Rose, who had a private pilot's licence, remained undeterred. They would fly to freedom. But Williams nearly spoiled everything by being banned from the workshop. He and Rose went on a recruiting drive and in Andrew Roger, who had beaten a swimming pool attendant to death with a crowbar, they found a willing partner with access to the workshops and useful experience in metalworking. Roger and Rose spent three weeks making a collapsible ladder out of central heating pipes. This they hid around the workshop in its constituent parts.

All three men were now showing signs of excitement and nervous exhaustion which they did everything to hide. The exercise gym was to be their first line of escape. It was a cold evening in January 1995, and the gym was unusually busy. This was the trio's opportunity. A nod from Williams set the escape rolling. Hidden behind the throng of exercising prisoners, he moved to the gym's back door and inserted his key in the lock and held his breath. The door opened.

Hearts pounding, but unrushed and feigning nonchalance, the three slipped out unnoticed, locking the door behind them. The gate behind the gym opened easily, as did the next one. Now they were in the workshop assembling the ladder. They worked in silence, collecting the pieces one by one from their hiding places. The pieces fitted perfectly and with the ladder assembled, the tension eased.

One of the trio grabbed a set of heavy-duty pliers which were lying conveniently on the worktop and silently they made their way with ladder and pliers across the yard to the fence, expecting searchlights and dogs and sirens. But there was nothing.

Piece by piece, the pliers cut through the wire fence and still the alarm remained silent, and in minutes they were at the foot of the perimeter wall. The ladder was leant gently and silently against the wall and Williams went up with a bundle of electric cable over his shoulder. Still no searchlights and no dogs. Williams tied one end of the cable to the ladder and let himself down on the other side. The two others followed. They were out.

Walking to the airfield would be time-consuming. And they knew the alarm would be raised any minute – if it had not happened already. A hurried, impatient and not altogether congenial discussion just yards from a main road concluded that three lads needing a ride on a cold night presented nothing out of the ordinary. So, they hailed a taxi and paid for the ride with money they had smuggled into the prison.

At the airfield they found two light aircraft parked on the grass. The first was without a battery, but the second showed more promise. Rose was quickly in the cockpit checking the controls, but when he switched it on nothing happened. He tried again, and still nothing. Williams and Roger were growing impatient; hearts were pounding and nerves were fraying. After several minutes Rose was forced to give up. It was not an aircraft he was familiar with, and with dissatisfied mutterings and tense, whispered recriminations, the airborne escape was abandoned.

The three fugitives were now cold, tired and hungry. Finding shelter in a disused garden shed, they settled down for the night, all the time expecting the dogs to sniff them out. Williams had brought his razor with him, and in the morning, Rose removed his beard, donned a cap and went shopping. In all, 250 police officers, dogs, a light aircraft and a helicopter found nothing and saw nothing. The inquiry into the escape was one of the most damning on prison security to have landed on the desk of a British Home Secretary.

For nearly a week, the three men lived in the garden shed undetected, until their resolve seemed to falter. It was that or a sense of being invincible after their week's freedom that prompted them to take to the main road.

It was the beginning of the end. An off-duty prison warder recognised Williams' distinctive gait and raised the alarm. A special constable was the first on the scene and apprehended Rose and Roger, but Williams ran off, little knowing he was running back towards the prison. When he got to the marina at Newport, he tried to steal a dinghy, but the ropes were tight and wet and his fingers were frozen.

And in no time, torches were approaching while overhead a helicopter was hovering with a searchlight and heat-seeking equipment. There was nothing for it but to get into the water.

It was cold, very cold. The helicopter had left but now there were torches at the marina's entrance. He could not risk moving. And the cold dug deeper. He was so cold he was losing his sight and hearing and could not feel his arms or legs. It was too much: he had to get out of the water.

As he did so, he was spotted by a police dog and its handler. Rather than give up, he chose the pain of the cold and slipped back into the water, but it was too late. Leaving his dog on the pontoon, the officer jumped in and pinned Williams to the side. The Channel had proved the final barrier to an escape that shamed the prison service and embarrassed the government of John Major. Parkhurst was downgraded and amalgamated with the island's two other prisons. Williams and Rose remain in prison: Roger was released in 2006.

⚓

On a farm at Godshill, ten minutes from the marina where Matthew Williams lost his freedom, a pop festival in 1968 was to set the trend for decades of festivals of music, mud, sex and drugs that became a rite of passage for teens and those in their early twenties the world over. It was still a year before the Woodstock Festival in America would attract a crowd of half a million, yet a small island in the English Channel was playing host to a pop festival that crowned the counter-culture movement of the sixties.

Organised by three brothers, local men in their twenties bent on raising funds for a municipal swimming pool, this first festival created a following that grew into hundreds of thousands by the following year.

The headline act in 1969 was to be Bob Dylan, deserting his home-grown Woodstock Festival to give his first live performance since his near-fatal motorcycle accident. This was a gathering of free-loving hippies, drawn to the island by their common interest in peace, pot and protest. They came as pilgrims with their guitars, tents and lengths of plastic sheeting and waited long into the night for their hero to take the stage. He had told a press conference earlier in the week that he had always wanted to perform on the island because it had been the home of the Victorian Poet Laureate, Alfred Lord Tennyson.

At 11 p.m., he made his entrance. Dressed in a white suit, white tie, white shoes and a yellow shirt, he took centre stage, his head just above the bank of microphones. There were screams from the crowd as it surged towards the stage and Dylan began his act with 'She Belongs to Me'. (Who 'she' was remains the subject of dispute among fans and music journalists.) He was on stage for just an hour, which seemed like short change for some members of the festival crowd.

The plans for a municipal swimming pool came to nothing, and the size of the crowd, which turned acres of the island into a rubbish tip, became an issue for the island's residents and their politicians. Nudity on the island's beaches further offended local sensibilities.

In spite of efforts to derail it, the 1970 festival went ahead the following year, with Hampshire's Chief Constable Douglas Osmond dressing up as a hippie and spending the day incognito among the crowd to satisfy himself that fears about public disorder were unfounded or at least exaggerated. He remarked afterwards that he saw more violence during a typical football weekend. Nevertheless, it was more than thirty years before the Isle of Wight hosted its next festival, by which time Glastonbury had stolen the show.

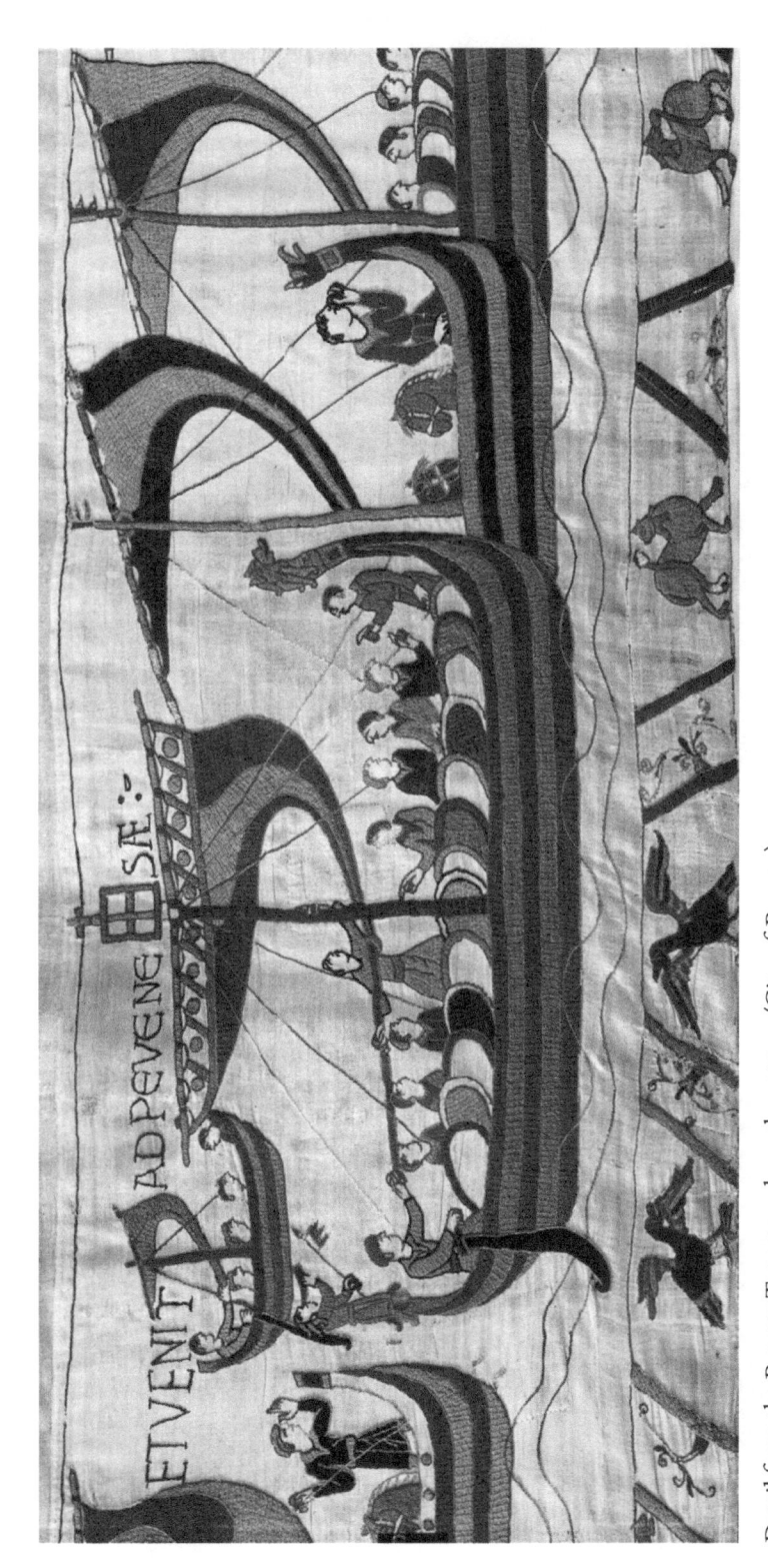

Detail from the Bayeux Tapestry, eleventh century. (City of Bayeux)

Frank Beken and his camera with its oral shutter system, 1924. (Courtesy Beken of Cowes)

At Dover, a car is slung onto a cross-Channel ferry, 1938. (Roger Hollingsworth/Alamy Stock Photo)

A First World War mobile pigeon loft.

Homing pigeons proved to be vital cross-Channel message carriers during both world wars. (IWM Collections)

Victor Hugo with his family at Hauteville House on Guernsey.

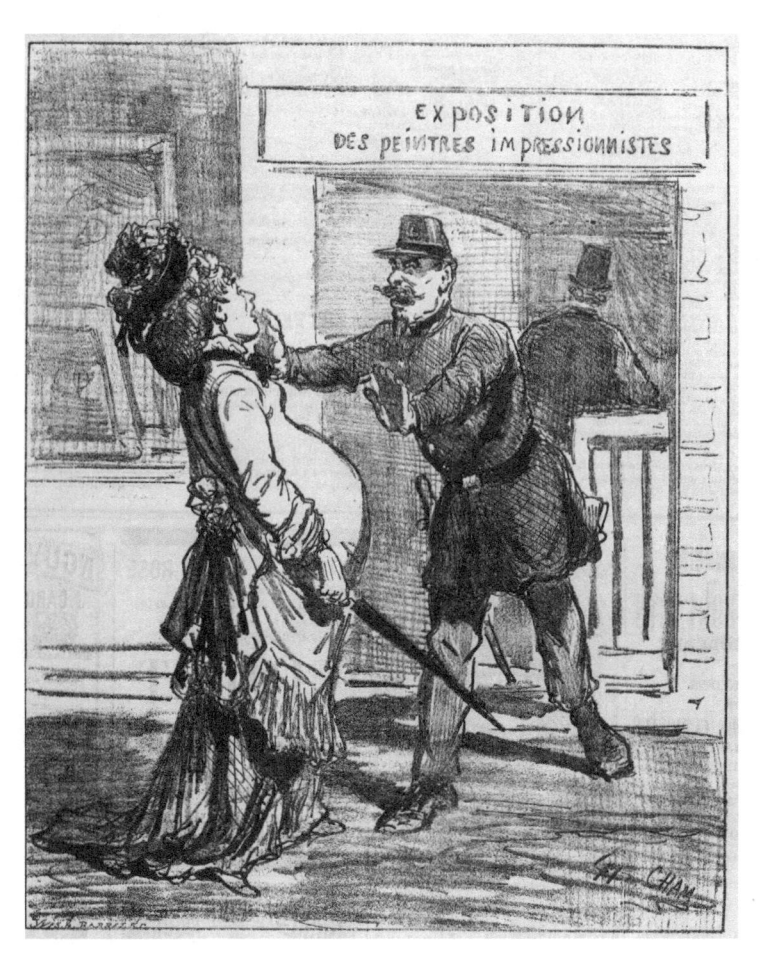

A cartoon in the French satirical magazine *Le Charivari* warns a pregnant woman not to view an exhibition of impressionist art. (Heritage Image Partnership Ltd/Alamy Stock Photo)

A Manet painting of Monet working on his floating studio ('Die Barke', 1874). (Bayerische Staatsgemäldesammlungen – Neue Pinakothek München. CC BY-SA 4.0 DEED, Attribution-ShareAlike 4.0 International)

Coco Chanel (right) outside her Deauville boutique.

Royal Marines Corporal George Tandy DSM. (IWM Collections)

A swordsman recruited by Henry VIII in Calais prepares to behead Anne Boleyn.

Cosette, a character in Victor Hugo's *Les Misérables*, depicted in the book's first edition by Émile Bayard.

An impression by an unknown artist of Louis Blanchard's balloon flight over the Channel in 1785.

The schooner *America* in the Channel mist, 1851.

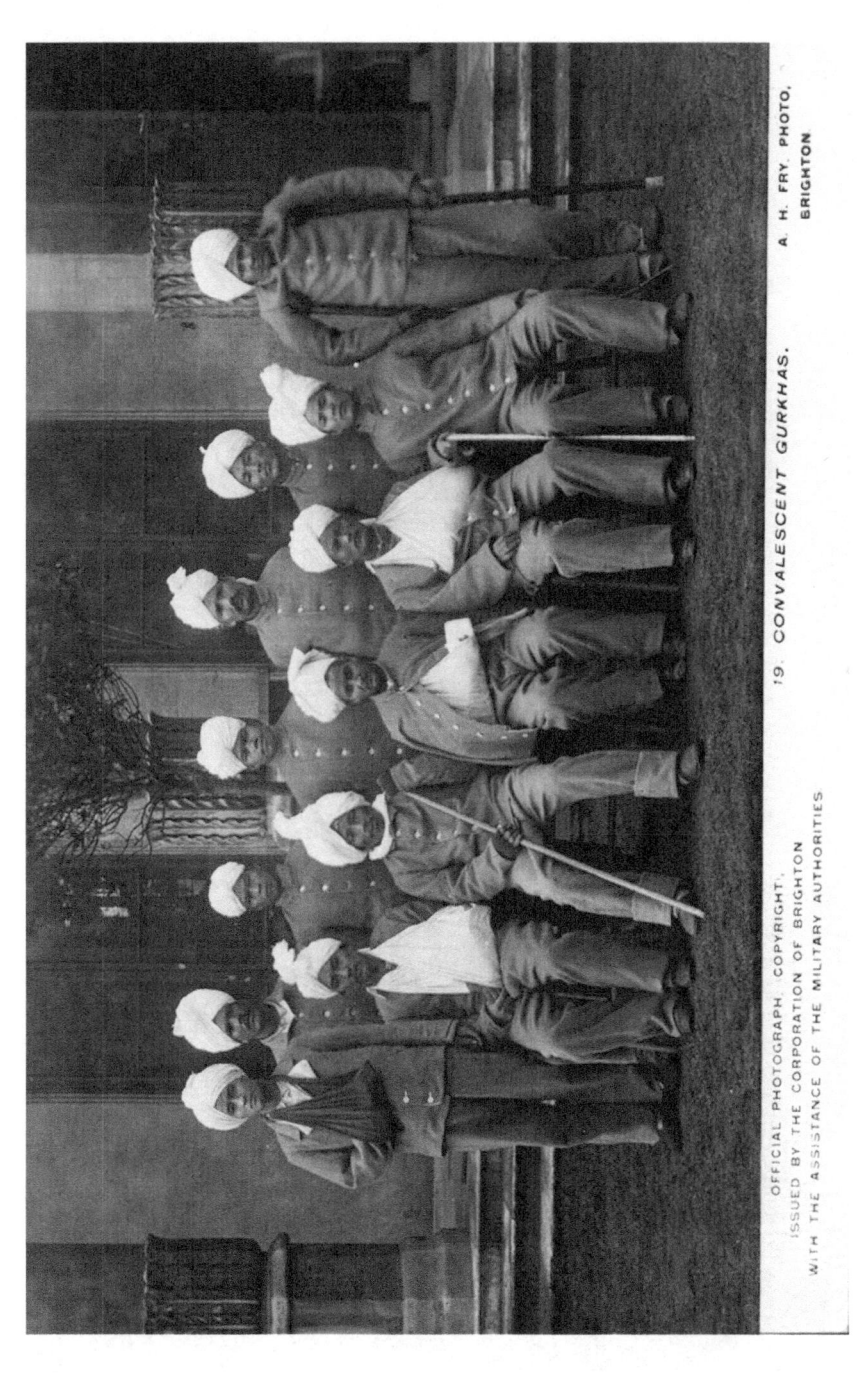

Wounded Indian troops outside the Brighton Pavilion, where they were being treated after returning from the Western Front during the First World War.

Henry Winstanley's Eddystone Lighthouse.

Louis Blériot after crash-landing beneath Dover Castle, having been the first to fly a plane across the Channel in 1909. (Library of Congress)

German occupation forces on Guernsey during the Second World War. (IWM Collections)

General Garibaldi visiting Alfred Lord Tennyson on the Isle of Wight, 1864. (Chronicle/Alamy Stock Photo)

Auguste Rodin's 'Burghers of Calais' presented to the city in 1889. (Gift of Iris and B. Gerald Cantor, 1989)

Beach photographer, Deauville *c.* 1900. (Brandstaetter Images/Hulton Archive via Getty Images)

Napoleon with his aides on the deck of HMS *Bellerophon* in Plymouth Sound, 1815.

CAPT. MATTHEW WEBB.
SWAM FROM DOVER, ENG. TO CALAIS, FRANCE
IN 21 H - 45 MIN.

Captain Matthew Webb, the first person to swim the Channel, 1875. (The Jefferson R. Burdick Collection)

The yacht *White Heather*, photographed by Frank Beken in the Channel in 1924. (Courtesy Beken of Cowes)

Waiting for the fishing boats to come in at the dockside in Boulogne.

6

The Channel at War

There were more lives lost and more boats destroyed in the Channel during the ten years of the two world wars than during the whole lifetime of this seaway. But there were, in all the mayhem, acts of heroism, courage and sacrifice, along with some examples of extraordinary maritime ingenuity.

It is a mistake to think of the Channel as a moat. Moats have drawbridges. When they are up, the moat presents a formidable barrier to an unfriendly foe. When they are down in times of peace, they facilitate the passage of people and goods across the water-filled ditch. Without a drawbridge, the Channel as a moat has put the population of England at a considerable disadvantage when it comes to travel and trade beyond its borders. Nor was the Channel wholly successful as a defensive moat; the Romans and the Normans made relatively light work of their cross-Channel invasions.

Come the First World War, with the enemy failing to reach the Channel coast of France, an invasion was never likely. But between August and September in 1914, and without the drawbridge, 176,000 officers and men, 1,389 nurses and civilians and 51,000 horses had to be shipped across the Channel from Southampton. With 2 million men on the Western Front, supplies of armaments, ammunition, food, tents and blankets, hospital equipment, trucks and more had to be shipped across. During the Battle of the Somme,

118,496 casualties were carried in uncomfortable converted hospital ships back to Southampton and Dover.

As a defensive seaway, the Channel did have a role to play, but not as a moat. The Dover Barrage – light steel nets anchored to the seabed with mines attached at various levels – ultimately proved successful in denying German U-boats passage through the Dover Straits into the Channel and on to the Atlantic. In 1916 and again in 1917, German torpedo boats attempted to break through the barrage but were fought off by British destroyers. But there was otherwise little action in the Channel during the years of the first war.

While the fighting did not reach the Channel, the bloodied results of the fighting did. Wounded soldiers from the Western Front arrived at the Channel ports in their thousands. A complex of three military hospitals swung into operation in Brighton to care, in the beginning, for the injured Indian soldiers (British soldiers were still being recruited and trained when war broke out, necessitating soldiers from the Empire being rushed to Europe). By the end of 1914, a full third of the British Expeditionary Force was made up of Indian troops. Many died and thousands of them were injured.

Trains with carriages converted into mobile operating theatres in which army surgeons struggled to steady their instruments against the lurching and swaying carried the injured to Boulogne and Le Havre. Here, they were transferred to hastily prepared hospital ships for the journey across the Channel.

Brighton's Royal Pavilion, the former summer residence of the royal family, was transformed into a hospital. The Great Kitchen became one of two operating theatres and 600 beds were installed in the State Rooms. X-ray equipment arrived in army trucks.

Muslims and Hindus were provided with their own water supplies, and nine separate kitchens were spread around the Pavilion grounds so that food could be prepared according to the different religious and caste traditions. An area of the eastern lawns was designated for Muslim prayers, while a tented gurdwara was erected for Sikhs.

Nor was the Pavilion the only Brighton landmark to become a hospital. The old workhouse in Elm Grove became the Kitchener General Indian Hospital and is today's Brighton General. York Place School, 38 Adelaide Crescent,

Hove Dispensary, Howard House in Sussex Square and No. 6 Third Avenue all became hospitals.

More than 12,000 wounded Indian soldiers passed through Brighton's First World War hospitals. Seventy-four of them did not survive their wounds. The Hindus and Sikhs were cremated on the hillside above the village of Patcham where the Chattri memorial serves as a place of remembrance. The Muslims were buried at Woking, the site of the Shah Jahan Mosque.

Towards the end of 1915, the Indian Army was redeployed to the Middle East and the Indian hospitals in Brighton had closed by 1916. But the Pavilion hospital was not idle for long. Having discharged its last Indian patient in January, it reopened as a specialist hospital for British amputees in April.

In total, 6,000 soldiers, sailors and airmen were treated here, with prosthetic limbs fitted to those most likely to benefit. A workshop was established to offer training in a range of new skills, from engineering to cine photography. Stoolball (cricket meets rounders meets baseball), which originated in Sussex back in the fifteenth century, became popular with the semi-mobile patients. The less mobile produced a monthly magazine which they called the *Pavilion Blues*. The hospital closed in the summer of 1919.

As the Second World War loomed, the Channel was regarded by many as a welcome part of the island's defences. In the eyes of the German High Command, it was a seaway that had to be under their absolute control for an invasion to take place. For the British, if the Channel provided a natural defence, without a drawbridge it also constituted a mighty obstacle in getting the British Expeditionary Force (BEF) across the Channel and into France, just as it had in 1914.

Under the command of General Lord Gort, the first of 390,000 men left Southampton in a variety of troop ships on 9 September 1939. In time, they would be supported by 300 tanks, armoured vehicles, trucks, armaments, tents, ambulances, field kitchens and all the paraphernalia of an army preparing for battle away from home. By 7 October, five weeks after the declaration of war, the Channel had seen colossal movements of men, 24,000 vehicles and 140,000 tons of supplies, all of which reached France without incident.

Mine barrages were laid across the Dover Straits and proved successful in interrupting the passage of U-boats between the Channel and the North Sea. And a fleet of ships based at Dover, led by the 'A'-class destroyer HMS *Codrington*, began patrolling the Channel.

The BEF was not in France for long. The Germans avoided the Maginot Line – the defences built by the French along their border with Germany after the First World War – and invaded France through the Low Countries. The British were unprepared and outplayed, and Gort decided to withdraw to the coast. Within nine months of making the journey across the Channel, the British force, with a sizeable French contingent, was on its way back.

Evacuating the British and French troops from Dunkirk – Operation Dynamo – was an exercise that required not just naval vessels but sailing barges, a London fireboat, nineteen lifeboats, paddle steamers, 160 fishing boats, cockle boats, Belgian mail packets, Thames cruisers, fifty steam-powered tugs and a Trinity House vessel called *Patricia*, which is now a restaurant in Stockholm. Dynamo was planned by Rear Admiral Bertram Ramsay, who was pulled out of retirement and worked in a tunnel underneath Dover Castle that once housed a giant dynamo. In all, some 700 little ships came to the aid of the Royal Navy, evacuating 338,226 British and French troops across a very full 'moat' that was constantly under attack from the air and from the occupied French shores.

The British warships in the Channel had been in action for several days before the command to commence Operation Dynamo was given during the evening of 26 May. There had already been heavy losses, and the crews who had survived were exhausted. But under the command of Admiral Ramsay, thirty-eight destroyers and thirty-six minesweepers were soon providing the backbone to the evacuation. They were constantly under attack from enemy aircraft and submarines.

The evacuation of the BEF along with French troops was a desperate but largely successful enterprise, but it came at a cost. By 4 June, the nine days of the evacuation had cost the lives of thousands of seamen and the loss or damage of a huge number of ships. While the majority of General Gort's BEF was home and preparing for the next stage of the war, there were senior ranking officers who believed his withdrawal was premature, even though King Leopold of the Belgians had surrendered, leaving the British force in what Gort believed was an untenable position.

Winston Churchill, then prime minister, believed Gort, a highly decorated officer with a Victoria Cross to his name, was not best suited to the job of field commander, and on his return from France, Gort was moved to a non-combatant role.

In the Channel, the destroyer *Codrington* became the warship of choice for VIPs crossing from Dover to Boulogne. Built at the Swan Hunter yard on Tyneside in 1929, HMS *Codrington* was named after Admiral Sir Edward Codrington, who served with Nelson at Trafalgar. (Christopher Codrington, a forebear, who had used slaves on his sugar plantations in Antigua and Barbados to amass a family fortune, gave his name to Hawksmoor's fine library at All Souls College in Oxford. He left the college £10,000 on his death in 1710. The college dropped the Codrington name from its library in 2021 in response to campaigns aimed at discrediting those who built their fortunes on the backs of slaves.)

The A-class destroyer was a full 350ft in length, had a complement of 185 officers and men and was armed with anti-aircraft guns and torpedoes along with five standard 4.7in guns, the extra one positioned between the two funnels. HMS *Codrington*'s most distinguished guest was the king himself (George VI), who sailed with Captain Casper Swinley from Dover to Boulogne in December 1939 to visit the Expeditionary Force, and then back again six days later. The First Lord of the Admiralty, Winston Churchill, twice made the crossing on *Codrington* in the New Year, the second time with Prime Minister Neville Chamberlain as they made their way to Paris for a War Council meeting.

In early May, the destroyer was again on royal duties when, in the face of the German advance through northern Europe, it rescued Prince Bernhard and Princess Juliana with their children from IJmuiden in the Netherlands and delivered them safely to Harwich. George Creasy had succeeded Captain Swinley and it was Creasy himself who carried Juliana's infant Princess Beatrix ashore.

Just days later, *Codrington* was back in the Channel to assist in the repatriation of the BEF from Dunkirk. Major General Montgomery was in command of the expedition's 3rd Division and was next to put his trust in *Codrington*. Having reached Dunkirk with his chief of staff and a wounded aide-de-camp, he and a bevy of senior officers, including his commanding officer, General Alan Brooke (later Field Marshal Viscount Alanbrooke), were taken on board *Codrington*. The ship remained at anchor with Brooke and Monty on board for several hours while German aircraft continued their attack on the British ships loading the retreating force. Brooke wrote in his diary, 'I am not very partial to being bombed on land, but I have no wish ever again to be bombed at sea.'

Montgomery remained perfectly calm throughout, speculating on the Luftwaffe's tactics rather than the possibility of his imminent death. In truth, he was a little distracted and smarting over his upbraiding from Brooke over a circular he had sent out to his men on the prevention of venereal disease. Both the Catholic and Church of England chaplains found Montgomery's language obscene and would have had the offending circular withdrawn, but Brooke let it stand.

Codrington received no direct hits and eventually delivered its precious cargo, which included 700 junior ranks, to the relative safety of Dover Harbour. There, it did a quick turnaround and, at the head of four destroyers, set out to cross the Channel in the cover of darkness. Together, the ships had brought home 64,500 men by dawn on 2 June.

However, HMS *Codrington*'s days were numbered. She had made eight round trips across the Channel at speeds of up to 31 knots (36mph), bringing home a total of 5,500 men. She was damaged while rescuing troops at Saint-Valery-en-Caux but was quickly repaired. Back in Dover for a refit, she was moored alongside HMS *Sandhurst* in the submarine basin, which had become a dangerously confined space.

On 27 July, the Luftwaffe launched a concerted attack on Dover with Messerschmidt fighter-bombers, which caught the port's defences off guard. One bomb narrowly missed *Codrington* but fell close enough to cause a powerful shockwave through the water. It broke the destroyer's back. There were three casualties. The attack on Dover and the sinking of *Codrington* was kept quiet until the end of the war when what was left of the destroyer was still visible in the harbour.

The patrol of the Channel continued without *Codrington*, destroyers and armed trawlers providing escorts for the convoys of merchant ships making the Channel run. The merchantmen – there were more than 3,000 merchant ships involved at the beginning of hostilities – brought coal from the mines in the north and the Midlands into the Channel, forming convoys with additional ships loaded with coal from the Welsh fields. Their cargoes supplied the power stations of the south coast, which alone needed 40,000 tons every week. The dockyards at Plymouth, Portsmouth and Southampton, the railways and vital industrial sites also needed their share of coal.

Some of the freighters made a dash through the Dover Straits, into the Thames Estuary and up the river to the docks of the capital. Escort duties were not popular, especially among the men of the naval reserve as the convoys made easy targets, and the escort ships were limited in their ability to defend them.

Whenever they could, convoys travelled at night, under cover of darkness. With navigation lights extinguished, there were wrecks to be spotted and negotiated. Compass readings were unreliable, the Dover Straits were heavily mined, and dimly lit buoys were hard to find in poor visibility – and there was plenty of that. To make matters worse, foghorns mingled with signals announcing a ship's intention to turn to port, starboard or go astern. When the ships of a convoy unavoidably bunched together collisions were frequent.

Hitler, meanwhile, made his intentions very clear, even if he did misread British resolve:

Since England, despite her hopeless military situation, shows no sign of being ready to come to terms, I have decided to prepare a landing operation. But first the English air force must be so disabled in spirit and physically that it cannot deliver any significant attack on the German crossing.

He gave his planned invasion the code name *Seelowe* (Sea Lion) and ordered the Channel crossing be a surprise. The south coast would be attacked from Ramsgate, in the east, to beyond the Isle of Wight, in the west.

Winston Churchill, by now Britain's prime minister, far from being intimidated, responded by telling Parliament:

> I expect that the Battle of Britain is about to begin. Upon this battle depends the survival of Christian civilization. Upon it depends our own British life, and the long continuity of our institutions and our Empire. The whole fury and might of the enemy must very soon be turned on us. Hitler knows that he will have to break us in this Island or lose the war. If we can stand up to him, all Europe may be free and the life of the world may move forward into broad, sunlit uplands.
>
> But if we fail, then the whole world, including the United States, including all that we have known and cared for, will sink into the abyss of a new Dark Age made more sinister, and perhaps more protracted, by the lights of perverted science. Let us therefore brace ourselves to our duties, and so bear ourselves that, if the British Empire and its Commonwealth last for a thousand years, men will still say, this was their finest hour.

The Battle of Britain was an air battle fought in part over the Channel. When it commenced during the third week of June in 1940, the first German aircraft to be shot down had got as far as Chelmsford. The Heinkel aircraft, on a reconnaissance flight, was shot down by an anti-aircraft battery and spiralled nose-first into the Bishop of Chelmsford's garden. Three of the four-man crew died. They were buried two days later with full military honours at a service conducted by the bishop himself. The fourth airman was captured later and held as a prisoner of war.

There followed, in the air above the Channel convoys, the battle that would put an end to Hitler's intentions. British Spitfires and Hurricanes took on the German Messerschmitts, Dorniers and Heinkels – the men of RAF Fighter Command, under Air Chief Marshal Sir Hugh Dowding, were pitted against the men of Göring's Luftwaffe. They were brave, these men, young too. Many of them had only a few hours of training under their belts, although a group of the German pilots had got in some practice with the Condor Legion, bombing Guernica for Franco during the Spanish Civil War. They all showed great courage.

Convoys of both sides were attacked, and the dogfights in the air were frantic and intense. But skill, determination and courage were sometimes not enough and hundreds of airmen from both sides went down with their planes into a Channel that Hitler believed he could command. The Americans thought he could too, believing Hitler would reach London before the end of the summer. It was not to be. On one day alone, thirty-one German aircraft were shot down. The Battle of Britain raged from early July until the autumn.

On 15 September, Hermann Göring, the Luftwaffe's not entirely competent commander-in-chief (he was responsible for building up the Luftwaffe but was unsure how best to use it), gave his gathered airmen a pep talk. He was sending them on a last-ditch mission to destroy the RAF and give his Führer an undefended Channel for the invasion.

One hundred German bombers, escorted by some 400 fighters took off from airfields in Holland, Belgium and along the French coast, from Calais in the north to Samer in the south. A menacing roar filled the skies over the Channel as this airborne armada began its journey towards London.

Aircrews at bases in the south of England were scrambled, and Spitfires and Hurricanes raced down uneven grassy runways to climb into the path of the enemy. Fifty-six German planes were shot down at a cost to the RAF of half that number. Yet, the bombers that did get through caused terrible damage before the British fighters forced them to turn around and head back across the Channel. A heavy price had been paid, but the RAF had been neither morally nor physically destroyed as had been Hitler's intention. And the German plan to present Hitler with an undefended Channel to facilitate his invasion had failed.

The Channel had proved to be the barricade that politicians and military strategists had long believed it to be. And as the social scientists and philosophers had long contended, it was also a narrow stretch of water between two widely differing cultures – two nations who had fought each other and distrusted each other through centuries of chronic antagonism.

The distrust continued. Communications between the British and French armies had been strained and at times, tested to breaking point. When France signed an armistice with Germany towards the end of June 1940, the British politely asked the French if they could have their warships. The request was, not the slightest bit politely, turned down. The French ships then lying in British ports were commandeered anyway and, determined that the Germans should be prevented from taking over French ships moored in Algerian ports across the Mediterranean, British forces destroyed the lot, and then made sure that additional ships of the French fleet taking refuge in Alexandria were left powerless. France broke off diplomatic relations with their erstwhile ally across the Channel – a move many British politicians and senior military figures, including Air Chief Marshal Dowding, considered a blessing.

General de Gaulle – by then in London – was using the airwaves of the BBC to call for support in France for his Free French forces. The French Resistance grew, and Free French naval forces fought alongside the ships of the Royal Navy. But old antagonisms fuelled an atmosphere of distrust between the two nations which lingered long after the war was over and lingers still today.

As the war progressed, the Channel was to witness a particularly shameful and humiliating episode that lives uncomfortably in the histories of both the Royal Navy and the RAF. Thanks to the work of Alan Turing and his colleagues at Bletchley Park, the British government's top-secret code-breaking establishment, the Germans' Enigma code had been broken. This provided the British defence establishment with much-needed intelligence on the movement of enemy ships.

Two mighty German battlecruisers, *Scharnhorst* and its sister ship *Gneisenau*, had been in the French harbour of Brest for repairs. Both, it seemed, along with the heavy cruiser *Prinz Eugen*, were to be moved from Brest to the German North Sea port of Wilhelmshaven. Rather than taking the lengthy Atlantic route and being confronted by British warships in Scapa Flow, Hitler, who could never resist taking control of each and every operation, decreed that his ships take the short route through the English Channel, where, he believed, a confrontation with the RAF would be a great deal easier to handle than an engagement with the Royal Navy. Not everyone in the German High Command agreed. The Luftwaffe even refused to provide air cover at the outset and only relented at the last minute.

On 11 February 1942, *Scharnhorst*, under the command of Kurt-Caesar Hoffmann, *Gneisenau*, commanded by Captain Otto Fein, and *Prinz Eugen* all steamed out of Brest under cover of darkness. They were given one of the most powerful escorts ever assembled by the German navy. Twenty destroyers were deployed to ensure safe passage for the three giant men of war after minesweepers had cleared a passage through the Dover Straits. In overall charge was Vice Admiral Ciliax, who had been promoted after a period commanding *Scharnhorst*. The Germans named the operation Cerberus: for the British, it was simply the Channel Dash. And it proved a costly and humiliating disaster.

Although the intelligence had been clear and reconnaissance aircraft had spotted the German convoy off the French coast, confusion and blunders on the British side failed to provide a planned blockade in the Channel. The German battle fleet continued to move eastwards along the Channel, hugging the French coast and encountering very little opposition. A final effort to sink the German ships fell to a squadron of outdated, slow, torpedo-carrying Swordfish biplanes of the Fleet Air Arm. Stationed at Manston on the Kent coast, the Swordfish squadron was led by Lieutenant Commander Eugene Esmonde, a fearless pilot who had been decorated for his part in the sinking of the *Bismarck*.

But even Esmonde was reluctant to undertake the mission without adequate cover from RAF fighters. When Esmonde finally took to the air at the head of five additional Swordfish aircraft, it was snowing, visibility was poor and the fighter cover had still not arrived. He and his squadron circled over Ramsgate, waiting for the Spitfires to turn up.

When they finally appeared, they set off together to find the German ships which, by now, had passed beyond Dover. Try as they might and with heroic determination, the Spitfires were no match for the Luftwaffe fighters and the accompanying flak from the convoy's escort destroyers. It left the Swordfish squadron with what must have already seemed like a suicidal mission. The following is an account of the Swordfish attack from the 'History of Manston Airfield':

> The first section of three Swordfish, led by Esmonde presses on past the screen of the destroyers. Expecting the greatest danger of all to their ships, a suicide attack, the German ships spot aircraft attacking at sea level. When they are 2,000 yards away, every flak gun in the fleet opens up on the slow aircraft.
>
> In the rear of Esmonde's aircraft, P/O Clinton fires his machine gun at the diving Luftwaffe planes that had now joined the attack. Gold tracer shells and flak bursting all around them, cannon shells from the enemy aircraft rips large holes in the fuselage and wing fabric. One of the Spitfire pilots, F/L Michael Crombie is shocked to see P/O Clinton climb out of his cockpit and crawl along the back of the fuselage to the tail where he beats out flames on his aircraft with his hands.
>
> Once past the destroyers, the eleven-inch guns of the battleships open up, creating smoke and flame whilst throwing spray into the aircraft. One shell bursts in front of Esmonde's aircraft and destroys the lower port wing. The shuddering aircraft dips, but Esmonde keeps it flying, whilst being badly wounded from wounds in his head and back, aiming for the *Prinz Eugen*. By this time, P/O Clinton and the Observer Lt Williams are dead from the last attack by an Fw190. In a last desperate effort, he raises the Swordfish's nose and releases his torpedo just before receiving a direct hit which blows his Swordfish to pieces. Spotting the torpedo, the *Prinz Eugen* took evasive action.

The German convoy made it to Wilhelmshaven, and *Scharnhorst* went on to do enormous damage to Allied shipping before she was finally sunk in the Barents Sea on Boxing Day in 1943. Of the crew of 2,000, thirty-six survived.

As the threat of a German invasion receded, operations in the Channel were reduced to regular patrols and the continued protection of the convoys of merchant ships. At the beginning of 1944, Allied commanders began to consider the liberation of Europe and an end to Nazi aggression. Limited planning for the invasion of occupied France had begun soon after the BEF had been brought home from Dunkirk. By July of 1943, Lieutenant General Frederick Morgan, together with British and Canadian officers, had already created detailed plans for the invasion. They included options for some convincing deception tactics to draw the German army away from Normandy.

Meanwhile, 9 million tons of supplies and equipment from North America had reached Britain and the number of US troops on British soil had reached 1.4 million. There was, in addition, a substantial Canadian force, as well as soldiers and airmen from Australia, Belgium, Czechoslovakia, Holland, France, Greece, New Zealand, Norway, Rhodesia and Poland.

General Eisenhower and his planners of the D-Day landings knew a great deal more about the tides in the Channel than either Julius Caesar or William the Conqueror. But their mechanism for determining the tides did not appear suited to the twentieth century. Rather, it had the look of a Heath Robinson contraption. Back in the 1870s, and building on the work done by two French mathematicians, Pierre-Simon Laplace and Joseph Fourier, the Scottish physicist William Thomson (Lord Kelvin) designed a mechanical analogue computer of gears and wires and pulleys to predict tidal data. He had his machine built in London in 1872. The Americans followed quickly with a machine that became known as 'Old Brass Brains'. Both worked remarkably well in 1944.

Field Marshal Rommel believed, reasonably, that an Allied invasion of occupied France would likely be launched during a high tide when the Allied troops would have less beach to cover under fire. He littered the Normandy beaches with all manner of underwater obstacles, many of them mined. But Allied aerial reconnaissance soon spotted Rommel's beach defences and plans for the landings were revised accordingly. Landing at low tide was now necessary so that teams of engineers could demolish the obstacles laid by Rommel's men and create safe passages through which the troops could negotiate the beach. And the tide had to be on the rise so the landing craft, having deposited their men, could withdraw without the risk of running aground.

The vital tide predictions backing the D-Day landings were in the hands of Arthur Doodson, a Liverpool University maths graduate, and six young women at the Liverpool Tidal Institute (the rest of his team had been redeployed). Doodson used both the Kelvin machine and 'Old Brass Brains'. The low-water times were different at each of the five Normandy beaches designated for the landings. There was an hour's difference between Utah in the west and Sword to the east, with about 100km between them. This meant that landing times would need to be staggered in line with the tide predictions for each beach.

Under cover of darkness, teams of Allied personnel in small boats and midget submarines set about measuring the tidal flows at each of the five enemy beaches. With this input, Doodson predicted that General Eisenhower had a window of just three days in the summer of 1944 to launch Operation Overlord: 5–7 June. But when the time came, bad weather forced Eisenhower to postpone the landings by a day.

On the other side of the Channel, Rommel thought the weather would put off the Allied operation and, still believing the invasion would come at high tide, opted to return to Berlin to spend time with his wife Lucia on her birthday. Love and war, it transpired, make uneasy bed fellows.

Prior to that first week in June, US commanders decided a rehearsal for Overlord was necessary. Slapton Sands in Devon resembled the beaches of Normandy and was selected for a practice run by the American rear admiral who was to command the landings on Utah beach; 3,000 residents were moved out of the area, and the Dart Estuary soon filled with naval craft.

The rehearsal itself, code-named Operation Tiger, took place on the night of 28 April 1944. A string of landing craft had made its way from Plymouth and was crossing Lyme Bay, each with a full complement of soldiers, armaments and fuel. HMS *Azalea*, a Royal Naval corvette, was in attendance as escort. As they crossed the bay, the lookout on *Azalea* was horrified to spot a group of nine German E-boats – fast, powerful torpedo craft – on the prowl for prime targets. The American landing craft were easy prey.

As the German boats closed in, *Azalea*'s wireless officer attempted to warn the crews of the landing craft. But there had been misunderstandings about radio frequencies and *Azalea* failed to make contact.

Mayhem ensued. The E-boats struck with deadly accuracy, quickly sinking two of the landing craft and damaging a third. The fuel reserves in one of the craft blew up, turning the water into a sea of flames. Bodies floated grotesquely in the burning water. The GIs who survived the initial onslaught found they were wearing their life jackets the wrong way round. The heavy back packs flipped them over, holding their heads and chests under water until they drowned. Through another dreadful mistake, the men who did make it ashore died as the intended live covering fire from a cruiser out at sea picked them off as they scrambled up the beach. In all, 946 US servicemen died that night. Their commander, Rear Admiral Don Moon, remained in command for the Normandy Landings proper but died by suicide two months later.

Rear Admiral Moon, along with other senior Allied commanders, had seemingly learned little from an equally disastrous exercise which had cost the lives of Canadian rather than American personnel. In the summer of 1942, a raid on the German-held Channel port of Dieppe involving troops from the Toronto Royal Regiment, the Calgary Regiment, the Cameron Highlanders and the South Saskatchewan Regiment proved a fatal disaster.

The raid originated with Vice Admiral Lord Mountbatten's Combined Operations. There was mounting pressure from the Soviets for the Allies to open a second front in western Europe; RAF Fighter Command was lobbying for a raid on one of the French Channel ports to bring the Luftwaffe out into the open; and the Canadians, who had been in England for two years, were itching to go into battle. Six beaches in front of and either side of Dieppe were to be stormed in a surprise operation, code-named Jubilee, to wrest Dieppe from the Germans.

Under darkness in the early hours of 19 August, 6,000 officers and men, the majority of them Canadians, set off across the Channel in 237 ships and landing craft and headed for the Dieppe shoreline. They were to be joined by tanks and given cover by RAF fighters and bombers. But things began to go wrong from the start.

The first of the landing craft ran into a German coastal convoy, which destroyed the intended element of surprise. Then the main body of the attack force reached the French coast late, after sunrise, exposing the hapless landing craft to the German gun emplacements above the beach. The tanks were delayed too and then got stuck in the shingle of the beach.

What little intelligence had been gathered about German positions and their strength turned out to be inaccurate and misleading. Dieppe remained firmly under German control. Nearly 1,000 Allied soldiers, most of them Canadian, lost their lives, their torn bodies littering the beach as their wounded colleagues retreated through the water onto their landing craft. Just under 2,000 men were taken prisoner. Of the twenty-seven tanks that left England, none made it back. Some lessons were clearly learned from both the Slapton Sands fiasco and the poorly executed Dieppe raid, lessons that were put in place as the D-Day landings were planned. But those lessons had come at a terrible price.

Operation Overlord began in the small hours of 6 June, five weeks after the Slapton Sands fiasco. Some 18,000 paratroopers were dropped into France, while members of the French Resistance and Britain's Special Operations Executive were sabotaging German positions. The ships carrying the invasion force – the naval element of Overlord was given the name Neptune – steamed out of Plymouth, Portland, Portsmouth, Southampton and the Isle of Wight. The first of the landing craft were on the two most westerly beaches – Utah and Omaha – by 6.30 that same morning.

Engineers working in the shallow waters and on the beaches were able to clear pathways for the troops behind them. They did so with a heavy loss of life, but the brass tide-calculating machines had done their job, and the landings went as well as could be expected. And they were witness to many acts of extraordinary fortitude and courage. Royal Navy Commander Kenneth Edwards recalls these incidents in his book *Operation Neptune: The Normandy Landings* (Fonthill 2014).

The first involved a Welsh medical orderly called Emlyn Jones on board one of the landing craft. They were leaving the beach with wounded men when they were hit by a shell below the waterline. The sick bay where Jones was working on the wounded began to flood. Seven of the men in his care were badly wounded and were at risk of drowning. These seven, Jones moved to safety and continued to treat their wounds. But some wounds were beyond treatment, so in a partially flooded landing craft still under fire, Jones, who had no training beyond the needs of an orderly, coolly amputated three crushed legs.

On a landing craft carrying tanks, George Wells, an able seaman twice hit by shells, had a shattered finger cut off with a pair of scissors. He was cracking jokes while the scissors were taking their time to cut through the bone.

Commander Edwards also recorded the actions of George Tandy, a corporal in the Royal Marines. Tandy's landing craft, No. 786, of which he was the coxswain, was critically damaged as it was being lowered into the water from the mothership, putting its steering mechanism out of action. Tandy, aged 19, lowered himself over the stern of the landing craft, putting one foot on the rudder and the other on the guard rail, while clinging on to a cleat. He steered the boat like this with his foot for the 7 miles it took to reach the beach, finding a path in the choppy waters through the mines and all the other landing obstacles, including the so-called Czech Hedgehog, a collection of interlocking steel beams with sharp, pointed spikes which could penetrate the toughest hulls.

Tandy managed all this and once his troops were ashore, he steered his landing craft back to the mothership in the same way. The blunt-nosed craft was pitching in the waves and made slow progress, throwing Tandy up one minute and dropping him down the next. The journey back against the wind took two and three-quarter hours. The exhausted and bruised helmsman had to be hauled out of the water and spent the next two days recovering in the sick bay. He was awarded the Distinguished Service Medal.

The Battle of Normandy, which followed the seaborne assault, paved the way for the advance across northern Europe and into Germany. By the end of August, the Nazis were in retreat, desperately attempting a counter-offensive in the Ardennes. The Battle of the Bulge took place in heavy snow and freezing temperatures, lasting five weeks and costing the lives of 120,000 Germans. The Allies were on the road to Berlin but were to suffer another 75,000 casualties on the way.

The tide machines that played such a crucial part in Operation Overlord would have been of little use had the officers and men undertaking the landings failed to show extraordinary acts of selfless duty and courage like those recalled by Commander Edwards. Yet, not a single Victoria Cross was awarded to members of the navy involved in the Neptune landings. Medals for gallantry were, however, awarded to thirty-two pigeons. While radio silence protected Overlord, one of those pigeons, known affectionately as Gustav by his handlers, flew news of the first Normandy landing at Sword Beach back across the Channel to Portsmouth. The message was in a container attached to its leg.

Royal Blue, one of the royal pigeons from the loft at Sandringham, was assigned to the RAF. Pigeons flew on every bombing mission and in October 1940, Royal Blue was on a Bristol Blenheim that was forced to make an emergency landing in Nazi-occupied Holland. The crew survived the landing, but their radio was shattered. Royal Blue, unhurt, flew straight back to Sandringham, alerting the Signals Corps to the plane's position.

Pigeons were already part of the wider war effort by the time Overlord was under way. MI14 was the pigeon section of British Intelligence; their trainers being required to sign the Official Secrets Act. Trained carrier pigeons were parachuted into France, and both Britain's Special Operations Executive and the French Resistance used them to relay messages from occupied northern Europe.

Aware of these winged spies, German marksmen on the French coast succeeded in shooting some of them down, but the greater obstacle to the birds' success came from British-based peregrine falcons, whose huge eyes, long, curved claws and a devastating attack speed were too much for the smaller carrier pigeons. To remove this obstacle to the birds' clandestine missions, a falcon-culling operation along the English Channel coast was ordered by the War Ministry. Without the falcons, fewer pigeons were lost.

The wartime National Pigeon Service claimed that 90 per cent of their birds delivered their various messages. Even without the falcons, that claim seems unreasonably high, but there remains little doubt that these wisest of old birds played an important but largely unsung role in defeating Hitler's army.

Peacetime races across the Channel attract the pigeons of thousands of fanciers whose take-off sites in Normandy stretch from Brest in the west to the Cherbourg peninsula, Saint-Malo, Lamballe and Carentan and along the Channel coast. With a following wind, the birds sustain speeds of 60mph, but a head wind forces a lower flight path, especially over the water. What has puzzled zoologists is the birds' tendency to fly a zig-zag course rather than follow a straight line.

Researchers have attached tiny GPS devices to birds, confirming their ability to recognise landmarks. In the *Journal of Theoretical Biology* (2004) Tim Guilford and his team at Oxford University reported one tagged bird that regularly 'flies along the road to the first roundabout, takes the third exit, follows the dual carriageway to the next roundabout and then leaves the road to fly across country'. But scientists are still trying to understand how the pigeons navigate over water. One theory is that they have an in-built compass system which logs onto the earth's magnetic field – tiny particles of magnetite have been found in the birds' beaks. Other theories point to the birds' use of the sun's position. Meanwhile, the racing pigeon fanciers remain less concerned with the how and more concerned with their bird's ability to outfly the competition.

Towards the end of the war, the Channel claimed one more life – of the man who made some of the most popular music of the age, music that set people dancing. It was a loss of life in which the enemy played no part. Ten days before Christmas in 1944, a single-engine high-wing Norseman C64 of the US Army Air Force prepared for take-off from the joint RAF/US Airbase at Twinford Farm near Bedford. At the controls was a 22-year-old, inexperienced pilot by the name of John Morgan.

Flying conditions that day were far from ideal. There was fog and snow, and Mr Morgan was not accustomed to flying on instruments. Nevertheless, he had lodged his flight plan for Paris and had run through the usual pre-flight checks. And his youthful confidence was probably overriding professional caution, an enthusiasm buoyed no doubt by the thought of a night or two in Paris.

Morgan was at this stage unaware of the identities of his two passengers. The first to climb aboard was Norman Baessell, an Army Air Corps colonel. Behind him came Major Glenn Miller, the King of Swing, who had turned the great American songbook into the music that filled dance halls from the mid-west to middle England. Baessell had business in Paris, and Miller needed to prepare his band for a Christmas show planned for US troops in the liberated French capital.

Morgan started the engine and made final pre-flight checks. Over the crackling radio, the Twinford control tower, a little haltingly, one imagines, gave him permission to taxi to the start of the concrete runway. At 1.45 p.m., he opened the throttle. The noise in the cabin was deafening as the aircraft sped down the runway and lifted into the wintry skies to be lost in a whirl of snow. The band leader was never seen again. Baessell and the pilot John Morgan were lost with him.

Countless theories have been propagated to explain the aircraft's disappearance. The most likely of them being that the plane simply ditched in the Channel. It was very cold, and the fuel pipes could easily have frozen. The plane was of a light build and would likely have broken up on impact.

Miller had not told anyone of his plans to fly with Baessell and was not missed until a couple of days later. But with any evidence at the bottom of the Channel swept away from the probable flight path by currents, the mystery remains unsolved. Swing did not die with Glenn Miller, but it did go into serious decline. And with the arrival of bebop and rock 'n' roll, the decline of the Big Band became terminal.

The war in the Channel finally came to an end with the liberation of the Channel Islands in May 1945. At St Peter Port on Guernsey, the destroyer HMS *Faulknor* took on board as prisoner Vice Admiral Friedrich Hüffmeier, the commanding officer of the German occupation force, and, with him, General Wolf, who had been in charge on Jersey. The German commanders were taken to Plymouth.

HMS *Faulknor* was back in the Channel Islands a month later in June, escorting a Royal Naval cruiser with King George VI on board for his visit to the islands as their hungry and weary people set to work on their road to recovery and normality.

7

Packets, Balloons and *Mal de Mer*

Caged in glass at a museum in Dover is a remarkable relic. Clearly the larger part of a boat's wooden hull, it was found under the well-trodden streets of the Kent port. The wood is oak and the binding withies are yew. Marine historians date the boat to around 1500 BCE and they believe it carried people and goods across the Channel.

Much later, in the first century CE, fishermen carried troops, merchants, animals and cargo across the Dover Straits in small, open boats. When the pious Edward the Confessor was on the throne, between 1042 and 1066, the royal mail became an additional element of cross-Channel shipments. The mail was carried between Dover and Wissant, a small fishing and farming community west of Calais, from where it was distributed on horseback across northern Europe. Domesday Book of 1086 records the king's messengers paying two pence to cross with a horse in summer and three pence to make the journey in winter. (There is no record of horses being sick during the crossing, but it cannot have been a pleasant experience for the animal or its rider.)

By the thirteenth century, Dover was one of the Cinque Ports, along with Sandwich, Hythe, New Romney and Hastings. There was no Royal Navy at the time, and the five coastal towns were obliged to provide ships and crews for the monarch in return for several attractive privileges. The ports paid no taxes or tolls and were given a degree of self-government. They could take it upon themselves to punish breachers of the peace and in the most serious cases, execute prisoners.

By the sixteenth century, shipbuilders on England's south coast were building 40-ton sloop-rigged craft to make the Channel crossings, not least to carry letters, messages and state documents from London to Europe's kings and princes. These collections of papers were known as *pacquettes*, and the boats carrying them adopted the name, which quickly became shortened and anglicised to 'packets'. By the seventeenth century, these packet boats, operating under sail, made regular crossings between Dover and Wissant, Calais, Boulogne and Dieppe, constrained as they were by both wind and tide.

They would drop anchor short of the beach and hand the sacks of mail to local boatmen in small craft who completed the journey. Transferring the horses of the king's messengers was a tricky business and not infrequently led to the horses taking an unplanned swim. The bags slung across the horses' backs – once they had reached dry land – carried the letters and royal papers. These bags were known as *malles*, the word that entered the English language as 'mail'.

On top of coping with nature's challenges, the packet boats were faced with marauding Dutch privateers – a legalised form of piracy – and beset by Dunkirkers, a large group of pirates who had come originally from north Africa and worked their way up through France to the Channel coast. The packet boat owners could ask for protection from armed ships, and with this extra security were able to make uninterrupted crossings and decent profits. But these crossings were not without mishap, which on one occasion served to illustrate the ever-strained relationship between England and France.

In his *History of Dover Harbour*, Alec Hasenson relates this telling story:

The Packet Express sailed from Dover to Calais with mails, and their [sic] being a heavy sea, Mr Pascall, the [English] mate of the Union Packet, then in Calais, thinking the Express could not enter the port, came out in a small boat, rowed by seven Frenchmen, to exchange mails at sea. The small boat was upset, and the seven Frenchmen drowned. Pascall contrived to get on the bottom of the capsized boat, but because the French were all drowned, the French soldiers would not allow a boat to go out and rescue Pascall, who was drowned within hail of hundreds of spectators.

The first steam packet, *Rob Roy*, arrived at Dover in 1820. She was a splendid vessel. With 80 tons of solid oak, sails, a 33 horsepower engine and two paddle wheels, she was the pride of the Dumbarton shipyards. William Denny built her, and David Napier in Glasgow put one of his new steam engines in her. She first worked the Glasgow–Belfast and Glasgow–Dublin routes, but her speed and some rebuilding of the passenger quarters soon had her on the south coast, lined up for the prestigious Dover–Calais run. And no sooner had she completed a series of successful Channel crossings than she was snapped up by the French Postal Administration, renamed the *Henri Quatre* and set to work under the French flag on a tight schedule, to carry mail and passengers between the Channel ports.

Henri Quatre was not the first of the Channel packets. Sailing boats had been doing the job since the middle of the sixteenth century, but she was the first powered vessel to provide the service and therefore a great deal more reliable than her predecessors, which needed a good wind, blowing in the right direction.

For passengers, seasickness on these steam-powered ships remained a deterrent to a relaxed or even manageable crossing. Buckets were handed out to passengers as they climbed aboard. On a bad day, the stench of vomit in the airless cabin was cloying and inescapable, and men and women – ragged, weak and pale – would stagger off the boat with little choice but to climb aboard the stagecoach, which would further challenge their equilibrium on bumpy journeys to Paris or London.

While many travellers were in dread of the coach journey, the writer and essayist Thomas de Quincey was happily addicted to mail coaches, just as he was to opium. His drug addiction began when, as a runaway teenager, he roamed the mean streets of London. *Confessions of an English Opium Eater*, published in 1822, was followed in 1849 by *The English Mail Coach*, an unfettered celebration of the stagecoach. He never ceased to delight in using them – he started to ride them in his student days to travel to and from Oxford – relishing the speed and sense of adventure. And when the railways arrived, De Quincey remained loyal to his mail coaches. Literary critics have found symbolism and metaphor throughout *English Mail*, with some finding the speed of the coaches echoing the speed of social change.

The Channel crossing and subsequent coach ride was a journey Queen Victoria's future husband made on a blustery February day in 1840. The packet paddle steamer *Ariel*, skippered by one Luke (later *Sir*) Smithett, Commodore of the Admiralty Packet Service, had brought Prince Albert to Dover for his marriage to Queen Victoria. Albert had never enjoyed the best of health and suffered dreadfully from seasickness. The Channel crossing, followed by a bumpy ride to London, must have been unpleasant and stressful in the extreme.

He was, however, fit enough to take Victoria to be his wife just four days later in the Chapel Royal of St James's Palace. It poured with rain on the day but was nevertheless a grand occasion, marred perhaps by Florence Nightingale's biting comment that Albert's uniform of a British field marshal had clearly been 'borrowed' for the occasion.

Stuart Townsend was born into a family of shipping merchants as Victoria's reign was coming to an end. A former artillery officer, he was not at first much interested in the family business. He preferred organising motoring holidays around Europe for British holidaymakers who were keen to explore a continent recovering from the ravages of the Great War. What soon became clear to Townsend was the travellers' preference to drive their own cars, which had a steering wheel in the right place and a gear stick where gear sticks ought to be.

The Townsend family business was not (yet) in the business of operating ferries, but this did not prevent young Stuart from chartering a coastal collier, cleaning out its insides and preparing it to operate as a car ferry. Cars were craned from the dockside into the well of the boat, secured and craned out again on the other side. Fifteen cars made the first journey in 1928.

And demand took off. Within two years, Townsend Bros was operating a regular cross-Channel car ferry service using two converted naval ships: a minesweeper and a frigate.

Lifting cars on and off ships was a time-consuming business and commercially inefficient. But a market for the cross-Channel car business had been well established and, after the Second World War, ferry companies began investigating alternative means of loading.

Stuart Townsend had bought a Bailey bridge in France and put it to use loading and off-loading at Calais in the summer of 1950. The system worked well and established the principle of motorists driving their cars on and off the Channel ferries. The Southern Railway arm of British Railways, now in fierce competition with Townsend Bros, was quick to follow suit and within two years launched their own drive-on/drive-off service in 1952.

Within ten years, total annual car crossings had reached 440,000, and within another ten years, the number had reached a million. Townsend Bros were holding their own against the publicly owned rail companies and in July 1956 decided to float their business on the London Stock Exchange. Their destiny, however, was taken out of their hands. On the day of the share launch, Egypt's President Nasser nationalised what he saw as *his* Suez Canal, and the markets crashed.

Stuart Townsend and his board had hoped for a flotation that would have attracted smaller investors who would let the Townsends carry on running the company. It was not to be. Monument Securities Ltd, with interests in property management, electricals and shipping, bought sufficient shares to gain control of the company, and promptly voted Stuart Townsend and his directors off the board. But the Townsend name was retained, and its operations were combined with those of the Norwegian Otto Thoresen.

Until, that is, the night of 6 March 1987. On that night, the *Herald of Free Enterprise*, sporting the Townsend Thoresen colours and with its logo on the funnel, capsized shortly after leaving Zeebrugge en route for Calais. The bow door had been left open, and the boat capsized within 90 seconds of leaving its berth. The assistant boatswain whose responsibility it was to close the door was asleep in his cabin; 193 people died.

Neither Stuart Townsend nor Otto Thoresen had anything to do with the building of the stricken ferry or its management, a management that Mr Justice

Sheen, who headed the public inquiry into the disaster, described as 'infected with a disease of sloppiness from top to bottom'. The directors, he said, 'did not have any proper comprehension of what their duties were'.

P&O, who had taken control of the company that bore the men's names, hastily painted over the names on the side of the ship and rebranded the ferry operations as P&O European Ferries. The two men who had pioneered car ferry services across the Channel were to be forgotten, and the broken ship towed to Taiwan and scrapped. Jeffrey Sterling, who had been Chairman of P&O for four years when the disaster occurred, was created a life peer in Margaret Thatcher's resignation honours list. P&O Ferries is now owned by the late Queen Elizabeth II's horse-racing friend, Mohammed bin Rashid al-Maktoum, the Dubai emir.

※

For sheer romance, in the years leading up to the outbreak of war in 1939, little could beat the night train from London to Paris. (In truth, the Orient Express that trundled from Paris to Istanbul through a Europe of kings and emperors was hard to match.) But the romance came at a cost.

You needed a first-class ticket, which bought you a berth in one of the wagons-lit coaches of the train. The service left platform 2 at Victoria Station at nine o'clock every evening, bound for the Gare du Nord. You dressed for the occasion. Lady passengers invariably wore hats. You had dinner and brandy in the dining car and by midnight were closeted in your en-suite cabin of polished wood, crisp white sheets and luxury soap which had none of the common carbolic smell of ordinary train toilets. But the air of romance was rudely interrupted when the train reached Dover.

The carriages were pushed and pulled, with disagreeable disregard for sleeping passengers, onto the ferry: to keep the ship balanced, half of the carriages were shunted onto the port side of the boat, the other half on the starboard side. Then heavy, clanky chains were used to fasten each carriage in position. The whole process took an age. Sleep required several brandies or a considerable amount of romantic exhilaration. And if a rough crossing brought on fits of *mal de mer*, any idea of an amorous crossing went out of the porthole. You could, if sleep eluded you altogether, take a stroll on deck to meet the night ferry coming in the other direction, and bear witness to ships passing in the night.

The same below-decks commotion would attend the pulling of the carriages off the boat at Dunkirk. Breakfast would be served as you passed through the fields of Normandy, and by nine in the morning, exactly twelve hours after leaving London, you would be in the heart of Paris. Not refreshed perhaps, but privileged to have crossed the Channel in noisy, bumpy luxury.

The first commercial passenger flight across the Channel was no less bumpy. The DH34A, a single-prop biplane of plywood wrapped in linen, had been developed by the De Havilland aircraft company for service during the First World War. On a wet August day in 1919, in the colours of Aircraft Transport & Travel Ltd, it took George Stevenson-Reece, a reporter from the *London Evening Standard*, on a bumpy two-and-a-half-hour journey from Hounslow Heath, on the western outskirts of London, to Le Bourget, in the north-eastern suburbs of Paris. The reporter's paper paid the fare of 21 guineas.

Lieutenant Eardley Lawford, an experienced De Havilland pilot who was at the controls, followed the railway line to Folkestone, bumped low across the Channel and tracked the Seine to Paris. With the reporter's luggage was a sack of grouse and some clotted cream from Cornwall for the British Ambassador.

In time, De Havilland adapted the plane to carry ten passengers. The fleet of eleven DH34 aircraft worked the cross-Channel London–Paris route, completing more than 8,000 flights in the first nine months of operation. The passengers were in an enclosed cabin, but the pilot and co-pilot sat up front in an open cockpit behind the engine. To refuel the aircraft, a mechanic carrying cans of petrol had to climb a ladder to reach the engine.

This first international passenger flight across the Channel evolved into the world's busiest air route, and that first commercial crossing was just ten years after Louis Blériot had wheeled the latest version of his own wooden aircraft, the *Blériot XI*, onto the beach at Les Baraques, just west of Calais, for his Channel hop.

It was on a Sunday in July 1909, with the sun barely up, that Blériot eased himself into the pilot's wooden bucket seat and signalled to his engineer (the engine had been developed by a manufacturer of motorcycles) to spin the rather beautiful walnut propeller. Lifting off at 4.41 a.m., Blériot climbed and banked over the water.

Claiming he was hopeless at navigating, he had not bothered with a compass, which proved a minor problem until the mist cleared and he spotted three ships he judged to be heading for the port of Dover. He followed them, and sure enough Dover Castle soon came into view. He flew over the tower of the castle's Saxon church and did a couple of circuits before landing on the slopes of Northfall Meadow, just below the castle itself. It was a heavy landing, and the *Blériot XI* smashed its undercarriage.

But the French aviator had become the first man to fly an aircraft across the Channel, and duly collected his £1,000 prize from the *Daily Mail*'s owner, Lord Northcliffe, a keen aviator and one-time editor of *Bicycling News*. Northcliffe had become the populist press baron who encouraged his journalists to concentrate on topics readers wanted to read about most: health, money and sex (which, with the exception of health, might have been difficult topics to get into a cycling magazine).

Even 70 years earlier, £1,000 looks a little modest against the £100,000 prize put up by the industrialist Henry Kremer for the first man-powered flight across the Channel. The prize was awarded to Dr Paul MacCready in 1979.

Dr MacCready, an American aeronautical engineer, had designed and built the pedal-powered *Gossamer Albatross*, which, with the long-distance cyclist Bryan Allen in the saddle, was the first aircraft of its kind to cross the Channel. The flight from Folkestone to the beach at Cap Gris-Nez took just under three hours, leaving Mr Allen dehydrated and exhausted – and Dr MacCready very rich.

Sadly, there was no such prize for John-Pierre Blanchard. Blériot may have been the first to achieve a powered crossing of the Channel, but he was not, by a long shot, the first to fly across. That honour belongs to Blanchard, who crossed from Dover in a hot air balloon in 1785. He had with him his Bostonian doctor friend, John Jeffries; and, considering a previous attempt at the crossing just months earlier had resulted in the death of both the pilot and his companion, their adventure was undertaken with a mixture of trepidation and courage.

Things did not go well. Blanchard had failed to allow for the degree of lift his balloon would need, and halfway across the Channel they began to lose height. They threw overboard the ballast but continued to drop. Next, their food and non-essential fittings were tossed into the sea. Then their coats and scarves (it was January and icy cold). And in a last desperate attempt to stay airborne, Blanchard took off his trousers and threw them out too. It seemed there was nothing else left. The two men faced disaster until Blanchard recalled they had both had plenty to drink before they took off. Could it just save the day and save them?

Mindful of the need to keep the balloon balanced, the two men urinated over opposite sides of the basket. This final shedding of weight did not halt their decline, but it did slow the rate of descent. They were by now flying dangerously close to the treetops of the forest at Guînes, yet somehow managed to find a clearing and, by grabbing branches, were able to ease themselves onto the ground. They were between Calais and Saint-Omer and people who had watched them come in over the coast and had followed their progress were quickly on hand with food and brandy and warm clothes – but no trousers. The flight had taken 150 minutes. If the balloonists had kept their trousers on and their bladders full, it might have ended much sooner.

A host of other adventurers and exhibitionists have chosen to make the Channel crossing on the water's surface rather than above it or in it. William Hoskins claimed to make the crossing in 1862 clutching a bale of straw. Bernard Thomas paddled across in a 5ft Welsh coracle; Richard Branson drove across in an amphibious Gibbs Aquada; Tim FitzHigham took nine hours to cross in a Victorian copper bathtub; and Bob Platten, a bank clerk, crossed on his grandmother's brass bedstead. He fastened it to two aircraft fuel tanks and stuck a 5hp outboard engine on the back.

Hilary Lister was not an exhibitionist. She was perhaps an adventurer, but above all, she was a woman of extraordinary determination and courage who refused to give in to her terrible illness. Miss Lister suffered from reflex sympathetic dystrophy, which left her paralysed from the neck down for all her adult life. She studied biochemistry at Oxford and was on a morphine drip when she sat her finals. She gained her degree, but she was forced to give up her PhD at the University of Kent when her arms stopped working.

The less she was able to do on her own, the more depressed she became and the more inclined to evaluate the quality of her life, which, in despair, she contemplated ending. Until she was taken sailing. The experience, she said, literally saved her life. And so it was, early on an August morning in 2005, with a light breeze sending gentle ripples across the water, that Hilary, by now 33, set off from Dover alone in her 27ft adapted sailing boat. Her support boat was not far behind, but she did not need it. In her mouth were two plastic straws. By inhaling and exhaling through the relevant straws she was able to control the sails and the rudder, her tongue covering the straw not in use. In just over six hours, she reached France.

The 'mad dream' she had a year earlier to sail across the English Channel had become a reality. Friends recall she was ecstatic. And she did not stop there. She sailed around the Isle of Wight, around Britain and across the Arabian Sea from Mumbai to Muscat. For her, water was a liberating place. 'It's where you feel free. Once on the boat I can sail it as well as anyone really. You just think about the next wave and the next puff of wind coming. The wheelchair is not there. I'm a sailor and it's a huge feeling of freedom,' she explained in *The Guardian* on 7 September 2018.

She started dreaming again. This time of a trip across the Atlantic. But it was not to happen. Hilary Lister died when she was 46.

Tribute after tribute recognised her courage and determination. Reflecting on her life, she would have seen it as an encouragement to others who suffer paralysis not to give in, and a plea to able-bodied people to reassess the way they related to those with physical disabilities. She knew she was determined but she did not own the courage that others saw: 'We do not need wrapping up in cotton wool and can go out and do silly or dangerous things if that's what we want to do.'

⚔

Abdulfatah may have been courageous, but did he not get across the Channel. Having got to the French coast, he was determined to complete the last part of his journey, oblivious to the risks involved.

Abdulfatah Hamdallah came from Sudan in an area close to war-ravaged Darfur. He fled the country to spend time with his brother in Libya, and then headed for France via Italy. But the French would not let him stay, so he decided to try his luck in England. He left Calais in the middle of the night with a friend in an inflatable dinghy using shovels as oars. They had not got far when the dinghy capsized. Abdulfatah's friend managed to swim back to the beach and was taken to hospital suffering from hypothermia. But Abdulfatah did not make it back, and his body was found on the beach at Sangatte.

According to Migration Watch UK and Home Office figures, in 2023 29,437 people crossed the Channel in small boats, compared with 45,774 in 2022. Nigel Farage, a right-wing British politician without a party, described the influx as 'a shocking invasion'. A total of 104,990 made the crossing between 2018 and 1 September 2023; sixty-four people drowned making the attempt.

The cross-Channel migrants were fleeing violence, war, poverty, repression and hunger, coming mostly from Syria, Eritrea, Iran, Sudan, Afghanistan and Yemen. Others came from Iraq, Nigeria and Pakistan.

Dr Yahya al-Rewi came from Yemen. *The Guardian* newspaper reported in July 2020 that the one-time senior civil servant (he had been Yemen's representative at the United Nations Economic and Social Commission for Western Asia Technology) arrived in Calais after travelling first to Switzerland. There he was beaten up by what he supposed were Yemeni government agents, who had tracked him down after he criticised the corruption of the government back home. The attack left him blind in one eye. Believing Switzerland to be no longer safe, Dr al-Rewi travelled to France by train and thence to Calais where he bought a dinghy and crossed safely to the UK.

The doctor was one of the lucky ones. A younger fellow Yemeni lost more than an eye. Newspaper reports told of a young male asylum seeker, who had escaped persecution at the hands of the Houthis, arriving in Libya only to fall into the clutches of an organ-trafficking gang who removed one of his kidneys. When he was strong enough, he made his way from Libya to Morocco, then Spain and finally to France, from where he crossed the Channel with others in a small boat. Arriving on the Kent coast, he was given medical attention, but chose not to say much about the organ traffickers for fear they would track him down.

In 2019, seventeen Iranian refugees broke into a fishing trawler at Boulogne for their escape to Dover. But they failed to start the engine. Five other Iranian asylum seekers took a small, overcrowded boat and perished. With them on the boat was a family of two adults and three children, Kurds from Sardasht, who had reportedly paid 5,000 euros for the trip, having sold gold and all the other family valuables. There were fifteen other refugees in the overcrowded boat when it sank.

The family from Sardasht drowned, and the body of their youngest child, a boy who was 15 months old, was never recovered. The other fifteen were rescued and taken to hospital. One Iranian who did make the crossing told newspaper reporters that for those who don't have the cash, the smugglers force them to join their smuggling operation in exchange for the free crossing. A young boy fleeing west Africa told a Save the Children official that he had fled because first his mother and then his father had been killed for being Christians.

Driven by the people-smugglers, refugees crossing the Channel say they do so because they have family or friends in the UK, because they speak or want to learn to speak English, because they believe the English will treat them better than the French, or because it is easier to find work in Britain. And they say the risks involved in the crossing are worth it.

In theory, the United Nations Convention obliges countries to open their doors to refugees and asylum seekers, and it forbids governments to send them back if they are at risk at home. The UK insists that refugees seeking asylum in Britain must only do so once they are on British soil.

Hundreds of inflatable dinghies crowd a piece of ring-fenced wasteland in Dover, testifying to the many thousands of asylum seekers who reached the UK and have asked to stay. The plight of these asylum seekers was the subject of Daljit Nagra's poem 'Look We Have Coming to Dover!' Nagra, the son of Punjabi parents who came to Britain in the 1960s, wrote his poem in 2004, starkly describing an immigrant's arrival and early days in England, and invites readers to reflect on Matthew Arnold's poem 'Dover Beach', penned more than 150 years earlier (see Chapter 8). In *The Guardian* on 7 October 2004, fellow poet Lavinia Greenlaw described it as a poem of 'contemporary resonance which engages playfully and powerfully with our literary heritage'. Only after years of adapting to their new world, the poem suggests, can immigrants see their world as optimistically as Arnold.

Desperate migrants, invaders, smugglers, adventurers, soldiers, merchants, exhibitionists, kings and queens have all made their crossings of the Channel. Most of them made the journey in boats of one kind or another; some flew and others swam. Some never made it. And then at last came a new way.

The idea of making the crossing under the water surfaced at the beginning of the nineteenth century. A French mining engineer, Albert Mathieu-Favier, envisaged a tunnel for horse-drawn coaches, lit by oil lamps and with an island in the middle where the horses would be changed. Chimneys along the route would provide the fresh air. He had been working on his scheme throughout the Revolution and in 1802 had the plans delivered to Napoleon at the Château de Malmaison, which Bonaparte shared with Josephine. The idea appealed in principle to the emperor but, lacking even the most basic of surveys, the plans gathered dust.

Some thirty years later, it was the turn of another French engineer and hydrographer, one Aimé Thomé de Gamond. De Gamond had been working on his plans for fifteen years, during which time he had done some serious, if basic and risky, geological surveys. These would entail taking a rowing boat out from the shore accompanied by his assistant and his daughter.

According to Thomas Whiteside, in his book *Tunnel Under the Channel*, de Gamond would drop over the side of the dinghy, weighted with bags of flints. He wore a safety rope around his waist and an additional red distress line attached to his left arm. He covered his ears with pads of buttered lint to protect them from the pressure of the water as he descended, but otherwise dived stark naked, apart from a belt of inflated pigs' bladders to aid his journey back to the surface. He had no breathing apparatus. Instead, he put a spoonful of olive oil in his mouth, which allowed him to expel air slowly from his lungs without letting water force itself in.

Convinced he had done enough research to stand his plans up, he took them to Napoleon III, who was as enthusiastic about the idea as his uncle had been with Mathieu-Favier's proposal fifty years earlier. (The omens were not good. A cartoon in the magazine *Puck* had shown a terrified English lion fleeing from a French cock emerging from a Channel tunnel.)

The emperor commissioned an appraisal of the scheme which proved positive, and a much encouraged de Gamond left France for England. There, he met with three established and respected engineers – Isambard Kingdom Brunel, who had already tunnelled under the Thames, Robert Stephenson and Joseph Locke – all of whom supported his scheme. And recognising that any permanent link between England and France would be as much a political challenge as it would be an engineering one, he secured an audience with Queen Victoria's consort, Prince Albert.

Albert was enthusiastic, as was the queen herself, not least because she was prone, like Albert, to getting seasick. But her prime minister, Lord Palmerston, had other ideas, and soon proved de Gamond right in his fears about the political challenges. Palmerston saw the Channel as a moat – and a narrow moat at that. Rather bluntly, he told Prince Albert that his enthusiasm for the link might have been tempered had he been born on the island of Britain. 'What!' Palmerston is reported to have exclaimed, 'You pretend to ask us to contribute to a work the object of which is to shorten a distance which we already find too short.'

For the French, England had long been *l'Albion perfide*. Palmerston had just given that belief renewed credence. He might also have recalled that Calais was English between the fourteenth and sixteenth centuries and Normandy was

English too during the fifteenth century. Back then, the Channel was a choppy route to the southernmost parts of the kingdom rather than a defensive barricade.

However, some twenty-five years after Palmerston's death, tunnelling did begin on both sides of the Channel. The British dug 2,000 yards, but the realisation that a link to the Continent had become a real possibility rekindled fears that an invasion route was in the making. *The Nineteenth Century*, a monthly literary magazine, carried a plea from writers and others:

> The undersigned, having had their attention called to certain proposals made by commercial companies for joining England to the Continent of Europe by a Railroad under the Channel, and feeling convinced that (notwithstanding any precautions against risk suggested by the projectors) such a Railroad would involve this country in military dangers and liabilities from which, as an island, it has hitherto been happily free hereby record their emphatic protest against the sanction or execution of any such work.

The signatories included Alfred Lord Tennyson, Robert Browning, T.H. Huxley, Herbert Spencer and the Archbishop of Canterbury. Senior military figures weighed in with their objections. The head of the British Army, Field Marshal Sir Garnet Wolseley, saw the Channel as a great wet ditch for England's protection. 'Surely,' he wrote, in a lengthy submission to the Board of Trade, 'John Bull will not endanger his birth-right, his property, in fact all that man can hold most dear ... simply in order that men and women may cross to and fro between England and France without running the risk of seasickness.' The project was dropped.

The political stance of successive British governments continued to be formed by these fears until ten years after the Second World War. Harold Macmillan, who was at the time Anthony Eden's Defence Secretary, was asked in the Commons if a Channel tunnel was still a risk *vis-à-vis* an invading army. His reply was brief. There was, he said, 'scarcely any military opposition to the Channel Tunnel'. This gave the green light to the railway companies, who were champing at the bit to get a tunnel under way.

But their enthusiasm, like that of de Gamond, was to be thwarted. It was not until the 1970s that a renewed enthusiasm emerged, but Harold Wilson's

Labour government claimed it could afford a tunnel or Concorde but not both. Concorde got the green light, and after ten years of planning and fourteen months of digging, the tunnel project was dropped – again.

Ten more years passed. Then, come the second Conservative government of Margaret Thatcher in 1984 and with François Mitterrand in the Élysée Palace, the Channel Tunnel project was given fresh impetus. The British and French governments agreed to invite bidders to present their plans. There were ten proposals from which four were shortlisted.

Thatcher, true to form, was adamant from the very beginning that the project be privately funded, a condition that the French were happy to go along with. And she favoured a tunnel for cars. But technical objections put paid to her vision of driving herself to France, and the chosen bid from a group of fifteen British and French construction companies and banks for a rail link won the day. Their scheme provided for a double rail tunnel – one for northbound trains and one for trains going south – separated by a service tunnel. The group, which became known as Eurotunnel, estimated the cost at £4.8 billion.

In the presence of Prime Minister Thatcher and President Mitterrand, the Treaty of Canterbury, which gave birth to the Channel Tunnel, was signed in the chapter house of the city's cathedral by Britain's Foreign Secretary Geoffrey Howe and his French counterpart, Roland Dumas. (Dumas narrowly escaped prison years later, having been convicted on corruption charges together with his lingerie model mistress, Christine Deviers-Joncour, and Geoffrey Howe found himself at the centre of an acrimonious parting from Margaret Thatcher's government.)

To fund the project, Eurotunnel went on the search for both equity and loans. There was plenty of interest in France but in Britain the shares proved unattractive to the point where lack of funding almost put an end to the project before it got started. Lord Pennock, a one-time senior ICI man with a green Rolls-Royce and a passion for tennis, who co-chaired Eurotunnel with André Bénard, resigned along with other senior executives on the British side. In his place came Alastair Morton, a tall, blue-eyed South African-born 'bruiser', who had just succeeded in turning around the failing merchant bank Guinness Peat.

Fluent in French and with a formidable intellect and a love of music, Morton was the son of a Scottish engineer and an Afrikaans mother. He grew up in

South Africa and went to university there before coming to England and studying law at Oxford. He was not a man to suffer fools gladly – the *Financial Times* described him as 'incendiary' – nor did he know much about the construction industry. But he managed on just five hours' sleep, and he did know how to bang heads together, especially when a project he believed in was at stake.

He talked to the Bank of England (at whose behest he had rescued Guinness Peat) and the bank persuaded the commercial banks and various corporations to buy his shares. The public offering did less well to start with but was fully underwritten in the end.

In all, 205 banks contributed to an initial loan and appointed technical watchdogs, but cost overruns – not uncommon with such large projects (according to contemporary reportage, an IT system for Britain's NHS went over budget by more than 400 per cent, and the Sydney Opera House overran its budget by more than 1,000 per cent) – meant Morton was soon asking for more. To some, the asking seemed more like bullying.

By contrast, Morton's co-chairman on the French side of things was more of a natural diplomat. André Bénard had studied engineering. He was an Anglophile, spoke fluent English, kept a flat in London and enjoyed golf. During the war, he joined the French Resistance in Morocco and was captured by the Germans. He escaped and was decorated for his bravery. After the war, he finished his studies before going to work for Shell, rising to become the company's boss in France. While Morton was banging heads together and twisting arms, Bénard was using his charm to persuade and cajole.

Their joint efforts – and observers have little doubt that without Morton and Bénard there would be no Channel Tunnel – led to work beginning at Coquelles on the outskirts of Calais in June 1988 and at Folkestone in December of the same year, with Eurotunnel given exclusive rights to operate the tunnel until 2042.

Not all went smoothly during the ensuing seven years. Ten workers died and others suffered serious injuries. Technical difficulties caused delays and further cost overruns. The excessive heat the trains would generate in the tunnel had been ignored, and an air-conditioning system involving refrigeration units at both ends of the tunnel had to be installed. The French tunnellers were allowed to smoke and enjoy 25ml of wine at lunchtime, while both pleasures were denied to workers on the English side. It was nine years in the building, delivered late and way over budget. A year after it opened, Eurostar announced a loss of £925 million.

Remarkably, it transpired that while the track gauge on both sides of the Channel was the same, the dimensions of the rolling stock were not. The rolling stock on the Continent is wider and higher than that in Britain. Adapted engines and carriages were ordered for the tunnel trains but were a year late in delivery, and the idea of being able to get on a night train in Manchester or Glasgow and be in Paris for breakfast was put on hold.

There were also one or two less-onerous distractions. A decision had to be made about the numbering of the passenger carriages: should carriage number one be at the British or French end of the trains? And Florent Longuépée, a right-wing Paris city councillor representing the 1st Arrondissement, wrote to the British government requesting Waterloo Station be renamed because it was upsetting for the French to be reminded of Napoleon's defeat when they arrived in London. His request was politely turned down (the later moving of the London terminal to St Pancras was about linking with high-speed track).

In 1996, two years after it opened, the tunnel and train operators suffered a serious setback when a lorry on a freight shuttle caught fire. The fire was spotted as the train entered the tunnel at Sangatte. Following emergency protocols, the train continued its journey so the fire could be brought under control in the open at Folkestone. But the automatic safety system detected a fault that suggested the train might derail. The train came to a halt halfway through the tunnel. It then lost power. Smoke from the burning lorry began to work its way along the tunnel and into the passenger carriage where the HGV drivers were eating their evening meal.

Firefighters could not at first locate the train, partly because markers on the tunnel wall were hidden by the smoke. When they did arrive, they found the lorry drivers and train crew already making their way to the service tunnel. While the French emergency services treated those suffering from smoke inhalation, British firefighters tackled the blaze, which took twelve hours to bring under control. The blaze reached temperatures of 1,000°C and took 300 firefighters sixteen hours to extinguish. Fourteen of the thirty-two lorry drivers suffered minor injuries. There was no loss of life.

In a BBC television interview during the construction of the tunnel, Alastair Morton had said, 'It is highly unlikely that there will ever be a serious incident.' There was another lorry fire in 2008 ... Morton moved to the Strategic Rail Authority and resigned at the collapse of Railtrack. He was knighted in 1992 and died following a heart attack in 2004 at the age of 66.

Over its first twenty-five years of operations, Eurostar carried, on average, 7.8 million passengers a year. Like all travel companies, it suffered a massive loss of business during the Covid pandemic, and in 2022 the company merged with the Franco-Belgian train operator Thalys. In 2023 Eurostar notched up 18.6 million passengers, and in 2024 it celebrates thirty years of journeys under our narrow sea.

In the broader sense, the tunnel broke the ferries' monopoly of cross-Channel traffic, pushing the ferry operators towards competitive pricing. The tunnel's future operations were, however, assured during the often-tense Brexit negotiations between the British government and the European Union.

8

Writers, Poets, Painters and Debussy

One way or another, the Channel has caught the eye of a broad family of artists and the imagination of an equally diverse worship of writers. Novelists and poets have drawn on its link between cultures; its role as defender of an island state; its unerring ability to reduce those crossing it to abject misery – and seized on its potential as an allegory for their political musings. And both writers and artists have been drawn by the ever-changing light imposing its moods on the pageant of nature's watery landscape.

William Shakespeare was well on his way by the 1580s, but it was not until circa 1591 that he wrote his first history play, *Henry VI*. In Part Two, which coincides with the end of the Hundred Years War between England and France, he introduces us to the Channel by name, the first time in literature 'Channel' had been used to describe the 'narrow sea'. The Duke of Suffolk has been banished for having had the Duke of Gloucester murdered. Pirates capture him off the Kent coast and set up a mock trial. His words put the pirate captain firmly in his place, and in so doing, refer to the Channel:

It is impossible that I should die
By such a lowly vassal as thyself. The words move rage not remorse in me:

> I go of Message from the Queene to France;
> I charge thee waft me safely crosse the Channell

His command, as it turned out, was to no avail and the duke was duly beheaded.

Shakespeare returns to the Kent coast and the Channel with *King Lear*. A blind Earl of Gloucester has become suicidal and asks to be led to the cliffs above Dover. His son, in disguise, takes him there and then tricks him into thinking he has leapt to his death and has had a miraculous escape. In *Richard II*, Shakespeare is more concerned with the Channel as part of an island's defence:

> This fortress built by Nature for herself
> Against infection and the hand of war,
> This precious stone set in the silver sea,
> Which serves it in the office of a wall,
> Or as a moat defensive of a house
> Against the envy of less happier lands;
> England, bound in with the triumphant sea,
> Whose rocky shore beats back the envious siege
> Of wat'ry Neptune, is now bound in with shame.

While Shakespeare used the Channel in his plots, other writers have found in the Channel a canvas for their prose and poetry and a platform for cultural comment. Most recently, the complex cultural relationship between the English and the French is explored through the stories of English men and women in France by the British novelist Julian Barnes in his collection of short stories, *Cross Channel*. Two centuries earlier, Charlotte Smith, the eighteenth-century romanticist, wrote her epic blank verse 'Beachy Head'. It was published shortly after her death in a volume entitled *Beachy Head With Other Poems*. Smith lived in a small cottage close to Beachy Head and takes her readers with her onto the clifftop:

> On thy stupendous summit, rock sublime!
> That o'er the channel rear'd, half way at sea
> The mariner at early morning hails,
> I would recline; while Fancy should go forth
> And represent the strange and awful hour
> Of vast concussion; when the Omnipotent
> Stretch'd forth his arm, and rent the solid hills,
> Bidding the impetuous main flood rush between
> The rifted shores, and from the continent
> Eternally divided this green isle.

Throughout her poem she rejoices in the nature that surrounds her, and observes the shipping that is already making the Channel a busy seaway:

> Advances now, with feathery silver touched
> The rippling tide of flood; glisten the sands,
> While, inmates of the chalky clefts that scar
> Thy sides precipitous, with shrill harsh cry,
> Their white wings glancing in the level beam,
> The terns and gulls and tarrocks seek their food.

From nature, she turns her attention to the fishing boats and freighters plying their trade:

> Afar off, and just emerging from the arch immense
> Where seem to part the elements, a fleet
> Of fishing vessels stretch their lesser sails;
> While more remote, and like a dubious spot
> Just hanging in the horizon, laden deep,
> The ship of commerce richly freighted, makes
> Her slower progress, on her distant voyage,
> Bound to the orient climates, where the sun
> Matures the spice within its odorous shell.

In common with too many writers, artists and composers, Charlotte Smith was destitute towards the end of her life. In her case, a rotten marriage had robbed her of the money her poems and novels had made. Her popularity had waned and, had it not been for the efforts of her family following her death, 'Beachy Head' might never have been published. She had been described as a 'genuine child of genius' but her work and that of other women authors went out of fashion in the Victorian era.

Some twentieth-century critics suggest she was trying to engage with her male contemporary, William Wordsworth. But what seems clear is that throughout her life, and for years afterwards, her gender was getting in the way of wider recognition and approval of her considerable talent. Wordsworth himself thought she wrote with 'true feeling for nature'. She was a lady, he said, 'to whom English verse is under greater obligations than are likely to be either acknowledged or remembered'.

For several years at the beginning of the nineteenth century, the poet and artist William Blake lived in a thatched ploughman's cottage in the Sussex coastal village of Felpham. It was here, with the salty Channel air blowing through the eves, that he wrote 'Jerusalem' as part of his epic work, *Milton*.

More than a century passed before Hubert Parry put 'Jerusalem' to music, turning it, controversially in some quarters, into a national anthem. It has been played on the last night of the BBC Promenade concerts since 1953, was adopted by the Women's Institute as their very own hymn and is sung at Twickenham when England plays rugby there – and it regularly tops the list of favourite hymns. It has not, however, been adopted by the Royal Yacht Squadron at Cowes, even though Parry was the first composer to be honoured by the club when it elected him a member in 1908. His two yachts, the yawl *The Latois* and the ketch *The Wanderer*, were regularly seen cruising in the Channel.

When William Wordsworth could tear himself away from the Lake District, he too found pleasure and took inspiration from the Channel and its coasts.

In 'Near Dover', he sees nature playing a political role, drawing France into 'frightful neighbourhood':

> Inland, within a hollow Vale, I stood;
> And saw, while sea was calm and air was clear,
> The Coast of France, the Coast of France how near!
> Drawn almost into frightful neighbourhood.
> I shrunk, for verily the barrier flood
> Was like a Lake, or River bright and fair,
> A span of waters; yet what power is there!
> What mightiness for evil and for good!
> Even so doth God protect us if we be
> Virtuous and wise: Winds blow, and Waters roll
> Strength to the brave, and Power, and Deity,
> Yet in themselves are nothing! One decree
> Spake laws to *them*, and said that by the Soul
> Only the Nations shall be great and free.

Wordsworth became interested in the ideas of the revolutionary forces at work in France but both he and his sister Dorothy were dreadfully seasick when crossing the Channel, which became a deterrent to their continental trips. And when Napoleon came to power at the end of the eighteenth century, the poet's views about the French did an about-turn and finally put an end to his cross-Channel adventures.

Charles Dickens, Wordsworth's junior by some forty years, was by contrast an avid traveller and was a regular on the cross-Channel paddle steamers. Not that Dickens found the passage any easier. In 'Crossing the Channel with Dickens', Dominic Rainsford, Professor of Literature in English at Aarhus University, asserts his belief that Dickens associated the crossing with a sense of 'trauma and dislocation'. He cites a letter from Dickens to his wife Catherine (whom he treated appallingly and later divorced after she had given him ten children), when he had made the crossing on one particularly rough day, 'I never knew anything like the sickness and misery of it. And besides that, I really was alarmed; the waves ran so very high, and the fast boat, going at that speed through the water, shipped such volumes of it.'

In 'The Calais Night-Mail', Dickens' loathing of the Channel crossing appears to colour his view of Calais, albeit after an admission of his indecision about the place. He begins by asking himself whether he shall benefit Calais in his will, 'I hate it so much and yet I am always so glad to see it'. However, in what must have caused the burghers of Calais and the French Tourist Board to bury their heads in their hands, he continues, 'Malignant Calais! Low-lying alligator, evading the eyesight and discouraging hope! [...] sneaking Calais, prone behind its bar, invites emetically to despair [...] Thrice accursed be that garrison town'. (The respect those burghers of Calais would have felt respect for their town, and indeed for themselves, was restored by Auguste Rodin, whose fine 1885 statue of local heroes outside the town hall commemorates the Siege of Calais during the Hundred Years War. And these men were true heroes prepared to die for their city, unlike the 'heroes' of a twenty-first-century need to boost national morale at every opportunity.)

Nor was Dickens' lacerating pen any more benign when it came to Dover. When he arrived early for the night packet – as he did on most occasions – waiting impatiently on board for the train from London to deliver the mail for Europe, his thoughts would turn to Dover itself:

> I particularly detest Dover for the self-complacency with which it goes to bed [...] the many gas eyes of the Marine Parade twinkle in an offensive manner, as if with derision. The distant dogs of Dover bark at me in my misshapen wrappers, as if I were Richard the Third.

However, by the time he sets foot in France he has forgiven Calais its failings. And the pain of the crossings is mostly forgotten in his novels. *A Tale of Two Cities*, his historical novel set in London and Paris, which might well dwell on the sea passage, mentions it hardly at all.

While Dickens was ambivalent about Calais, he had a soft spot for Boulogne, spending three summers there with the family. 'Our French Watering-Place' was written in praise of the town and published in the weekly journal *Household Words*. Dickens has nothing but compliments for the place:

> If this was but 300 miles further off, how the English would rave about it! I do assure you that there are picturesque people, and town, and country about this place, that quite fill up the eye and fancy. It is more picturesque and quaint than half the innocent places which tourists, following their leader like sheep, have made impostor of.

Elisabeth Jay, an Oxford literature professor, gives Thomas Carlyle the prize for the most uncomfortable account of the nineteenth-century Channel crossing. Carlyle, the eminent Scottish writer and historian, pays 7*s* 6*d* for a passport at the Foreign Office, and from the Reform Club obtains times of the packet-boat crossings and connecting trains. He has taken the precaution of sailing with Robert and Elizabeth Browning, both experienced travellers who would show him the ropes.

They meet at London Bridge Station and travel by train to Newhaven. The fare to Paris is 22*s*. On board the steam packet that awaited their arrival in the small harbour, the passengers, many of them French, are in a jovial mood. The Brownings have their child and the child's nanny with them.

The jollity soon recedes as the boat encounters the angry sea beyond the harbour wall. There are benches on deck covered with painted canvas hoods and plenty of stewards busying themselves about the place. As the boat starts to pitch, nearly all the passengers sink into the general sordid torpor of seasickness accompanied by its miserable guttural retching.

Carlyle, not feeling on top form himself, sees Browning being sick and bears witness to continuing groaning and vomiting all around him. One elegantly dressed Frenchman has wrapped himself in a blanket and huddled into a corner with his head between his knees. Regardless of the wind and the spray, Carlyle manages to light a cigar, which takes his mind off the pitiful mass of humanity gathered wherever he chances to look.

It takes eight hours to reach Dieppe. Carlyle and the Brownings are in a wretched state but manage to stagger into a quayside hostelry to find warm tea and cold coffee. Everyone pretends not to be thinking about the return journey, but everyone is doing just that.

Professor Jay, in claiming Thomas Carlyle to have best expressed the horrors of Channel crossings, may not have come across the work of Hans Ostrom. Ostrom is a Professor of English at the University of Puget Sound at Tacoma in Washington State. Ostrom made the trip on an especially rough day and pulls no punches in recording the stupor on the unhappy tossing ship. His poem 'Channel Crossing' is not recommended reading for anyone about to make the journey. He describes a passenger standing near him throwing up and sending 'wet pink pebbles' his way, while below decks he finds 'a dense-macabre of vomiters'.

Back in nineteenth-century England, and sharing this scourge of seasickness, was an inventive engineer who pioneered a cost-effective process for manufacturing steel, and in so doing, put Sheffield firmly on the industrial map. Henry Bessemer – later *Sir* Henry and a fellow of the Royal Society – was a design engineer with limitless energy. While the production of steel was his main interest, his energy and inventive skills led him in some unlikely directions. One of these paths of discovery resulted in the invention of a ship for people who suffered from seasickness. SS *Bessemer*, based on a design by the naval architect Sir Edward Reed, had a saloon which rested on giant gimbals that kept the saloon on one plane while the ship rolled from side to side. Unfortunately, Bessemer did not pay as much attention as he might to the ship's navigation.

On its first crossing from Dover, it collided with the pier at Calais, demolishing a significant part of the structure and causing a great deal of damage to his new ship. Whether it was human error or a design fault was never made clear, but the incident dashed the hopes of regular Channel crossers who once again faced the indignity of throwing up every time the sea tossed and turned.

Matthew Arnold, Dickens' junior by just ten years, was born on Christmas Eve in 1822 in the pleasant Thameside village of Laleham, nestling between Staines and Chertsey. By the time he was 6, his father, Dr Thomas Arnold, had assumed the headship at Rugby School, where he put religious and moral principle at the top of the agenda, followed by gentlemanly conduct and academic ability.

Arnold was to follow his father into education, becoming a schools' inspector. He was at the same time a vocal social critic: the aristocracy were 'Barbarians' and the commercial middle classes no better than 'Philistines'.

Arnold's strident views were, for some, tempered by his striking good looks. His hair was thick and wavy and parted down the middle, so the parting created a continuous line with his long, narrow nose. Arthur Clough, whom he had met at Rugby, used to tease his friend by saying, 'his hair is guiltless of English scissors'. Clough, too, was an educationalist and, like Arnold, a poet.

But work and his social criticism left Arnold little time to write his poetry. Yet what he did write was ranked with the best of the romanticists' work of the early nineteenth century.

'Dover Beach' was no exception. Arnold visited the Channel coast with his wife Frances soon after they were married, moving some critics to describe the work as a honeymoon poem. It seems very likely that it was written in 1851, although it was not published until much later in 1867.

The poem describes the waves rushing to and fro over the pebbles, evoking the rise and fall of human misery. There is grief, too, for the rapid decline in religious faith. The poem suggests that Sophocles heard the same sounds as the waves washed over the pebbles of the Aegean. (In truth, this is a little unlikely as there are virtually no tides in the Aegean, so the waves would rarely wash up the beach in the way they do in the Channel.)

'Dover Beach' gives Matthew Arnold his place among the great writers who have found inspiration in the waters of the Channel:

The sea is calm tonight,
The tide is full, the moon lies fair
Upon the straits; on the French coast the light
Gleams and is gone; the cliffs of England stand
Glimmering and vast, out in the tranquil bay.
Come to the window, sweet is the night air!
Only, from the long line of spray
Where the sea meets the moon-blanched land,
Listen! You hear the grating roar
Of pebbles which the waves draw back and fling
At their return, up the high strand,
Begin, and cease, and then again begin,
With tremulous cadence slow, and bring
The eternal note of sadness in.

Sophocles long ago
Heard it on the Aegean, and it brought
Into his mind the turbid ebb and flow
Of human misery; we
Find also in the sound a thought,
Hearing it by this distant northern sea.

The Sea of Faith
Was once, too, at the full, and round earth's shore
Lay like the folds of a bright girdle furled.
But now I only hear
Its melancholy, long withdrawing roar,
Retreating to the breath
Of the night-wind, down the vast edges drear
And naked shingles of the world.

It was the waters of the Channel that likewise influenced the writings of François-René de Chateaubriand, the precursor of French romanticism. Born in Saint-Malo on the Brittany coast in 1768, Chateaubriand had a distinguished career as a diplomat before becoming pre-eminent among French writers at the beginning of the nineteenth century.

The son of a slave trader, he was born during a storm, and his mother told him that his cries at birth had been drowned out by the waves crashing on the shore. As a boy growing up, he liked to challenge those waves, and in his writings associated the sea with the transient nature of life. His best-known work is his autobiography, *Mémoires d'Outre-Tombe*, designed originally to be published, as the title suggests, long after his death. It has frequent references to the sea. His tomb, on a tiny islet just off Saint-Malo, is surrounded by water when the tide is in.

Unlike Chateaubriand, François-Marie Arouet was not greatly interested in the Channel itself, rather he was intrigued by the differences in culture enjoyed by the two nations it separated. Arouet used the pen name Voltaire, and no Frenchman could have written more flatteringly about the English. His enthusiasm for England's liberal ideals were summed up by his biographer Evelyn Hall in her book *Friends of Voltaire*, 'I disapprove of what you say, but I will defend to the death your right to say it'. The quote, often wrongly ascribed to Voltaire himself, was how Hall illustrated Voltaire's belief in free speech.

However, the thoughts and beliefs he himself set out in his *Letters Concerning the English Nation* did not go down well with his compatriots. He maintained that England's success in the sciences and business and in their wars against his countrymen came from the country's tradition of personal freedoms. When it was translated into French and published as *Lettres Philosophiques*, it received widespread and hostile criticism and nearly landed him in prison. Copies of the book were burned in front of the Palais de Justice.

Comparing the philosophical ideas of Descartes with those of Isaac Newton, he claimed that Descartes' ideas constituted a sketch while Newton's were masterly. Voltaire had made it clear that 'when it comes to matters of the mind, the English Channel is far wider than the Atlantic'.

The French did not like it one little bit. Voltaire was driven out of Paris into exile and into the arms of Émilie du Châtelet, a married mother of three, who was twelve years his junior yet already a match for his intellect. Sharing Voltaire's interest in science, she had translated Isaac Newton's *Principia*, a translation still used today. Émilie became Voltaire's long-term companion and together they created a fully functioning laboratory at her château in the Haute-Marne, where she indulged her scientific pursuits. When they both submitted treatises to the Paris Academy on the nature of fire, Émilie became the first woman to have a paper published by the Academy.

Two centuries later, Albert Camus was to take his place in the long line of distinguished French writer/philosophers. Camus believed the human situation was absurd and devoid of purpose. This belief was echoed by a collection of American and European writers, including the Irish novelist and playwright Samuel Beckett, best known for his work for the theatre *Waiting for Godot*, which opened in Paris in 1953. Beckett, in common with the writers and painters before him who had challenged the conventions of their art, flouted theatrical convention. He and Camus both won the Nobel Prize for Literature. (Beckett was also awarded the *Croix de Guerre* for his work with the French Resistance.)

The Irishman spent much of his life in Paris, and it was there that he met the American heiress and art collector Peggy Guggenheim, with whom he had a brief and fiery affair. The affair fizzled out (Guggenheim had countless lovers during her life) and Beckett began what turned out to be a lifelong relationship with the pianist Suzanne Déchevaux-Dumesnil, although for many of those years Beckett was having an affair with a BBC script editor called Barbara Bray.

But it was Suzanne rather than Barbara who was to inherit the rights to the author's work and, to ensure that happened, he needed to marry her – in

England. In March 1961, and intent on keeping the marriage under the radar, Beckett flew from Le Touquet to Lydd and checked in at the Hotel Bristol in Folkestone using his middle name, Barclay. In the Channel port there was a whiff of something mysterious going on, prompting journalists to start asking questions. But the wedding plans remained a secret, with the writer's agent spinning a tale about Beckett being on holiday in Africa. The wedding took place at the town's registry office on 25 March before the couple returned to France.

Voltaire had railed against his country's tyranny over freedom of thought and speech – freedoms he so admired about the English. And in his writings, he was always challenging authority which continually seemed immune to change. A century later, a young French artist was challenging the authority, tyranny even, of the French art establishment, represented by the members of the Académie des Beaux-Arts.

Claude Monet grew up in Le Havre, the son of a local merchant. There he met Eugène Boudin, who recognised the young man's talent for painting and persuaded him to become a landscape artist. Monet took up the challenge but with some very particular ideas of his own.

His new ideas did not go down well at the Académie, but the Channel coast of France was witnessing the birth of impressionism. In the 1860s, this new art form challenged the orthodoxy of the history painters and the narrow vision of the judges at the Académie des Beaux-Arts in Paris. It was these judges who were responsible for choosing the works for their annual exhibition, the Salon, where inclusion had been the apogee of an artist's endeavours for centuries past.

The impressionists, led by Claude Monet, working with flat, filbert-tipped, coarse animal-hair brushes (hog's hair was commonly used), were no longer interested in history painting, nor were they constrained to work in their studios. In Britain, John Constable had ventured beyond the studio, setting up his easel within the landscapes he was painting for the past forty years, but the idea was new to the French.

For the impressionists, the world outside their studios and the everyday lives of ordinary people was what interested them – that and the quality of natural light. It was their determination to capture light as it brought new life to the land and particularly to the water. Nor did they feel constrained to hide the brush strokes that carried the light into their paintings. Rather, they used generous dabs of paint to create the effect of light, especially when it fell on water (their oils were in tubes, small containers invented by the British artist John Rand for painters working away from their studios). Painting outside and without making the necessary sketches for studio work helped to define this new approach.

Art historians have also suggested that photography, which originated in France, had released painters from the shackles of faithfully recording life with as much realism as their tools allowed. It was in 1838 that Louis Daguerre launched the new medium with his photograph of the Boulevard du Temple in Paris. This new art form, which 'never lied', gave painters the opportunity to branch out and find new subjects they could interpret in new ways.

Yet this revolutionary style of painting did not go down well with the self-important arbiters of fine art at the Académie des Beaux-Arts. History painting had been considered important. The critics said these new works were mere sketches – impressions at best.

These 'impressions' – created by Monet, Courbet, Boudin, Degas, Pissarro and Manet – were rejected time and again when submitted to the Salon and were ridiculed among the population at large. A cartoon of 1877 in French satirical magazine *Le Charivari* shows a country constable warning a pregnant woman not to enter an exhibition of impressionist art lest the works do her psychological harm.

Yet this much-maligned art form was, in time, to flourish like no other. And it all began on the Channel coast of France.

Eugène Boudin was a son of Honfleur, the pretty fishing village on the estuary of the Seine that was miraculously untouched by the ravages of the Second World War. He was among the first landscape painters to work outside, and he loved to paint the estuary and the sea in all its changing colours. 'Learn to draw well and appreciate the sea, the light and the sky,' he advised his friend, Claude Monet.

Monet took his advice, and then in the summer of 1870 married his lover of five years Camille Doncieux. With their son, Jean (his father painted him on a horse tricycle wearing a dress and a pretty bonnet), the Monets came to Trouville for their honeymoon, during which the bridegroom could not resist setting up his easel. He painted the beach scene – a painting that has tiny grains of sand that blew in the wind stuck on the surface – which included figures indulging in the new craze for sea bathing. His 'Boardwalk at Trouville' recorded the promenading of the middle classes, and back in Le Havre, having to cling on to his canvas in the wind, he managed to capture the angry white sea buffeting the jetty.

It was in Le Havre in finer weather that Monet made his first uniquely impressionist work, 'Impression: Sunrise'. Sitting at his bedroom window, and using rapid brush strokes, he caught the essence of the early morning as life began in the Channel port. And, although he did not realise it at the time, he had sparked the beginning of the impressionist movement.

'A landscape is only an impression, instantaneous, hence the label they've given us – all because of me, for that matter,' Monet remarked:

> I'd submitted something done out of my window at Le Havre, sunlight in the mist with a few masts in the foreground jutting up from the ships below. They wanted a title for the catalogue; it couldn't really pass as a view of Le Havre, so I answered: 'Put down *Impression*'. Out of that they got impressionism, and the jokes proliferated.

The painting was not well received and was predictably rejected by the judges at the Salon. Two years later, Monet entered it at the first combined show of the Cooperative and Anonymous Association of Painters, Sculptors and Engravers held in the Paris studio of the photographer Felix Nadar. There were works by Renoir, Degas, Pissarro and others.

'Sunrise' got a bad press. Singled out by the critics, reviewers and the public at large, the painting attracted vitriolic criticism. Visitors to the exhibition expressed disgust and outrage, complaining they could not even recognise what they were looking at. Monet toughed it out, for which the art world will be eternally grateful.

His friends were at work further along the coast. Gustave Courbet, who was a witness at Monet's wedding, worked at Étretat. (It was here at Étretat that Monet, getting too close to his subject, was swept off a rock by an unexpected wave. His palette smacked into his face leaving his beard coloured in blue and yellow.) Manet was at Boulogne, Degas at Valery-en-Caux, Pissarro at Dieppe. Eugène Boudin himself painted at Trouville and Le Havre.

And then, some 150 years later, along comes David Hockney. More interested in trees than the sea, Hockney arrived in the autumn of 2018 and settled into a farmhouse with a disused cider press that he converted into a studio. 'The French,' he told *The Times* art critic, 'they love painting. It seems to flow in their blood.' The apple orchard with its river bordered by tall poplars looked, he said, like a Monet, 'And I feel like Monet ... And I am still a smoker like Monet.'

⚉

While the impressionists were at work on the French side of the Channel, the Pre-Raphaelites in England were creating their own challenge to the canon of the day. While their loose community of artists, the Pre-Raphaelite Brotherhood, emphasised the importance of painting from nature and of recording nature in all its wondrous detail, they were not drawn to the changing light over the Channel in the same way as the French impressionists had been. But their landscapes were nevertheless complemented by the waters of the narrow sea.

John Brett sailed around the south-west coast, stopping to paint when the inspiration took hold. His 'The British Channel Seen from the Dorsetshire Cliffs', which he painted in 1871, is a fine example of his work; and 'Britannia's Realm', which he completed nine years later, also received generous praise from the critics.

John Ruskin had set his fellow Pre-Raphaelites objectives that were proving hard to meet. He had limited insights into the practicalities of painting outside beyond the studio. Using oils in the studio was one thing, using oils outside for paintings that took weeks or months to complete in the face of all sorts of weather conditions was something else.

John Inchbold used graphite and watercolours for his 1861 painting of Tintagel, and William Holman Hunt also used watercolours for 'Asparagus Island', which he painted in 1860 from the Lizard Peninsula. But his 'Sheep on a Hillside Above the Channel' was painted in 1862 in oils and was both full of colour and painstaking in its attention to detail in true Pre-Raphaelite style, as ordained by Ruskin, 'Go to nature in all singleness of heart, and walk with her laboriously and trustingly, having no other thoughts but how best to penetrate her meaning, and remember her instructions; rejecting nothing, selecting nothing, and scorning nothing'.

Hunt worked on the painting from Lovers' Seat on the cliffs overlooking Covehurst Bay, near Hastings. It was one of his finest works, yet, ironically, was very nearly lost to a force of nature. Hunt explained at the time:

For two or three days Val [Prinsep] and I remained working on the cliffs. My drawing was on a block [of paper] of which the sun had gradually drawn up one corner; this warped surface did not seriously interfere with my progress until one day a sudden gust of wind compelled me to put my hand on brushes in danger of going to perdition, when, turning round on my saddle seat I saw my nearly completed picture circling about among the gulls in the abyss below. Luckily a fresh gust of wind bore it aloft, until the paper was caught by a tuft of grass at the brink of the precipice. It proved to be in reach of my umbrella which fixed it to the spot until, with the help of my friend, I was able to rescue the flighty thing for completion.

The Channel was also the setting for one of the most poignant pictures of the period. Ford Madox Brown was never a paid-up member of the Pre-Raphaelite Brotherhood but his friendship with Rossetti and other members of the group associated him with the movement in the minds of the critics. His 'The Last of England' was an oil-on-panel painting, completed in 1856 and currently in the collection of Birmingham's Museum and Art Gallery. (There is a watercolour replica painted by George Rae in the Hungarian State Gallery in Budapest.)

The picture shows a couple and their child in an open boat in the English Channel, leaving England on their way to Australia as emigrants. Madox

Brown himself was thinking of emigrating to India, and the Pre-Raphaelite sculptor Thomas Woolmer had already left for Australia.

The models for the work were Madox Brown himself and his second wife, Emma. The man's expression reflects his anxiety about the voyage while his wife remains serene, reflecting her trust in her husband. Madox Brown adhered to the Pre-Raphaelite doctrine of being true to nature and painted the picture outside, having persuaded his wife to sit in all weathers, although not actually in the water. Accompanying the couple in the boat are what he described as 'an honest family of the green-grocer kind'. The White Cliffs of Dover are just visible in the background, and the cabbages around the boat are intended to suggest a lengthy voyage.

Hot on the heels of the impressionists and Pre-Raphaelites, and their challenge to the existing order in the art world, came a composer who would similarly challenge the accepted forms of composition, in his case, within the world of music. The composer was Claude Debussy.

Three giants of twentieth-century music – Pierre Boulez, Daniel Barenboim and the American pianist and jazz-band leader Duke Ellington – all attested to Debussy's influence across a huge range of modern musical forms, leading to what one music critic described as 'a velvet revolution overturning the extant order without upheaval'.

La Mer was part of that revolution. He began to write the work in three sections in Burgundy in 1903 and was quick to admit that 'the ocean doesn't exactly wash the hillsides of Burgundy'. Debussy loved paintings and might well have become a painter himself had the talent been there, but instead drew inspiration from the work of others. Monet and J.M.W. Turner and their paintings of the Channel were of particular influence. *La Mer*'s subtitle 'Three Symphonic Sketches' suggests the work is pictorial in nature.

Debussy loved the sea even though he could not swim, and although he began the work in Burgundy, he moved to Pourville on the French coast, just west of Dieppe, where Monet painted the beach. To finish the work, he moved to Eastbourne where he took a sea-view room with a balcony at the Grand Hotel. 'The sea,' he wrote to his French publisher Durand, 'unfurls itself with an utterly British correctness.' *La Mer*, says the music writer and broadcaster Rob Cowan in *The Gramophone* of September 2018, is 'a cross-Channel triple-tier seascape'. The final violent movement is subtitled 'Dialogue of the wind and the sea' and clearly evokes Turner's 'Fishermen and Sea', a moonlit picture of the Needles off the Isle of Wight with a fishing boat riding the violent waves.

Turner, inspired by sunlight, storm, rain and fog, found it all in the Channel. He produced his Channel sketchbook and pictures of Le Havre, the pier at Calais, Dover Castle overlooking the Channel and pictures of Folkestone Harbour. Yet it was neither a Turner nor a Monet that Debussy chose to front the score of *La Mer*. Instead, he went for Hokusai's woodblock print 'The Great Wave off Kanagawa'.

La Mer was premiered in Paris in 1905, when it perplexed its audience and greatly disappointed the critics. One wrote, 'I do not hear, I do not see, I do not smell the sea.' Another suggested the audience was not served with an ocean but 'with some agitated water in a saucer'.

Fortunes changed for *La Mer* and Debussy when he conducted the work himself in London on 1 February 1908 at the Queen's Hall, then home of the BBC Symphony Orchestra. (The hall stood within a stone's lob from what is now Broadcasting House but was destroyed by an incendiary bomb during the war.) Although reviews were lukewarm, the audience clearly loved the piece.

In the same *Gramophone* issue, the pianist Sviatoslav Richter called *La Mer* 'a piece I rank alongside the "St Matthew Passion" and the "Ring Cycle" as one of my favourite works'. And when the New York Philharmonic reviewed its concert performances since 1917, it found that Toscanini, John Barbirolli, Leonard Bernstein and Pierre Monteux had all included it in their various programmes. Pierre Boulez chose to include it in his last appearance with the orchestra.

The two years Debussy had spent writing *La Mer* had been troubled ones. The biographer François Lesure had described Debussy as 'withdrawn, unsociable, distant and shy', and his fame had become unsettling. His marriage to the fashion model Lilly Texier was falling apart and the composer was having an affair with the singer Emma Bardac, the wife of a Parisian banker who was already having an affair with Gabriel Fauré. Bardac ended that affair and was now carrying Debussy's baby. (Keep up.) On learning this, Lilly tried several times to take her own life and finally shot herself in the stomach. She survived when the bullet stuck in a vertebra.

Bardac married Debussy and was with him in Jersey and then at Eastbourne when he completed *La Mer*. Having received the initial downbeat reception, *La Mer* became one of the most popular pieces in the classical repertoire.

9

Just Another Day at the Office

The writer and poet John Masefield loved the sea and he loved books. Born in 1878, he was orphaned by the time he was 10, disliked school and left when he was 13 to join the naval training ship HMS *Conway* on the Mersey. He said that he went to sea to cure himself of his obsession with books. (Matthew Webb, the first man to swim the Channel, preceded him on the training vessel.)

After three years, Masefield was ready for his first voyage and joined the four-masted barque *Gilcruix*, bound for Chile. Far from curing his obsession with books, he found himself below decks devouring Dickens, Chaucer, Keats, Kipling, Alexandre Dumas and Robert Louis Stevenson. His reading was interrupted by bouts of violent seasickness, and he suffered from sunstroke soon after arriving in Santiago. The young, disillusioned Masefield returned home as a passenger.

Back in England, he took things easily until he recovered his sea legs and set sail again on a cargo ship crossing the Atlantic to New York. He did not enjoy that trip much either and jumped ship to spend a couple of years roaming across America, before getting a job at a carpet factory in Yonkers, Ella Fitzgerald's hometown in New York State. (It happened also to be the hometown of Elisha Otis, the engineer and entrepreneur who gave the world a lift, and there were rumours that Ella Fitzgerald's rendering of the song 'Miss Otis Regrets' was in some way connected. It wasn't. Cole Porter wrote the song in 1934, recording the lynching of a society woman who murdered her lover.)

Back to Masefield, and a carpet factory was not the start in life you might expect of a man who would go on to succeed Alfred Lord Tennyson as King George V's Poet Laureate. Yet carpets failed to smother Masefield's skills as a writer. On his appointment as Laureate, *The Times* declared, 'His poetry could touch to beauty the plain speech of everyday life'.

In 1902 he published 'Sea Fever', written in the first person and telling of his love for life at sea. Then, a year later, and still in his early twenties, he wrote 'Cargoes', a short poem that quickly became one of his best loved:

> Quinquireme of Nineveh from distant Ophir,
> Rowing home to haven in sunny Palestine,
> With a cargo of ivory,
> And apes and peacocks,
> Sandalwood, cedarwood, and sweet white wine.
>
> Stately Spanish galleon coming from the Isthmus,
> Dipping through the Tropics by the palm-green shores,
> With a cargo of diamonds,
> Emeralds, amethysts,
> Topazes, and cinnamon, and gold moidores.
>
> Dirty British coaster with a salt-caked smoke stack,
> Butting through the Channel in the mad March days,
> With a cargo of Tyne coal,
> Road-rails, pig-lead,
> Firewood, iron-ore, and cheap tin trays.

Today, the iron ore shipped through the Channel will likely be on its way to the steelworks in Germany. There will be tankers as long as two football pitches carrying crude oil from Russia, the Gulf and Mexico. Bulk carriers will be bringing grain from America. There will be cars from Japan and South Korea, and top-end European makes on their way to the Americas. Ships piled high

Just Another Day at the Office

with containers full of everything made in China – telephones, toys, t-shirts, toothbrushes, shoes, suitcases, sunglasses, cosmetics, bicycles, tyres, machinery, chairs, tables and even Covid-testing kits – hurry towards ports in the UK, Holland, Germany and Scandinavia.

Ships with refrigerated containers bring pineapples, melons, bananas, oranges, avocados, dates, peaches and kiwi fruit from sunny climes; lamb and venison from New Zealand; dates, almonds and chickpeas from north Africa. Cruise liners bring together the people who like glitzy hotels that float. Ferries shuttle cars and their passengers between the coasts of England, France and Belgium. British and French trawlers ply their trade. Six hundred freighters have been known to pass through the Dover Straits in a single day. The Channel has become a trading highway like no other.

Russell Smith, who captains a car ferry on the Dover–Dunkirk run, compares the passage to a pedestrian crossing a motorway in rush hour. Smith has been at sea for more than twenty years, and if the conflation of shipping – the tankers, container ships, coastal vessels, fishing boats, yachts and overcrowded inflatables – on the two-hour passage causes the occasional stress and a rush of adrenalin, it does not show. There are no furrows on his brow, nor a hair on his head – the latter, he adds quickly, having nothing to do with stress. Nor is there a watch on his wrist. He takes his position on the bridge and runs through the passage plan for the crossing with his deck officers. He knows the direction and force of the wind and the strength of the prevailing tidal currents. And he has had a look on the electronic charts at the traffic moving through the Channel and across his potential path.

The officers of Dover Port Control, situated at their control room on the end of the harbour's eastern wall, give Smith permission to depart on schedule. The quartermaster checks all the navigation systems and everything else that could undermine the ship's safety. Smith signs the pre-departure check list (additional checks were added for ferries after the Zeebrugge disaster when a ferry left the Belgian port with its bow doors open) and notifies his duty engineer.

While loading the cars and trucks – the ship's capacity is 100 lorries and 250 cars – a clever balancing system automatically moves water between tanks on either side of the ship to keep it on an even keel. The ship has bow and stern thrusters, small engines that greatly improve the ship's manoeuvrability in port: these are now turned on. And unlike most other ships, the drive shaft from the engine only rotates one way, so to go astern, the pitch of the propeller blades is changed. The mechanism that achieves this is carefully tested.

Smith and his first officer now move from the controls in the centre of the bridge to the controls on the port wing of the bridge. From here, they can see through 180 degrees and are able to keep an eye on the stern of the ship as they move off the mooring.

Smith gives the order to cast off. The lines are hauled back onto the ship, while the bow and stern thrusters are employed to ease the ship off its berth. Away from the quay, Smith turns the ship to port, heading for the harbour's eastern entrance. The traffic lights at the end of the harbour wall are green, confirming his exit passage is clear.

Out in the open sea, the quartermaster takes the wheel and, a mile clear of the port, engages the autopilot on a bearing of 090 – due east – as the second officer takes command from his captain. Smith is now free to catch up on paperwork in his office immediately below the bridge.

The ship is already approaching the Goodwin Sands, that shifting mass of Channel sand that has claimed so many seafaring lives. But the south-west Goodwin Light is still to the north as the ferry increases its speed to 17 knots. Now comes the first real challenge for the team on the bridge. The ship must turn to starboard to cross the south-west lane of the traffic-separation scheme.

The ship's radar and real-time chart plotters show a stream of ships coming down the lane, and a decision must be made when to cross between them. They have their schedules, just as the ferries do, and for them, the less deviation in speed and direction the better. The ferry's navigation equipment shows the speed of each ship. But the other ships are not the only consideration. The strength and direction of the wind and the speed of the tidal currents come into play. And the separation channel must be crossed at right angles. A ship may deviate from this angle of approach only to avoid a collision.

There is a nautical mile between the southbound ships – sometimes less. Given their speed and limited manoeuvrability, a mile is not much. The gap for the ferry must be carefully chosen. Right now, a less-experienced bridge team might feel their white shirts beginning to stick to their backs. But on this ferry, there is an air of professional calm. Calculations have been made and through their binoculars, the second officer and quartermaster have made visual contact with the line of ships – a tanker, coasters, a container ship and more coasters.

The decision is made to reduce speed and slip through the gap between the third and fourth southbound ships. The bridge team keep a careful eye on the stream of traffic, watching for any change of speed or direction. Past the CS4 light buoy, and the quartermaster turns the ship to starboard to make the transit.

There is a visible change in the body language on the bridge when the ferry is safely through the unending stream of ships. So far so good. Now the bridge team prepares to negotiate the shipping in the north-east channel of the separation scheme.

There is less traffic heading north this morning, so the transit on this side of the separation scheme is relatively straightforward. 'But it is dangerous to be complacent at any stage of the journey,' warns Captain Smith, talking partly to himself and with a nod to his crew.

Through the lane, and there is a turn to port onto an easterly heading again. But it is not yet time to relax, because now the crew must find a passage through the ships at anchor off the French coast. Port officers at Dunkirk are called up on the radio, giving an hour's notice prior to arrival. The French officials ask for the number of crew members and check the captain has a pilot's exemption certificate that qualifies him to bring his ship into port.

They pass south of the orange solar-powered Dyck Buoy, which marks the beginning of the area requiring a pilot, and a member of the bridge team slips out onto the deck to raise a French courtesy flag – a common gesture observed by all ships entering the waters of a foreign country. The red ensign continues to fly at the stern.

The ship is now on course for Dunkirk's west port, the autopilot is switched off and the quartermaster takes the wheel. Catching the crew's attention on the ferry's port bow is the brand-new container ship *Montmartre* – all 400m of her (more than four football pitches). She has just departed Dunkirk West on route for Algeciras on Spain's Andalusian coast, across the bay from Gibraltar. Its bow is painted green, denoting it is powered by liquified natural gas. 'Saving the planet,' quips the first officer. It is piled high with containers – not quite to its 23,000 capacity but not far off. It has its own recycling unit and a swimming pool. It will turn the heads of ancient mariners as it makes its way along the Channel.

As *Montmartre* picks up speed, Captain Smith reduces his and radios the Dunkirk controllers asking permission to enter the port. This he is given, and he takes over control from the second officer. Ten minutes pass and the ship is through the outer breakwaters.

The aesthetics are not entirely pleasing. To the left is the liquified natural gas terminal and to the right, the most northerly of France's Channel nuclear power stations. The beaches of the wartime evacuation are to the north. There is little to please the eye here. Passengers imagining the delights of rural France must surely tread a little warily to their cars.

Captain Smith has moved from the centre of the bridge to the starboard wing controls, where he has clear sight of the quay. Down on the bow, a deck officer with a radio is talking him into the berth. Slowly, slowly, down to just 2 knots. Mooring lines are thrown over the side – three from the bow and three at the stern. With the pitch of the propeller blades reversed, a touch of power brings the ship to a standstill. The mooring lines are made fast. More safety checks, and the order is given to open the bow doors.

It takes about an hour to turn a Channel ferry around. The quartermaster remains on the bridge, keeping watch while the ship is in port. Here in Dunkirk, he might spot a small inflatable dinghy being launched into the canal that runs into the inner port. People smugglers, hindered by an increase in security measures further down the French coast, have moved their operations into the area around Dunkirk. 'It has,' says Captain Smith, 'become a real concern for us. The migrants launch at night, show no navigation lights, are often overloaded and tend not to show up on the radar.' The French authorities put a beam across the canal but that was quickly smashed by refugees desperate to get across to England.

If Smith spots migrants at sea, he reports their position to the French Channel controllers, or to the Dover Coastguard if he is in British waters:

We realised recently that our estimates of people on board these craft was not realistic: we could see the people sitting on the side but not those clustered in the well of the craft. Having seen them at close quarters here in Dunkirk we have doubled our estimates.

If the passengers in these flimsy craft – the ferry crew have seen beach lilos being used – are in distress, the ferry crew would be obliged to render assistance. But in practical terms, the high-sided ferry is not best suited to the task. British and French Search and Rescue vessels that patrol the Channel do the job more efficiently.

The return journey to Dover begins with the captain running the ship astern from its berth until there is space to turn. Then he hands the wheel to his quartermaster and passes control of the ship to the second officer. Smith, who did nautical studies at GCSE and joined the Royal Marine cadets, says there is no other job he would rather do.

Just as landing is the tricky part of flying, so docking is the tricky part of managing a ship. While the skipper on a long-haul voyage might dock four times in a month, Smith docks his ferry four times in a day – in all manner of weather conditions. And he enjoys the challenge.

Ferry captains like Russell Smith combine to make 100 crossings of the Channel every twenty-four hours, sharing the cross-Channel traffic with the tunnel on a 50/50 basis. It was nearly 100 years ago that Stuart Townsend winched and craned fifteen cars into the well of his converted collier to launch a cross-Channel ferry service.

A roll-on/roll-off service began in the 1950s, followed in July 1959 by Christopher Cockerell's first hovercraft crossing – fifty years to the day after Louis Blériot flew across the Channel. The first hovering car ferry crossed from Dover to Boulogne on a summer's day in 1968. It took just twenty-five minutes, reaching speeds of 50 knots. These speedy hovercraft became known, with unfortunate prescience, as the Concordes of the Sea. Like the supersonic aircraft, the passenger hovercraft enjoyed a glamorous but short-lived existence. Although not through any dreadful accident. Rather, with competition from the Channel Tunnel and duty-free sales no longer supplementing ticket sales, the Channel hovercraft ceased to be commercially viable and were withdrawn from service in 2000.

With no sign of Channel car ferries becoming redundant, Captain Smith is looking forward to another twenty years at sea. He wonders what nautical

The Busy Narrow Sea

innovations might be brought to Channel shipping before he steps down from the bridge.

The sun has yet to rise on an inclement morning at Brixham, on the south Devon coast. About 4 miles offshore, the captain of a very large crude oil carrier has requested a pilot to assist his passage through the Channel and the congested Dover Straits. The tanker is on its way to Rotterdam with 2 million barrels of crude oil. In Brixham Harbour, an easterly wind snaps halyards against the masts of the yachts moored along the pontoons. In the blackness, there comes a hint of salt and fish on the wind.

At the nearby pilot's flat, the phone rings. The caller reports there is just one hour to go before the tanker reaches the agreed rendezvous point. Kim Sykes adjusts his uniform, puts his decaffeinated coffee and his milk chocolate bars into his bag, and checks his GPS and electronic chart plotter. He pulls on his safety boots and grabs his high-viz jacket from the hook on the back of the door.

Walking briskly around the harbour, he passes lobster pots and fishing nets piled on the quay. The pilot cutter is waiting for him, restless in the breeze. He climbs aboard, and down below, gets a cool welcome with complaints about the early hour. A hot mug of coffee would have been nice. The talk among the crew, in their slow Devon drawl, is of fishing.

The boatman casts off, and they move slowly from their mooring towards the harbour entrance, passing an incoming trawler with which they exchange greetings. The lighthouse is flashing its position to yachtsmen who may be seeking refuge. Brixham is sheltered from the prevailing westerlies, but this morning's wind from the east is disturbing the usually calm water within the harbour walls. It does not bode well for the state of the sea beyond, and the tanker is 4 miles away.

Having cleared the harbour, the pilot cutter has increased its speed, and the sea, whipped up by the easterly wind, is lifting it on its 8ft waves one minute and throwing it down the next. It is not a comfortable ride. Any hot coffee would not have stayed in its mug.

The skipper has taken note of the wind direction and swell and asked the tanker to position accordingly. The cutter will come alongside to transfer the pilot, a tricky exercise at the best of times, made more challenging by a swell and turbulent seas.

After thirty minutes, the lights of the tanker are just discernible. Sykes prepares himself. The tanker has slowed to 6 knots: any slower and the cutter would have difficulty holding her course; any faster and the interaction with the ship's low-pressure wave would suck the cutter alongside and capsize her.

The cutter manoeuvres to a position along the tanker's leeward side, adopting precisely the same speed, and the two vessels proceed side by side, close enough for the cutter's boatman to touch the tanker's hull – 325m of super tanker and 15m of pilot cutter, like a whale cub clinging to its mother.

Sykes comes up onto the cutter's deck, checking his life jacket. The boatman comes with him. The skipper must keep his boat exactly in tune with the larger ship. A rope ladder is thrown down the side of the tanker's hull. But the cutter is moving up and down on the waves and Sykes must be patient. Quietly, and partly to himself, he says, 'This is where I need to be really careful: it does not do to let this bit become routine. I always do a visual check of the ladder.'

There is a pause as he concentrates, and then, 'This is the hazardous bit', in what sounds like an understatement. And this morning it is.

Sykes is not wearing gloves. 'I like to feel the rope,' he says quickly. He looks behind him to make sure his bag is ready to be hauled up after him. The cutter is still in place, doing exactly 6 knots in time with the tanker. But the rise and fall of the sea takes the ladder out of reach and back again in fractions of a minute. The waves are 2m high.

Sykes pauses to judge the rhythm. Had the ship been rolling in a swell, the ladder would be swinging away from the tanker's side, making the necessary ascent even more difficult. Sykes waits for the ladder to be in reach, grabs it and puts his left foot quickly on the first rung. He has done this so many times, but he must not let the comfort of routine make him complacent.

He is clear of the pilot boat now and, viewed from afar, he will resemble a lizard clinging to the side of a towering wall. He has 6m of ladder to climb to reach the main deck. When a tanker has delivered its cargo, it rides much higher in the water, and an extra ladder will be needed when a pilot climbs on board for the return journey.

He makes the climb safely and is greeted by one of the ship's officers as his bag is hauled up and the cutter eases away for its return to Brixham. His ID is checked and a visitor's permit issued. Visitors' permits might be a twenty-first-century irritation for landlubbers, but in the middle of the Channel? Still, with permit in hand, it is up to the bridge to meet the captain, who turns out to be a flamboyant Russian – not a language familiar to Sykes – with just enough English to brief the pilot about the ship and its peculiarities.

Together, they go over the passage plan. Sykes would rarely take issue with a prepared passage plan unless there had been recent wrecks or if he has knowledge of unusually strong tides and currents. That done, and the ship now on course for the Isle of Wight, Sykes leaves the crew to it and hunkers down for a couple of hours' rest.

What first alarms a visitor to a modern ship's bridge is the absence of a wheel. It would be foolish to expect a huge varnished-wood wheel with handles, the kind Nelson would have stood behind, but there being nothing vaguely resembling a wheel is, well, alarming. However, a more studied examination of all the screens and dials reveals what looks like the yoke in the cockpit of an aircraft. This apparently controls the dynamic positioning system. That's modern ships for you.

When the tanker reaches a point due south of the Isle of Wight, Sykes rejoins the officers on the bridge. The ship must now turn to starboard and head for the start of the Channel's eastbound traffic-separation lane.

'The job of piloting has changed,' says Sykes as he scans the radar:

With the electronic navigation aids available today, the job of steering a ship through unfamiliar waters is so much easier than it was even twenty years ago. What we are mostly here for is to relieve fatigue. A ship's master, who has had responsibility for his vessel for days or weeks, gets tired. And that's bad news when a ship must negotiate a difficult passage at the end of a long voyage. And the English Channel can be difficult.

They wait for a gap to cross the stream of ships travelling west. Dawn has broken and Sykes can use his binoculars as well as the electronic equipment to scan the water. It is the same with most men of the sea. 'Navigation aids are a godsend,' they say, 'but God gave us eyes.'

Once on the French side of the Channel, Sykes decides when to turn the ship to port and begin the north-east passage towards the Dover Straits. It is a short run before they are level with the Greenwich Lightvessel, a floating, red-hulled weather station sitting on the Greenwich Meridian and the spot that marks the starting point for the traffic separation scheme which puts a no-go zone between the eastbound and westbound shipping.

Their passage must be precise because this huge tanker is restricted by its draught. Three red lights in a vertical line on the ship's mast warn other ships about the depth of water the tanker needs, and the limitations this imposes on altering direction.

Once on course, there is a noticeable drop in tension on the bridge. Ferries, says Sykes, are obliged to give way. 'Fishing boats,' and this is a private incantation, 'fishing boats somehow always manage to gather where they give us the most problems.' They can be asked to move, he says, but can't be relied on to do so. However, this morning there are thankfully no fishing boats nor any other unlikely hazards.

His relationship with the ship's master is congenial, and it triggers memories of the Japanese master of a cargo ship who was the 'perfect English gentleman', immaculately dressed and polite to the point of deference. And there was Captain Oh Yoonki, the Korean master of a huge car transporter. His team of officers consisted mostly of women, who would gather on the bridge for coffee and a gossip every morning at 10. 'Unusual,' according to Sykes, 'to find women crewing cargo ships.'

Now approaching the Dover Straits, the ship must report to both the Dover Coastguard and the French Channel controllers with details of the ship's cargo, its draught and the number of people on board. That done, the ship is now on course for the deep but narrow passage of water that avoids the Sandettie Sandbanks, which present a hazard just beyond the north-east corner of the Dover Straits. The Sandettie Lightvessel Automatic, managed by Trinity House, warns ships to stay clear.

For reasons that appear obscure, Poet Laureate Simon Armitage has chosen to title his collection of commissioned work *Sandettie Light Vessel Automatic*. He attempts an explanation in the book's introduction, saying he has been pursuing a kind of writing that views poetry as a trade and as a craft. 'Like the craft in this book's title, perhaps, fulfilling a practical service as a navigational device in the English Channel while offering imaginative possibilities as part of the hypnotic litany of the Maritime and Coastguard Agency's Shipping Forecast.' The light vessel has yet to feature in the shipping forecast, but Mr Armitage is the Poet Laureate.

Once through the Sandettie passage and beyond the poet's metaphorical light ship, the final run into Rotterdam presents Sykes with few challenges. And it is not his job to take the tanker into port anyway; that is down to two more specialist pilots who arrive by helicopter. Sykes bids his captain farewell and takes advantage of an airborne lift ashore, his work done for the day.

Not all ships passing through the English Channel choose to have a pilot on board. There is no legal requirement to do so, and some captains take the suggestion of using a pilot as an affront to their ability to skipper their own ship. British pilots are licensed by Trinity House. On average, some 600 ships pass through the Channel every day. That does not include the fishing boats, ferries or yachts, or the swimmers' support boats. A pool of French and British pilots is on call every day of the year and in all weathers to help the masters of the giant tankers and container ships navigate a safe passage on this final stage of a journey that has taken weeks, or even months, at sea.

According to official reports, the car carrier *Tricolor* did not have a pilot on board. It did not need one. Its captain knew the waters of the North Sea and Channel well. It was a dank, miserable day on the Belgian coast at Zeebrugge. Those of a superstitious nature would later point out that it was Friday 13th (December 2002).

A light mist hung over the ranks of factory-fresh cars lined up in their thousands on the quayside. The sea beyond the port lay still and grey. Dockside drivers in their blue overalls were loading spanking new luxury vehicles onto the car carrier, bound this evening for Southampton and onward to the United

States. There were Saabs, BMWs and Volvos. Some had protective covering over their roofs and bonnets.

Headlights blazing, they filed up the ramps and into the six car decks. With the cars parked, the drivers were collected in people carriers and taken back onto the quayside to collect their next vehicle. When they had finished, 2,781 cars, worth about £30 million were safely stowed. Most of them had customers waiting for them. This was not a particularly large consignment; there are car carriers that can accommodate 8,000 vehicles on fourteen decks.

As darkness descended, the mist thickened to a heavy fog. Undeterred, *Tricolor*'s crew made the ship ready for departure. The long-serving Norwegian skipper took his place on the bridge, and before the port clock on St Donatian's Cathedral struck midnight, the ship slipped its hawsers and eased out into the fog that enveloped the port.

Once in the North Sea proper, the captain set his ship on a course that would take it to a point some 20 miles off the French coast, then through the Dover Straits, past the Varne Lightship, across to the westbound shipping lane of the Channel and into Southampton. The ship's radar would see through the fog.

The first leg of the journey was uneventful. But by two o'clock in the morning, a train of events was in motion that would change this routine passage into a mariner's nightmare. From various accounts and from the reports of court proceedings, it became clear that as the *Tricolor* was on course for Southampton, a Bahamian-flagged container ship, *Kariba*, was steaming at about 16 knots ahead of her in a section of the Channel's traffic-separation scheme. *Tricolor* was half a mile to starboard, steaming at 18 knots and in the process of overtaking. *Kariba*'s navigation lights were just visible through the fog.

It was then that both ships registered radar contact with a third vessel, a Singapore-registered bulk carrier, *Clary*. She was about 5 miles distant, steaming towards them.

It was *Clary* that occupied the attention of the master and second officer on the bridge of *Kariba*. *Clary*'s intentions were unclear to the officers on the bridge of *Kariba*, as the bulk carrier continued to steam towards them at 13 knots. Remarkably, none of the vessels communicated with any of the other ships by radio. Nor did they sound their fog horns.

Kariba's master held her course and speed, assuming the approaching vessel would turn to starboard to pass behind her. But *Clary* held her course too and ploughed on. *Kariba*'s master and second officer were on their ship's bridge, and on perceiving no change in *Clary*'s heading and, fearing a collision, turned hard to starboard through 30 degrees. On so doing, they saw, to their horror, the lights of *Tricolor* and pushed the rudder still further in an attempt to avoid a collision by passing behind. But it was too late. Still steaming at 16 knots, *Kariba* smashed into *Tricolor*'s port beam.

The car carrier capsized within minutes and sank to lie on its side in the mud 30m down. Miraculously, none of the twenty-four crew – the captain, the Swedish cargo superintendent and a crew of twenty-two Filipinos – were injured. Some were picked up by a tugboat that happened to be in the area, the others by a lifeboat launched by *Kariba*.

There was now an urgent need to warn other shipping of the danger posed by the sunken vessel, which broke the surface at low tide but was beneath the waves when the tide was full. Radio alerts were issued and a UK Coastguard vessel joined a French maritime police patrol boat to guard the wreck. Warning buoys were dropped into place as a further precaution. But none of this prevented the Dutch cargo vessel *Nicola* from running onto the sunken hull of the *Tricolor* and needing tugs to pull her clear.

The number of guard buoys was increased, but just over two weeks later, on New Year's Day, *Vicky*, a Turkish-registered vessel bound for New York with 2 million gallons of kerosene, also ran onto the wreck. It lost none of its cargo and was floated off at high tide. Yet this maritime disaster was still not over. Three weeks later, a salvage tug knocked a safety valve on *Tricolor*, resulting in a massive oil spill.

The salvage operation took over a year. The hull was cut into nine sections using diamond-encrusted wire and lifted to the surface by floating cranes. All 2,781 cars were written off. Their metal components were sent for recycling. Insurance companies haggled in the courts, and that fateful night served, and continues to serve, as a stark reminder of the challenges for crews in the congested waters of the English Channel.

The Channel has been the divider between two nations who cannot help but harbour differences and disagreements that go back centuries. And antagonism is never far from the surface when nation speaks to nation about fish.

Few fishermen in Britain or France will have been surprised that the UK's exit deal from the European Union was nearly scuppered at the last minute by arguments over fishing rights. Fishing contributes relatively small amounts to both nations' economies but creates a disproportionately large amount of cross-border angst.

Ever since humankind has walked this planet, delineation of borders has been inextricably linked to the need for sources of food, water, shelter and energy. As the Cambridge historian Renaud Morieux puts it, 'Were fish, wherever they were found, the common property of all mankind? Or did the territorial logic also apply to this roaming manna of the sea?'

The question has vexed governments and their fishing communities for centuries and has set Channel fishermen at each other's throats since they first dragged their nets through this narrow sea. There was a time when ownership of the seas was determined by the distance a cannon ball would travel when fired from the shore. But that did not necessarily confer rights to the *contents* of the sea.

Seaweed was a prime example of the problem. Long used as a source of fertiliser and fodder – and even as a source of soda-ash for making soap – it posed the problem of who it belonged to. It was one thing when found floating on the open sea but quite another when washed up on the shore.

Throughout the eighteenth century, the *guerre des algues* (the war of the weeds) set the farmers, fishermen and kelp burners of Normandy and Brittany against one another. At the same time, French fishermen were looking for fish along England's southern coastline, while English fishermen were doing the same along the coastlines of Brittany and Normandy. A recipe for trouble was in the making.

Shellfish created a particular problem because they live on the seabed where they are creatures of the sea while the tide is in, but creatures of the land when the tide is out. As creatures of the sea, does that make them legally available to all comers or, as creatures of the land, are they the property of the landowner?

The scallop, which thrives in the Channel, has claimed more than its fair share of trouble-making. During an otherwise quiet night in August 2018, five British scallop dredgers from Brixham and Peterhead were fishing legally in the bay of the River Seine between Barfleur and Le Havre. They were surrounded by forty French boats from ports along the Normandy coast who maintained the British were fishing out of season. Rocks were hurled, smoke bombs thrown and boats rammed, and the British boats were forced to retreat.

Sure enough, the Brexit deal triggered more French aggression in May 2021 when an armada of boats threatened to blockade the harbour at St Helier on Jersey. The island, claimed the French, was imposing unfair and arbitrary rules. It took some old-fashioned gunboat diplomacy to see off the French intruders.

Yet, while the jealousies and distrust simmer, Devon's Channel trawlermen remain the most productive in British waters. Take the helm Adam Cowan-Dickie.

At the start of 2019, Mr Cowan-Dickie's fame was vicarious – his son Luke was playing rugby for England in the World Cup. By the end of the year, fame was his own. Adam and his five-man trawler crew landed the most valuable catch Brixham had ever seen. The haul of Dover sole – the crown jewel for Devon fishermen – sold at the town's auction for £125,000. Of Mr Cowan-Dickie, more later.

While fishing contributes a small percentage to the GDPs of both Britain and France, it has in the past been a vital industry for towns on both sides of the Channel. In the western Channel, shoals of pilchards (sardines longer than 15cm) swimming near the surface could be spotted by look-outs on the cliffs, who would signal to fishermen below. Small-mesh cotton nets were deployed in the shape of a horseshoe around the shoal and gathered in by teams of fishermen in low-sided open boats.

In 1587, Thomas Bodley, an Exeter-born scholar, parliamentarian (he represented Portsmouth) and diplomat, married a Devonian widow called Ann Ball, whose late husband had made a fortune selling Channel pilchards to the French. Ten years into their marriage, Mr and Mrs Bodley spent a hefty portion of Ann's inheritance on a library that became the centre of learning for the English-speaking world.

The Bodleian Library at Oxford University, which was to become one of the world's great libraries, had been built on the backs of a widow's pilchards. The work was completed in 1602 and Bodley was knighted two years later.

Pilchards do not, however, appear on college high tables, not even in occasional celebration of the founding of the great library. What on earth would the university's wine masters select to accompany pilchards?

By the eighteenth century, the notion that territorial waters were defined by the distance a cannon shot would travel had been replaced by an (albeit unwritten) rule that sovereignty extended a nautical league (3.5 miles) from the shore. This convention of 1793 held, more or less, for eleven years, when five British fishing vessels were escorted by French cruisers into the port at Dieppe for having fished a no-go area of oyster beds between Jersey and the French coast. In Parliament, the Foreign Secretary of the day mused that the oysters had gained sound intelligence about the no-go fishing zone and used this knowledge to establish a safe colony. He further assured the house that communications from his man in Dieppe confirmed the British ships were to be released imminently.

Oysters had been popular among both rich and poor since the fifteenth century, a food encouraged by the Church, which designated days of the week when eating the flesh of animals and birds was forbidden. Fridays and Saturdays were fish days.

Come the seventeenth century, the notion that it was unhealthy to eat fish and meat together was disappearing, while at the same time the Church's edicts were being followed less religiously. Sausages of pork or mutton had oysters added, while turkey stuffed with oysters and dressed with oyster sauce was a treat on a rich man's table. But in time, polluted waters led to a shortage of the mollusc and a challenge for fishermen creating oyster farms in the Channel along the coasts of Sussex and Brittany.

In Victorian times, while grand banquets invariably included a course of fine, fresh oysters, pickled oysters were a staple in poorer households. 'Poverty and oysters always seem to go together,' remarks the astute Sam Weller in Dickens' *Pickwick Papers*. In more recent times, the English palate has eschewed the oyster, while the French continue to knock them back with undiminished relish.

Does this appetite for oysters do anything for the French libido? The shellfish is rich in zinc, essential for testosterone production and for the maintenance of healthy sperm. And the oyster has been shown to boost dopamine, a hormone known to increase libido. Giacomo Casanova, the Venetian-born polymath,

The Busy Narrow Sea

was said to eat fifty oysters for breakfast. *The Story of my Life*, his erotic memoir, written in French, certainly reflects an energetic sex life, but as yet there is no scientific evidence to support the oyster lovers' claim that shellfish boosts sexual arousal in men – or women. But that has not prevented French women saying they would rather their men ate oysters than downed little blue pills.

In the estuary of the River Fal on Cornwall's Channel coast, today's oyster fishermen work in much the same way as they did in Dickens' time – and do nothing to discourage the perceived connection between the little mollusc and an appetite for sex. The sail-powered wooden boats are allowed to fish between 9 a.m. and 3 p.m. on weekdays, and between 9 a.m. and 1 p.m. on Saturdays. There is no fishing on Sundays. The season opens on 1 October and closes on 31 March. Boats' engines must not be used while fishing, and no mechanical gear used to pull the nets in.

These gaff-rigged sailing craft – a four-sided sail aft of the mast, with a jib and foresail forward of the mast – crewed by two or three fishermen, drift downstream on the tide and then sail back up again on the wind for the next run. A decent wind gets them back upstream in forty-five minutes. Powering the boats and relaxing the hours for fishing would lead to overfishing and an end to those healthy oyster beds which benefit from a mixture of Atlantic and brackish waters.

Chris Ranger skippers the 25ft *Alf Smythers*. With his long, straggly beard, light-blue woolly hat and blue-and-yellow waders, he and his traditional oyster boat – oiled pitch pine on an oak frame with a rust-coloured mainsail – are at one. What you don't see are Ranger's special waterproof socks – a vital part of his layers of winter fishing gear.

In a good week, Ranger will collect 2,000 oysters from the bed of the Fal Estuary. They are at their optimum size after four or five years; smaller ones that slip through a measuring ring must be thrown back. British waters are not the cleanest in the world and the oysters must be purified in seawater tanks treated with ultraviolet light to kill off dangerous microbes.

Ranger sells his cleansed oysters for 80p a throw to upmarket restaurants in London. When it comes to oysters, food writers have adopted the wine writers' desperate and flowery hyperbole. My favourite line describes the wild oysters of the Fal as 'blooming with notes of melon and cucumber before easing into a metallic finish'. The metallic finish apparently comes from Cornwall's tin mines. The mines produced millions of tons of tin in their heyday but have been closed since 1998. Bretons describe their farmed rock oysters as having 'a touch of hazelnut'. Codswallop!

Ranger, who lives alone in a cottage decorated with pictures and paintings of his traditional oyster boat, has yet to find a pearl. He thinks it unlikely that the constantly moving waters of his estuary would allow an irritant like tiny particles of sand or a parasite enough time to settle in the lining on the inside of the oyster shell. The irritant needs to remain in place long enough for nacre, layers of aragonite and conchiolin, to encase it and protect the mollusc from it – it is this tiny ball of nacre that becomes the pearl. (The economies of Bahrain and Qatar relied on pearls until the oil arrived. The abundance of warm-water parasites and the quiet seas of the Persian Gulf provide an ideal environment for pearl-producing oysters. Georges Bizet's opera *The Pearl Fishers* was set in what was Ceylon. South-east Asia and Australia have been and remain important producers of oyster pearls.)

Ranger is the first to recognise that this small mollusc is more than the food of love. It is, he says, no less than an ecosystem's engineer, cleaning water, absorbing CO_2, and helping a diversity of seawater species survive. 'The oyster is capable of making huge changes to our marine environment,' confirms the Zoological Society of London. And Ranger, the former computer marketing man, is at the heart of Cornish attempts to make this happen.

The herring is altogether less romantic – lots of bones, no pearls and no increase in libido. It had been fished in the North Sea and in the Channel for centuries until 1977 when its depletion in stocks brought about a five-year ban (although some local trawlers defied the ban and continued to fish small quantities of spawn herring). The fish is rich in protein and vitamin D. It becomes a kipper when smoked, and pickled and rolled it becomes a rollmop.

It spawns in the eastern reaches of the Channel between November and January when an average-sized female will deposit upwards of 40,000 eggs on rocks and seaweed.

The fish remains popular with the French, and the Herring Festival is a high point on Boulogne's calendar. But to the English, it has largely lost its appeal (and its value), and today's fishermen put their nets out for mackerel, mullet, plaice, turbot, a variety of shellfish, and Dover sole.

Adam Cowan-Dickie's record catch still stands. Mr Cowan-Dickie skippers one of the twenty beam trawlers that operate out of Brixham on Devon's southern coast. Fishing is in the family. He has five brothers, four of whom are at sea. His own crew of four includes two of his sons, Zac, who is 22, and his elder brother, Rico, 32. Channel fish are top quality, says Cowan-Dickie senior, 'These waters provide a good breeding ground, and the water quality is good too. You don't see sick fish in the Channel. No fish with worms in them.'

Their feeding ground is less predictable. 'You never know what the sea might throw at you, so you need to know your boat well. Twice a day I'm down checking on the engine. You must keep dry: your hands and feet especially must stay dry.' In his modern cabin there is air conditioning and radiators. On deck in winter, he wears thermals, two shirts, oilskins and thermal boots.

The first and last time he was seasick was when he was 14, but days on end of rolling and pitching are still exhausting, even if they no longer cause him to throw up. It is not, he says, a job for women. Yes, he knows it sounds sexist, but it is very hard work for seven days non-stop, and anyway, there is no room below to accommodate a woman with the appropriate privacy.

Adam is cook as well as captain and engineer. On the first evening of a trip, he serves up a light meal for the sake of everyone's stomach (stomachs get used to the sea after that). Then it's fresh fish for breakfast every morning, and an evening meal at 7.30 sharp. A fridge and a freezer carry plenty of food for the trip. But there's no alcohol, not even a tot of naval rum.

Cowan-Dickie and his crew take their 30m beam trawler (beam trawlers shoot their nets out over both beams instead of over the stern) to sea for seven days at a stretch. Then it is back for a night off before setting out again for another seven days. Two weeks on and one week off. He and the crew work six-hour shifts, day and night. They trawl for ninety minutes, haul the nets in, empty them and shoot them back out again.

The skipper plots a grid for the boat to cover, the length and breadth depending on what the echo sounders are telling them about the likely shoals underneath them. 'When you bring the nets in you never know what you will find in them. You win some, you lose some.'

Mr Cowan-Dickie wins more than he loses. He is an experienced and respected trawlerman and knows his Channel well. He is fishing for Dover sole, turbot, plaice, brill and lemon sole and in the winter months, cuttlefish. The tide, wind and temperature all play a part in filling the nets. A one-degree change in the water temperature will influence the fish he catches.

Historical catches play a big part in being at the right place at the right time. A detailed diary records what fish he has caught where and when, and under what climatic conditions. Cowan-Dickie says he finds top-quality sole, turbot, mullet and monkfish around Hurd's Deep, the deepest part of the Channel, north-west of Guernsey. But there is a downside to fishing here. The Deep is on the fringe of the eastbound lane of the Channel's traffic-separation scheme. It is here that fully laden tankers and giant container ships vie for space. They travel one behind the other at speeds of 20 knots (23mph) and need a lot of persuading to change course.

'We can see all the ships on our electronic real-time chart plotter,' explains Cowan-Dickie. The chart plots every ship's progress, records its name and shows where it is heading:

> We can call them up if we feel threatened, but half the time there is no one on watch so we don't get an answer. So we call the coastguard and ask them to call on our behalf. But that does not guarantee a response from the advancing vessel.

Cowan-Dickie keeps one eye on his plotter and one eye trained across the sea through his binoculars. 'It's a race track in this part of the Channel, and you never know if these giant ships will change course.'

Most often, it is a case of give and take, and collisions in the Channel involving fishing boats are relatively rare. In 2001, a Breton trawler sank after colliding with a Norwegian tanker, and in 2011, a French whelk fishing boat was sliced in half by a high-speed ferry, killing the fishing boat's captain. In November 2020, a Brixham scalloper sank off Newhaven with the loss of two crew members.

Cowan-Dickie is clear that safety comes at the top of a skipper's responsibilities. His rather unusual crew of a first mate, his two sons and two deck hands from the Philippines make a good team that generates a healthy camaraderie. He says local youths are no longer interested in a life at sea, and those who do show an interest don't have the stamina to last the course. Cowan-Dickie does not need new crew members with experience, rather he is happy to mould a man from scratch. He is looking for team players and people with a good work ethic. He gets both from his sons and his Filipino deck hands, who spend eight months of the year fishing with him and four at home in Manila.

Cowan-Dickie, bespectacled, with a strong beard but little hair and a tattooed right arm, looks very much at home in his wheelhouse full of screens and all the other electronic paraphernalia of a modern trawler. He exudes the kind of quiet confidence of a man who knows the sea, his crew and his boat – and has respect for all three. The boat, he says, has a place in his heart, and you can hear these are not just pretty words. His cabin behind the wheelhouse has everything he needs, including a television and a telephone. A John Grisham novel is next to his bunk. At home, he gets his exercise on a bicycle and his fun on a sexy Suzuki Bandit motorbike.

There are twenty beam trawlers like his at Brixham, but local crews are becoming harder to find. He was for Brexit, but says he and his fellow trawlermen are now worse off. 'I would never vote Conservative ever again.' The words of a disillusioned, angry man.

Since the end of the Second World War, Britain has been served by sixteen prime ministers, nine of them Conservatives. During the same period, the Fourth and Fifth Republics in France have seen ten presidents in the Élysée Palace. But none of them have managed to put an end to the cross-Channel antagonisms which continue to sour relations between the two countries. Yet, the last time an invader made it across the Channel was in 1066.

Just Another Day at the Office

The historian Andrew Roberts points out that it was not the Channel alone that saved Britain from invasion, 'What finally scuppered Napoleon's Europe was of course the fatal combination of the English Channel and the Russian winter; the same unlikely partnership that also did for Hitler's Europe.'

Centuries earlier, William Shakespeare had not envisaged any frozen parts of Europe coming to the defence of England. His King Richard II believed the Channel could do the job on its own, serving 'in the office of a wall/Or as a moat defensive to a house'.

The English Channel continues to witness ugly scenes involving French and British fishermen, but rather than serving as a defensive moat, it has become one step on the road to Europe, *une étape* on the road to Britain. Ferries carry millions of travellers across Shakespeare's silver sea. More choose to cross beneath the waves in trains. Above the waves, up to 600 vessels – freighters, cruise ships, trawlers and naval vessels – ply their trade between the North Sea and the Atlantic Ocean every twenty-four hours. On average, just forty ships transit the Suez Canal in the same period. And crossing the paths of the giant freighters and ferries come the small craft, mostly cheap rubber dinghies, full of desperate migrants beholden to criminal smuggling gangs and risking their lives on the final stage of a journey to a better and safer life.

The Channel has attracted the creative minds of writers and poets; inspired the creative talents of artists; challenged the spirits of countless adventurers and swimmers; served as a defensive moat; introduced Victorian bathers to the sea and provided settings for the birth of seaside resorts. Its waters have been the scenes of bitter battles – as have the skies above them.

It has been, and still is, a productive fishing ground, while the winds which once drove the 130 ships of the Spanish Armada up the Channel now power 100 turbines off the Sussex coast. The Channel's islands and islanders have between them endured occupation during the Second World War; created tax havens; nurtured one of the world's most successful cattle breeds; developed an early and popular potato; housed some of the country's most violent and dangerous criminals and promoted a pop festival that crowned the counterculture of the 1960s. Above the waves, the Channel provides a racetrack for yachtsmen and pigeons; below, tunnellers produced one of the great engineering feats of the twentieth century.

It is a seaway without parallel, linking and separating two nations that find it so hard to get on. And it separates an island state, not just from its immediate cross-Channel neighbour but also from the wider continent of Europe. An English Channel reflecting the islanders' ambivalent feelings about being Europeans.

Acknowledgements

Many people contributed thoughts and ideas for this book, and I am grateful to all of them.

Michael Delahaye took time to read the first draft and make corrections and suggestions where appropriate, while providing much-needed encouragement. It would have been a lesser book without his involvement.

My thanks to Jesper Christensen at the ferry operator DFDS for giving me access to one of his cross-Channel ferries. Russell Smith, Kim Sykes, Adam Cowan-Dickie and Chris Ranger were all generous with their time in explaining their various working practices on boats in the Channel. And Sarah-Jane Sterling was wonderfully open and honest about her cross-Channel swim.

Mary, who chooses to have her surname withheld, generously shared her family's experiences on the island of Guernsey during the Nazi occupation.

My wife Aileen was always patiently on hand with support, and with her knowledge and understanding of grammar that is infinitely better than mine.

In addition, I am grateful to the administrators of John Masefield's estate for allowing us to reproduce his poem 'Cargoes', and I am also grateful to Kenneth Beken for allowing us to use pictures from his family archive. And my thanks to Amy Rigg and Chrissy McMorris at The History Press for guiding the book through the publishing process.

Every effort has been made to trace copyright holders and to obtain their permission to use copyright material. I apologise for any errors or omissions, and on learning of such will ensure that corrections are incorporated in any further reprints or editions of this book.

Bibliography

Aldersey-Williams, Hugh, *Tide: The Science and Lore of the Greatest Force on Earth* (Penguin, 2017).
Armitage, Simon, *Sandettie Light Vessel Automatic* (Faber, 2019).
Arnold, Matthew, *'Dover Beach'* (1867).
Barnes, Alison, *Henry Winstanley* (Tourist Information, Saffron Walden, 2003).
Barnes, Julian, *Cross Channel* (Vintage, 2009).
Briggs, Asa, *The Channel Islands: Occupation and Liberation* (B.T. Batsford, 1995).
Calder, Nigel, *The English Channel* (Penguin, 1987).
Chateaubriand, François-René de, *Memoirs from Beyond the Tomb* [*Mémoires d'Outre-Tombe*] (Penguin Classics, reprint 2014).
Cicero, Marcus, 'Letters from Marcus Cicero to his brother, Quintus' on en.wikisource.org
Cunliffe, Tom, *The Shell Channel Pilot* (Imray, 2010).
Defoe, Daniel, *A Tour Thro' the Whole Island of Great Britain* (1724/The Folio Society, 1983).
Dickens, Charles, 'The Calais Night-Mail' in *The Uncommercial Traveller* (Oxford University Press, 2021).
George, Rose, *Ninety Percent of Everything: Inside Shipping, the Invisible Industry ...* (Picador, 2014).
Gibson, Robert, *Best of Enemies: Anglo-French Relations Since the Norman Conquest* (Impress Books, 1995).
Greene, Graham, *Brighton Rock* (Vintage, 2004).
Hall, Evelyn, *Friends of Voltaire* (Smith, Elder & Co., 1906).
Halley, Edmond, *Correspondence and Papers* (Arno Press, 1975).
Hasenson, Alec, *The History of Dover Harbour* (Aurum Special Editions, 1980).
'History of Manston Airfield' on manstonhistory.org.uk.
Kipling, Rudyard, 'A Smuggler's Song' in *Puck of Pook's Hill* (1906).
Leroy, Maxime, *Charles Dickens and Europe* (Cambridge Scholars Publishing, 2013).
Lesure, François, *Claude Debussy: A Critical Biography* (Fayard, 2003).

Maitland, Rear Admiral Sir Frederick (KCB), *The Surrender of Napoleon* (William Blackwood and Sons, 1904).
Morieux, Renaud, *The Channel: England, France and the Construction of a Maritime Border in the Eighteenth Century* (Cambridge University Press, 2016).
Northern France (Insight Guides, 2011).
Rainsford, Dominic, 'Crossing the Channel with Dickens', in Anny Sadrin (ed.), *Dickens, Europe and the New Worlds* (St Martin's Press, 1999).
Rattenbury, John, *Memoirs of a Smuggler* (John Harvey, Sidmouth, undated).
Roberts, Andrew, *Napoleon the Great* (Penguin Books, 2015).
Rousmaniere, John, *The America's Cup 1851–1983* (Pelham Books, 1983).
Russell, William Clark, *History of Britain and Ireland* (British Library Editions, 1889).
Shakespeare, William, *Complete Works* (Collins Clear-Type Press, 1923).
Sharp, Nigel, *Dunkirk Little Ships* (Amberley Publishing, 2015).
Smith, Charlotte, *The Poems of Charlotte Smith/Beachy Head with Other Poems* (1807).
Smith, Peter C., *Hold the Narrow Sea: Naval Warfare in the English Channel, 1939–45* (Moorland Publishing, 1984).
Tombs, Robert, *This Sovereign Isle: Britain In and Out of Europe* (Allen Lane, 2021).
Trevelyan, G.M., *History of England* (Longman, 1926).
Waugh, Mary, *Smugglers in Kent and Sussex* (Countryside Books, 1985).
Whiteside, Thomas, *The Tunnel under the Channel* (Rupert Hart-Davis, 1962).
Whitfield, David John, *Marconi Radio Officer: The Convoys of a WW2 'Sparks'* (CreateSpace, 2015).
Wordsworth, William, *The Poems of William Wordsworth*, en.wikisource.org.

Additional Sources

Financial Times.
The Guardian.
Hansard, 18 June 1940.
Illustrated London News.
Journals of Queen Victoria (made public by Queen Elizabeth II in 2012).
kiplingsociety.co.uk.
Le Monde.
New York Times.
Report of Mr Justice Sheen's Public Inquiry into the Zeebrugge Disaster.
The Times.
Victor Hugo in Guernsey Society.

Index

A Smuggler's Song 70
Addington, Dr Anthony 36
Aga Khan, the 41
Agricola 15
Alanbrooke, Field Marshal Viscount 134
Albert, Prince 56, 94–9, 152, 162
Alderney 15, 27, 68, 77, 87, 89, 90
Alfonso XIII, King 41, 42
America (yacht) 98–100
Antipodes 28
Ariel, packet paddle steamer 152
Armada, Spanish 28, 57, 211
Armitage, Simon 199, 200
Arnold, Matthew 161, 177, 178
Arnold, Samuel 55
Associated Press 42
Audley End House 19
Aurora (yacht) 99

Barclay brothers, the 91, 92
Barnes, Alison 19
Barr, Charlie 100
Baude, Patrice 49
Bay of Fundy 14
Bay Psalm Book, the 58
Bayeux tapestry 16
BBC 141, 87, 88, 138, 167, 172, 180, 187
Beagle 14, 65, 66

Beckett, Samuel 180, 181
Beken, photographers 100–2
Bellerophon, HMS 59–65
Benard, Andre 164, 165
Benedictine monks 50, 51
Bernhard, Prince 133
Bertrand, Fanny 64
Bertrand, General Henri 62, 63
Bessemer, Henry 176
Betjeman, John 46, 47
Blake, William 172
Blanchard, Jean-Pierre 8, 157
Bleriot, Louis 8, 155–7, 195
Bodley, Thomas 204
Boleyn, Anne 54
Bonaparte, Napoleon 48, 60–5, 82, 83, 161, 173
Bonham, Edward, British Consul in Calais 56
Boudin, Eugene 30, 37, 181, 182, 184
Boulogne 7, 8, 15, 48–50, 52, 130, 133, 150, 160, 175, 184, 195, 208
Brecqhou 78, 92
Brett, Jacob 29
Brett, John 184
Brighton 7, 8, 35–45, 94, 130, 131
Brighton Rock 43
Bristol Channel, the 14

217

Britain, Battle of 136, 137
British Expeditionary Force 130, 131
Brittany 13, 18, 179, 203, 205
Brixham 196, 198, 204, 208–10
Brotherhood, the Pre-Raphaelite 184, 185
Browning, Robert 163, 175, 176
Brunel, Isambard Kingdom 162
Burchett, Josiah 16, 18
Burnley Miners' Social Club, the 51

Cabourg 16
Caligula 30, 48
Cameron, Julia Margaret 96
Cap de la Hague 15
Careme, Antonin 40
'Cargoes', Masefield poem 190
Carlyle, Thomas 175, 176
Casanova, Giacomo 205
Chamberlain, Neville 133
Chanel, Coco 44
'Channel Firing', poem 47
Charles I, King 103
Charles III, King 77
Charlestown 25
Chateaubriand, François-René de 179
Chaucer, Geoffrey 16, 189
Chelmsford, Bishop of 136
Cherbourg 15, 68, 146
Chichester, Francis 66
China, People's Republic of 28, 191
Churchill, Winston 39, 87, 133, 136
Cicero, Marcus 15
Cinque Ports 75, 149
Clary, bulk carrier 201, 202
Clinton, P/O 140
Clouded Yellows (butterflies) 30
Cockerell, Christopher 195
cod 49, 51, 78, 85, 91
'cod houses' 85, 86
Colman the Younger, George 55
Condor Legion 137
Constant, cargo ship 20
Conway, HMS 189
Cornwall 24, 71, 155, 206

Courbet, Gustave 182, 184
Cowes 94, 95, 97–100, 172
Creasy, Capt. George 134
Cubitt, Thomas 94
Cumberland, Duke of 39, 42
Codrington, HMS 132–5
Cowan-Dickie, Adam 204, 208–10
Cawsand 57–60, 66–8
Calais 7, 14, 52–6, 70, 83, 137, 149–51, 153, 156, 157, 159, 160, 162, 165, 174–6, 187
Croizon, Philippe 9, 30
Cap Gris-Nez 13, 49, 156
Caesar, Julius 15, 16, 48, 141

D-Day 8, 89, 141, 142, 144
Daguerre, Louis 182
Darwin, Charles 65, 66
de Gamond, Aime Thome 161–3
de Gaulle, General Charles 56, 87, 138
de la Haye, Hugh 84, 85
de Maupassant, Guy 50
de Morny, Duc 37–40, 46
de Quincey, Thomas 151
de Saint-Pierre, Eustache 52
de Vienne, Jean 52
Deal 73
Deauville 7, 3035, 38, 40–6, 50
Deborgher, Pascal 49
Debussy, Claude 186–8
Defoe, Daniel 21, 71, 73
Deptford 18
Devon 29, 68, 69, 142, 196, 204, 208
DH34A bi-plane 155
Dickens, Charles 48, 173–7, 189, 206
Doodson, Arthur 142
Dover 12–15, 27, 29, 31, 49, 52, 130–5, 139, 149, 150–61, 170, 173, 174, 176, 186, 187, 191, 194–6, 199, 201
Dover sole 28, 49, 204, 208, 209
Dowding, Air Chief Marshal Sir Hugh 136, 138
Drake, Sir Francis 57, 78
Dubroca, Jean-Louis 12

Index

Dumas, Alexandre 37, 189
Dunkirk 83, 132, 134, 141, 150, 155, 193, 194, 195
Dylan, Bob 108

Eddystone 7, 20–2, 25
EDF, French energy company 14
Edward III, King 52, 55, 86
Edward the Confessor 149
Elizabeth I, Queen 54, 78, 79
Élysée Palace 164, 210
English lace 55, 56
Esmonde, Lt Commander Eugene 139, 140
Eurotunnel 164, 165

Farouk, King 41, 42
Fécamp 49–51
Firebrace, Henry 103, 104
Fitzgerald, Ella 189
Fitzherbert, Mrs Maria 40
Fitzroy, Robert 14, 65, 66
flounders 28

Gare Saint-Lazare 39
Garibaldi, Giuseppe 95, 96
geometry, chair of 19
George III, King 36, 39, 45
Gibbard, Professor Philip 11, 12
Goliath, paddle steamer 79
Göring, Hermann 136, 137
Gort, General Lord 131–3
Gossamer Albatross 156
Gravelines, *ville des smoglers* 83
Green, Winifred 87
Greene, Graham 43
Greenpeace 13
Gregory, Tom 30
Guernica 137
Guernsey 77–82, 86–8, 93, 99, 148, 209
Guerre des Algues (War of the Seaweed) 203
Guilford, Tim 146
Guise, Duke of 55

Gulliver, Isaac, smuggler 45
GUNS (Guernsey Underground News Service) 87
Gupta, Sanjeev 12

Hall, Henry 23
Halley, Edmond 7, 16–19, 28
Hamburg 28
Hamdallah, Abdulfatah 159
Hardy, Thomas 47, 48
Harris, Jersey Lieutenant-Governor Edward 82
Hastings, Battle of 16
Hathaway, Dame Sybil 92
Hawkhurst smugglers, the 68
Henry VIII, King 54, 96
Herald of Free Enterprise 153
herring 8, 28, 49
Hitler, Adolf 48, 87, 88, 90, 135–7, 139, 146
Holford, Harvey Leo 43
Holland, Henry 40
Honfleur 30, 36, 37, 182
Hong Kong 91
Hôtel du Golf 41
Huffmeier, Vice Admiral Friedrich 148
Hugo, Victor 77, 80, 81, 94
Hundred Years War 52, 75, 169, 174
Hunt, William Holman 185
Hurd's Deep 27, 209

Imperial College 12
Impressionism 30, 181, 183
Irving, Sir Henry 46
Isle of Wight 18, 94, 95, 97–100, 102, 103, 108, 136, 144, 158, 187, 198

Jersey 51, 67, 77, 80, 82–8, 148, 188, 204, 205
Jersey Royal (potatoes) 84, 85
Jethou, island of 77, 93
Jones, Emlyn 144
Juliana, Princess 133, 134

Kanaal, Engelse 13
Kariba, container ship 201, 202
Khan, Ambassador Mirza Abul Hassan 58, 59
Kipling, Rudyard 48, 70, 74

L'Illustration, illustrated Paris weekly 37
La Manche 7, 13, 33
Land's End 13, 71
Lander, Edward 56
Le Chatillon 49
Le Druillenec, Harold 87
Le Grand, Alexandre 50
Le Havre 13, 39, 98, 130, 181, 183, 184, 187, 204
Leopold, King 132
Les Misérables 7, 77, 80
Lister, Hilary 158
Liverpool Tidal Institute 142
Lizard, the 18, 57, 185
longitude 17
Louis XIV, King 20
Lynn, Vera 29

MacCready, Dr Paul 156
Mackenzie, Compton 93
Mackerel 28, 49, 208
Maginot Line 132
Maitland, Captain Frederick 60–5
mal de mer 155
Malles 150
Manny, Sir Walter 52
Maritime and Coastguard Agency 14, 200
Mary, Queen 55, 79
Masefield, John 189, 190
Mathieu-Favier, Albert 261, 262
Mayfield gang, the 72
Mayflower 57
Meteorological Office 14
Meuse, river 11
migrants, cross-Channel 33, 159, 161, 194, 211
Miller, Glenn 147, 148
Minas Basin 14
Monet, Claude 7, 30, 37, 39, 181–4, 186, 187

Mont-Louis 13
Montgomery, Major General Bernard 134
Montmartre, container ship 193, 194
Monument Securities Ltd 153
Moon, Rear Admiral Don 143
Morse, Samuel 29
Morton, Alastair 164, 165, 167
Mountbatten, Vice Admiral Lord 143
Mozin, Charles Louis 36, 37

Napoleon III 41, 162
Negra, Daljit 161
Nelson, Admiral Lord 29, 73, 133, 198
Newfoundland fishing grounds 51, 78, 85
Newton, Isaac 19, 180
Nicholas, Tsar 40
Normandy, hotel 41
Normans, the 130
North Sea 12, 132, 138, 200, 201, 207, 211
Northcliffe, Lord 48, 156
Nova Scotia 14

Odo, Bishop 16
'Old Brass Brains' 141
Olliffe, Dr Joseph 38
Operation Dynamo 132
Operation Tiger 142
Osborne House 94–7, 101, 102
Osmond, Chief Constable Douglas 108
Ostrum, Hans 176
Otis, Elisha 189
Overlord, Operation 142, 144–6
owlers 72
oysters 91, 205

packet boats 70, 132, 150–2, 174, 175
Palmerston, Lord 56, 162, 163
Pantcheff, Captain Theodore 89, 90
Paramour 17
Parkhurst, prison 102, 104, 107
Parry, Hubert 172
Pavilion Blues, magazine 131
Peel, Robert 94
Pevensey 16
Philippa, wife of King Edward III 53

Index

Pickles, Lord Eric 90
Pickwick Papers 205
pigeons 8, 30, 99, 145–7, 212
pigs' bladders 67, 162
Pissarro, Camille 7, 30, 182–4
Pitt, William, the Younger 70, 76, 83
Plymouth 14, 20–2, 33, 57, 58, 60, 64–6, 135, 142, 144, 148
Portland 14, 18, 24, 47, 144
prince regent 40, 61, 65
privateers 20, 78, 150

Railway, London–Brighton 38, 43
Ramsay, Rear Admiral Bertram 132
Ranger, Chris 206, 207
Rashleigh, Charles 24, 25
Rattenbury, John 68–70
Remigius, Bishop of Lincoln 50
Rennes 14
Rennie, John 38, 39
Renoir, Pierre-Auguste 81, 183
Rhine, river 11
Riga 13
Rob Roy, steam packet 151
Rodin, Auguste 53, 174
Rogers, John 74
Romans 7, 94, 129
Rommel, Field Marshal 141, 142
Romney marshes 75
Rotterdam 28, 196, 200
Royal Guernsey Light Infantry 86
Royal Pavilion, Brighton 8, 130
Royal Society 19, 176
Royal Yacht Squadron 97–9
Rudyard, John 23
Russell, Dr Richard 35, 36, 39, 41

Sadek, Narriman 41
Saffron Walden 19
Sandettie Lightvessel Automatic 199, 200
Sark 77, 87, 88, 91–3
scallops 28, 49
Scandinavia 28, 191
Scharnhorst 138–40
Scheldt, river 11

'Sea Fever' 190
Sea Lion, Operation 136
seaweed, source of fertiliser and fodder 29, 85, 203, 207
Seely, Charles, Liberal MP 95
Shakespeare, William 7, 27, 169, 170, 211
Shelley, Percy Bysshe 46
Shelton, Major 88
Sisley, Alfred 30
Slapton Sands 142, 144
Smeaton, John 24, 25
Smith, Capt. Russell 191–5
Smith, Charlotte 7, 170, 172
smuggling 45, 63, 66–78, 82, 160
soldiers, Indian 130, 131
Somme, battle of the 129
South Korea 28, 190
Soviet Union 13
'spotsman' 67
St Helena 17, 64, 65
St-Malo 7, 14, 51, 77, 88, 94, 146, 179
St-Valery-en-Caux 16, 134
Stevens, John Cox 98
Stevenson, Robert Louis 46, 188
Stevenson-Reece, George 155
Stirling, Sarah-Jane 31
Strabo, Greek historian 15
Straits, Dover of 12–14, 130, 132, 135, 139, 149, 191, 199, 201
Surrender of Calais, Opera 55
Swinburne, Charles 50
Swordfish bi-plane 139, 140
Sykes, Kim 196–200

Tacitus, Cornelius 15
Tandy, George 145
Tapps-Gervis, Sir George 45
Taylor, Elizabeth 20
Tennyson, Albert Lord 95, 96, 108, 163, 190
Thames, River 11, 15, 18, 71, 82, 131, 135, 162
Thatched House Tavern, the 97
Thatcher, Margaret 44, 154, 164
Thomas, Sarah 30
Thomson, William (Lord Kelvin) 141

tides 14–18, 28, 46, 57, 141, 177, 198
Tombs, Robert 12
Tomkins, Gabriel 71–5
Townsend, Stuart 9, 152, 153, 195
traffic separation scheme 192, 193, 199, 201, 209
Tregonwell, Lewis 45
Tricolor, car transporter 200–02
Trinity House 20, 24, 132
Trouville 37–9, 183, 184

U-boats, German 130, 132
Utah Beach 142
Utrecht, Treaty of 51

Vendroux, Yvonne 56
Victoria Cross 133, 145
Victoria, Queen 8, 40, 42, 56, 80, 94–7, 99, 101, 102, 152, 162
Victory, HMS 29
Vierge 13

Vincelli, Dom Bernardo 50
Voltaire 179–81
Von Mensdorff-Pouilly-Dietrechstein, Count Albert Joseph Michael 46

Walker, Charles 29
Webb, Matthew 9, 30
West Polmear 24, 25
Western Front 129
Whiteside, Thomas 162
Whitfield, Walter 20
Whittington, Dick (Sir Richard) 7
Wight, Isle of 8, 18, 92, 94, 95, 97–100, 102, 103, 108, 136, 144, 158, 187, 198
Wilde, Oscar 39
William I, King 7, 16, 50, 141
Wilson, Harold 163
Winstanley, Henry 7, 19–21, 23, 24
Wolffe, Jabez 31
Wool 53, 54, 72–5
Wordsworth, William 7, 95, 172, 173

Also by the Author …

COCONUT
HOW THE SHY FRUIT SHAPED OUR WORLD

ROBIN LAURANCE

978 0 7509 9061 5

'A bounty of a book.' – Sir Peter Stothard, former editor of *The Times*

Robin Laurance looks beyond the oils and health drinks to uncover the unexpected, often surprising, and vital roles played by the coconut palm and its nut in times past and present.

The History Press

The destination for history
www.thehistorypress.co.uk

Save Me, Stranger

Also by Erika Krouse

Tell Me Everything: The Story of a Private Investigation
Contenders
Come Up and See Me Sometime

Save Me, Stranger

Stories

ERIKA KROUSE

FLATIRON
BOOKS
NEW YORK

This is a work of fiction. All of the characters, organizations, and events portrayed in this collection are either products of the author's imagination or are used fictitiously.

SAVE ME, STRANGER. Copyright © 2024 by Erika Krouse. All rights reserved. Printed in the United States of America. For information, address Flatiron Books, 120 Broadway, New York, NY 10271.

Stories from *Save Me, Stranger* have appeared in a slightly different form in the following:

"The Pole of Cold" in *One Story*, "The Piano" in *Alaska Quarterly Review*, "North of Dodge" in *Glimmer Train*, "Eat My Moose" in *Conjunctions*, "Save Me, Stranger" in *Boulevard* under the title "Lotus," "When in Bangkok" in the *Kenyon Review*, "The Standing Man" in the *Iowa Review*, "Jude" in *Colorado Review*, "Fear Me as You Fear God" in the *Southern Review*, "The Blue Hole" in *Glamour*, and "Wounds of the Heart and Great Vessels" in *Crazyhorse*.

www.flatironbooks.com

Designed by Susan Walsh

Library of Congress Cataloging-in-Publication Data

Names: Krouse, Erika, author.
Title: Save me, stranger : stories / Erika Krouse.
Description: First edition. | New York : Flatiron Books, 2025.
Identifiers: LCCN 2024010597 | ISBN 9781250240330 (hardcover) |
    ISBN 9781250240347 (ebook)
Subjects: LCGFT: Short stories.
Classification: LCC PS3561.R68 S28 2024 | DDC 813/.6—dc23/eng/20240315
LC record available at https://lccn.loc.gov/2024010597

Our books may be purchased in bulk for promotional, educational, or business use. Please contact your local bookseller or the Macmillan Corporate and Premium Sales Department at 1-800-221-7945, extension 5442, or by email at MacmillanSpecialMarkets@macmillan.com.

First Edition: 2025

10 9 8 7 6 5 4 3 2 1

For JD and K

*Stories are the only enchantment possible,
for when we begin to see our suffering as a story,
we are saved.*

—Anaïs Nin

Contents

The Pole of Cold   1

The Piano   23

North of Dodge   33

Eat My Moose   54

Save Me, Stranger   73

When in Bangkok   93

The Standing Man   111

Jude   127

Fear Me as You Fear God   139

I Feel Like I Could Stand Here with You All Night and It Would Be the Worst Night of My Life   168

The Blue Hole   177

Wounds of the Heart and Great Vessels   194

Acknowledgments   207

Save Me, Stranger

The Pole of Cold

I live in the coldest town on earth.

You may have heard some debate about it. Our Siberian village, Oymyakon, reached minus 71.2 degrees Celsius in 1924, but they had to guess at the temperature because the thermometer froze. Another Far East town tried to claim the title, but then Oymyakon erected a sign in Russian that said THE POLE OF COLD, and it was settled. Signs can do that.

The mountains surrounding the Oymyakon Valley create a natural inversion; chilled air suspends in place while heat flees the earth. Birds freeze to death in midflight. Hair sticks to the bed, and eyeglasses fuse to the face. You can hammer a nail with a banana. You can hear a voice outside for over a mile. To find a person, you can follow the trail of breath as it hovers behind.

Boiling water thrown from my kettle crystallizes into a cloud of powder midair. Vodka freezes, salt water hardens, plastic shatters, steel splinters. You can't touch a doorknob with your bare hand. You can't turn off your car and expect it to start again. Everything breaks. Even sap sometimes turns to ice, and trees explode.

There is one east-west road spanning Siberia, and without it, we would die. No one knows how many gulag prisoners perished building the road; some say one million. Their bodies became part of the Kolyma Highway itself, Stalin's Road of Bones, and the only way here is to drive over them.

Like our neighbors across the Bering Strait, we worship the same bears who try to eat us every spring. About eight hundred people live in my village, almost all of us native Sakha. I'm a twenty-two-year-old woman and the mayor of Oymyakon. I always imagined when my mother returned for me someday, we would leave this place together and I would never look back.

This winter the weather scientists didn't arrive until early February. For money, my aunt Lyuda and I often host one or two of them for a week. We have Papa's empty room, I speak English, and the money helps.

We need the scientists, but I don't like them in my home. They talk and flirt. Trying to impress me, one of them once told me he had helped with the Alpine excavation of Ötzi the mummy, a.k.a. "The Iceman." Ötzi was a prehistoric herder who left his village in Anatolia. Outside his territory, between two peaks in the Alps, four men ambushed and killed him with an arrow and blows to the head. Ötzi lay facedown inside a glacier for five thousand years, perfectly preserved with the blood evidence of his murderers, never to return home. The Iceman found the courage to leave his known world for an unknown one, to grow beyond his borders, even if he had to leave his family behind. I wondered how my mother had found that courage. Or if it was courage.

"The American will be here soon. His name is An-der-son," Lyuda now said. "This one isn't a scientist, just a tourist. Go meet him, Vera; my feet hurt."

Out the window, I watched my neighbor Aytal's truck separate from the convoy. You have to drive here in convoys, because if your truck breaks down or you get a flat tire, you could freeze to death before another truck comes along. To pass a stranded vehicle without offering help is murder here, and it's illegal. Aytal always drives the scientists, now that his father is older. His truck began heading our way, so I pulled on my coat and walked outside.

It was almost minus 60 degrees Celsius, and I had to cover my mouth with a wool muffler to breathe. Aytal braked the truck, rolled down his window, and smiled shyly, the frostbite marks jumping on his cheeks. We went to school together; he'd proposed to me seven times, but I always said no. "Vera," he said, like my name was made of sugar, but I just nodded and wiped my nose.

The truck door opened. A lanky man stumbled out, encased in bright orange nylon.

"Mr. Anderson?" I asked in English.

"Urrgrrth," the man said.

"I'm Vera. Welcome to the Pole of Cold." I picked up his enormous bag and tugged on his arm. "Come. Please." The man walked like a robot. I had to tow him and his belongings from the road to our house.

Once inside, he stood beside the door, a mute orange mass. He was as tall as a Norwegian. His bag was bigger than our table and dominated the living room. "Sit, sit," Lyuda said in Sakha, and I pushed him into a chair.

"You need to take off your coat so the warm air can get in." I peeled off layers of artificial fabrics until a human body emerged. He was about my age. I pinched the ice from his eyelashes, and he blinked his green eyes. His cheeks were a just-slapped pink above a reddish-gold beard that was soft when my hand accidentally brushed it.

"It's cold," he finally stammered in English.

"You'll get used to it."

Lyuda handed him a cup of tea. He hunched over it. I took off my own coat, muffler, and hat, and the man gasped when I turned back around.

I forgot to mention I am beautiful.

I have heard all the words for "beauty" in Sakha and Russian, and the village boys and men used to look up words in French and Italian, trying to outdo each other. Every year since I was fifteen, the Miss Sakha Republic pageant asks me to compete, but I throw their letters away. I always imagined if I lived in a place where I wasn't covered in skins all day, my beauty might have purpose. Here, you have to muck stables and gut fish no matter what you look like.

When the man went to Papa's room to change, Aunt Lyuda wiggled her eyebrows. "Your age *and* handsome," she said in Sakha.

"He'll hear you."

"He doesn't speak anything."

The American emerged from Papa's room. He had taken off his knitted hat. His hair was wavy and a glinty shade of orange and gold, a little lighter than his beard. It looked soft, too, like he had brushed it.

Lyuda clunked bowls onto the table. "Stop staring and eat," she told the man in Sakha, and lit a cigarette while we sat.

"*Zdravstvuyte, spasibo,*" he said in Russian so bad it was nearly gibberish. His nose twitched. "What's in the bowls?" he asked me in English.

"That's reindeer jelly and tongue. And *khaan*—horse blood sausage. Salad is here," I said.

"Salad?" He frowned. "It looks like frozen meat."

"Frozen fish. We call it *stroganina*. Salad."

"I'm a vegan," he said. "I don't eat meat."

I translated this for Lyuda.

"How can he live if he doesn't eat?" She poked me. "Ask him."

"Meat," he specified. "I don't eat animal products."

"That's our food," I said. "Reindeer meat. Blood. Yogurt. Horse." He wrinkled his nose again, as if I had said something disgusting. I pushed the *stroganina* so it rocked his plate. "Here, this is just fish."

"You don't eat any vegetables? Local plants? Grains?"

I pointed out the window. "You want some lichen?"

At four o'clock, the sun was already down, and fog covered the rising moon. "I guess I'm eating meat then," the American said. He took a bite of intestines and smiled, food in his teeth. Lyuda poured him a glass of *kymys*, fermented mare's milk we save for special occasions, which he glugged, choked on, and then sipped. "Vera, please tell your mother it's delicious."

"She's my aunt."

"Oh," he said. "Where are your parents?"

"I don't know, where are yours?" I snapped.

My mother, Starkova Tuyaara Zaharovna, left with one of the weather scientists when I was a baby. She crawled into the truck with him, and Aytal's father drove it away. That was his job. He last saw her in Yakutsk, boarding a plane to Moscow with the rest of the scientists. But maybe she went farther than that, maybe even New York. Those scientists were from New York. I didn't know who gave her the money for her ticket. Regardless, my mother was still married to my father. I thought she should know her husband was dead, and I was still alive, waiting for her to return and take me away.

I don't remember my mother, and aside from muttered comments, nobody but Aytal had ever told me anything important about her. She left nothing behind, no clothes or furs or jewelry. When I try to imagine her, sometimes I see my own face, and sometimes I see only white space. Like white snow, the white arms of larch trees, my own white breath hovering in the white air. My mother is everywhere.

"Your English is amazing," the American now said. "I hadn't expected someone so fluent. Where did you study?"

"My papa taught me."

"How did he learn?"

"He went to college in Yakutsk. He made me read the whole English dictionary and books by Ernest Hemingway, Herman Melville. When people visit the city, they bring back English books for me."

"You're absolutely beautiful," he told me. "Do you know that?"

My aunt nodded at her plate, uncomprehending, while this American flirted in English in front of her. It gave me that prickly feeling you get when you come inside and the blood returns to your face. "You never properly introduced yourself." I leaned back in my chair. "That's rude in our culture."

"It's rude in every culture. I apologize." He pressed his palm to his chest. "I'm Theodore Anderson. Most people call me Theo."

Green eyes, red hair, Theo, *Thyo*. I recognized this name. I recognized this face.

"Are you okay?" he asked.

I pushed away from the table and retreated to my room. I grabbed my wooden treasure box from under my bed. The door to my room didn't shut all the way, so I leaned my back against it.

"Vera," my aunt called sharply and tried to push the door open. Then she returned to the kitchen and said to the man who couldn't understand, "I'm sorry about her. Eat. You're not hungry?"

I lifted the gold locket from my box. The metal stung my skin. I clicked open the clasp, pulled out the tiny slip of paper I had tucked into the center, and read what I had written seven years ago: *Thyo*. The picture inside the locket was faded, but I could just make out two flecks of green in the teenage boy's eyes and that reddish-gold hair, lighter in the photo. I memorized the tiny photograph all over again, with the door against my back. That boy was in our kitchen right now, grown into a man.

"Get him out of here," I yelled to Lyuda in Sakha. "He can stay in town with the scientists."

"No way," she yelled back. "He's paying us double." I heard murmuring, and Lyuda again yelling to me, "Take him to the store tomorrow. He keeps saying something called 'vegetables.'"

I lost my father in the plane crash here, seven years ago. I was fifteen. Maybe you read about it. He was out with the reindeer. I was currying snow from the horses' coats. The temperature was fifty-five below Celsius. I couldn't feel my face, and I liked that feeling; it was safe and cozy and reminded me of how warm my arms were in my reindeer coat. And then my favorite horse, Omtoon, looked up.

A plane going down is a memorable sound. It's a powerful drone, so loud it hurts. Metal screeches against itself. Shrieks snuffed out. Then silence and snow, which has its own sound.

I was screaming for my father. That has its own sound, too. Before the plane hit him, he was running in the middle of the field, waving at me, waving me back. Over the noise, he shouted, "Stay where you are." The reindeer had already fled from him. He didn't even look back at the plane, which grew so big, so fast upon his small body. It aimed straight for him, as if trying to find the one person in the vast taiga to kill.

After the plane crashed, I ran through the smoke to the fire, shouting for my father, my bucket clanging against my legs until I remembered to drop it. I ran around the plane in widening circles, searching the snow, but he was nowhere. The putrid fumes congealed in the air, and the plane itself moaned. Screams inside. I pulled on the door, and people fell out.

A woman caught me by the ankle as she hit the snow. Her face was half-black and burned off. She grabbed for my hand and pressed something into my mitten, something metal. It was a necklace, a locket. Then she said, "Theo," and fell still.

Later, I learned that the plane was carrying scientists from America, a married couple. They were on one of those expeditions to check out our weather instruments and feel the temperatures for themselves. But they didn't drive from Yakutsk for three days on the Road of Bones like the other scientists, the poor ones from universities. These ones were famous and rich, and they bribed a Yakutsk pilot two thousand dollars to fly them here.

Airplanes are not supposed to fly over this valley in winter. Even our own birds can't. The plane that killed my father dug deep into the permafrost, burned, and then iced over. It took chain saws to detach what body parts they could from the plane and ship them home to America.

The locket that lady gave me was gold and heavy, the only piece of metal jewelry I have ever owned. I didn't realize it opened until days later. It held a picture of a boy inside. I wrote down the English word the lady said, as I had heard it—*Thyo*—and tucked that scrap of paper into the locket. I didn't know the name, and the word meant nothing in any of our dictionaries.

They are a burden, our dead. In old times, we performed sky burials. We would wrap the body in canvas and perch it high in a larch tree. After the Russians conquered Sakha, they made us bury the bodies in the ground, even in winter. So we build a bonfire. It melts a few centimeters of permafrost, and we dig it up. Then we build another bonfire on top of it, and then dig under that fire, and so on for three days, until we have a hole deep enough for the corpse. Even then, the permafrost rejects the bodies. It pushes their bones back to the surface over time, so sometimes we have to do it all over again. This is how we care for the souls of the dead—with a fire and digging.

But my papa. He's still under the plane.

At the general store, our neighbor Yegor clapped my shoulder and said, "Vera, nice foreigner you've got. Amerikanetz?" The American circled the store, rubbing his mittens together and inspecting the snapshots tacked to the wooden paneling.

"From New York," I said in Sakha. "He won't eat our food. What news is there?"

"I'm out of food," Yegor said. He's Oymyakon's heating engineer and hasn't been paid by the Russian government in over a year. Without him and adequate firewood, everyone in our town would die within five hours.

I gave Yegor part of the money we got from the American and told him, "We've got some reindeer fat. And I'll contact the government again."

"How would I survive without you?" Yegor said, his face exploding in wrinkles. Yegor came here from Moscow to slip the authorities. Whatever he did there, he doesn't do it here. Many of our ancestors once escaped from something—their pasts, the law, famine, the Great Terror. Some say Oymyakon was founded on escape, when Mongol horsemen ran from Genghis Khan. It's safe to hide where nobody dares follow you.

The American touched items: frozen lingonberries, milk, vodka. Finally, he paid for a sack of flour, a package of macaroni, three Bounty bars, and a long can of foreign potato chips left over from the Christmas wares. He walked in a stiff-legged waddle.

"I discovered the hard way I can't digest meat anymore, and that outhouse is pretty far," he said. "Maybe I should have prepared more for this trip." I didn't think preparation was his problem. That morning he had put on seven layers—his parka over a jacket over a vest over a sweater over a shirt over a T-shirt over a silk undershirt. He looked like an orange stuffed bear. "At least the store carries these twelve-dollar potato chips." He pulled off the top and offered me the can.

I had never tried them before. I picked out a chip and ate it. "It tastes like nothing and salt."

He read the label. "I think those are the first two ingredients." We ate the chips together in the store, our teeth crunching the cardboard food. Neighbors touched Theo's arm in passing, welcoming him, and he made friendly noises.

"Listen," he said, turning to me. "I know you don't like me. I'm sorry for whatever I did. I'm not used to the culture here."

It was hard to stay angry at this man for what his parents did to my life. He was orphaned, too. But friendliness felt disloyal, so I escaped outdoors. The air cooled my face and calmed me.

Theo followed me outside and gulped. "Yup. Still nippy."

"We're getting near the record, I think. It's at least sixty below."

"The same temperature as Mars," he said.

"Mars doesn't have air like this. Listen." I exhaled, and the vapor turned to ice crystals, followed by the tinkling sound as they collided and fell. "We call that *shepot zvezd* in Russian, 'the whisper of the stars.'"

Theo breathed out himself and listened, grinning. Then he said, "You know, I could tell you things, too. Things that don't revolve around temperature. I could teach you how to hail a cab or make a real salad or customize your iPhone. There's more to life than subsistence—" He was interrupted by a crunching sound, the potato chip sound, but nothing was in his mouth. We stared at each other and then at his coat.

It was shriveling. Before our eyes, the coat cracked into long orange strips. Feathers dumped out and drifted to the ground. Then his snowpants began to do the same.

Fear saturated Theo's face. "This gear is GORE-TEX and eight-hundred-fill goose down. It's made for Mount Everest."

We had spent too long outside for foolish American clothing. "Come." I pulled his sleeve, and part of it came off in my mitten.

There was another hissing sound, and his plastic boots split their tops like two smiles. "My feet," he said. We ran home, his broken boots squeaking. It took only a few minutes, but in temperatures like these, that can be enough to kill a man without fur. Theo was stumbling by the time I shoved him inside our door and slammed it.

I pushed him into a chair in front of the woodstove and pulled off the remains of his boots. I wrapped a blanket around his feet and another around his knees, the body part most vulnerable to freezing. He rubbed his arms and chest until he could speak. "I don't have any other gear." He glanced around our house at our fireplace, our little television. "I'll have to stay here until summer. Cut off from the world."

"This is the world, too." I lit the burner under the kettle for spruce tip tea and a hot-water bottle, and used my knife to wedge loose oakum back into a drafty gap in the wall by his head.

"I'm the Iceman." Theo's voice was flat. "Oh my god. The Iceman stayeth." Feathers still leaked from his parka. He looked like a shattered tropical bird. "I'm going to lose my feet and my cheeks and die here."

"Calm down. You just need some fur," I said.

When Papa was buried under the plane, he was wearing his new fur coat, still under mortgage. The one that was now hanging in his closet was the old reindeer coat I remember him in. It was so worn the stiff hide had become almost supple. The fur lining was bald in the seams and under the arms. It had a stain on it from where I once spilled a bowl of blood gravy all over him. It still smelled like my father—sweat and food and mare milk and hair.

I pulled the coat off its hanger and handed it to Theo, along with Papa's suede undergarments and my grandfather's decorated reindeer boots, a family heirloom. "This is what we wear here," I said.

"I'm opposed to fur," he said. I pushed his hand through the armholes of the long coat anyway. The coat fit, as did the boots after I removed the balls of wool from the toes. I wedged Papa's old arctic fox hat onto Theo's head, and it puffed out on all sides. "I feel ridiculous," he said. But his cheeks grew pink, and he smiled. "I'll bet my head looks pretty big now."

"It looked big when you stepped out of the truck." I brushed off the coat, and he blushed.

I didn't know how I felt about giving my father's coat to the son of his killers. But if I'd let Theo die in his torn plastic clothing, I would have been a murderer. I tugged hard on both sleeves of my father's coat and said, "This isn't a gift."

I was sleepless all night and didn't stumble into the kitchen until almost lunch, after the winter sun had already risen. When there are only four and a half hours of daylight, we hoard that time, and Lyuda was mad. "I fed and milked the animals," she reproached. "The American helped. Sort of." Theo blushed at the sight of me. Lyuda handed me tea, and I drank it slowly to let Theo get used to me again.

He cleared his throat. "Please show me the airplane. The one that crashed here seven years ago."

"Too far," I said. "You'll freeze."

"They told me it was right there, in that field next to yours." He pointed out the window to our pasture. "That's why I'm staying with you."

"What does he want?" Lyuda asked. "Does he want to try ice fishing? It's too cold, I think."

"He wants to see the airplane."

"Then show him the airplane."

I made an annoyed sound, but that didn't sway the American. He touched my arm. "Please," he said. "Take me there, or take me to someone who will. I'll pay."

There was no way this man was going to my papa's grave without me. So after I finished the remaining chores, we left together.

Trudging toward the field, the American looked up at the bright, milky sky. "What do people do here? What do you do?"

"I help teach school. I do languages: English, Russian, and Sakha. School closes when it's minus fifty-five Celsius or more, like now, so I have the day off." Maybe I could ask him for a donation for the kids. He was rich enough to come here. "I'm also the mayor."

"Mayor?" Theo almost stopped walking, but it was too cold. "Aren't you a little young for that job?"

"My papa was the mayor of Oymyakon. After he died, no one had the heart to replace him, so I took over. I was fifteen. It was supposed to be temporary."

"I'm impressed."

"It's no big deal. Mostly I nag the government for salaries and run the town meetings." I didn't mention that I also sign certificates for tourists to celebrate their bravery for spending a few days in the place where we live all the time. Or that we built a post office and the new school with indoor plumbing, and hired two new teachers from Yakutsk and one from Moscow. We're not drowning at the bottom of a collective vodka bottle, like some of the other towns in Sakha. We're living.

I stopped walking, and Theo did, too. The airplane lay before us, no more than a snowy mound. The actual plane is visible only in summer. I used to spend hours there after the thaw, standing in puddle grass, mosquitoes feasting on me while I waited against hope for movement.

After the plane crashed, the town tried to retrieve my father's body. But the plane was too big and the ground too hard, and they didn't know where he was, exactly. After the thaw, they lit fires to dig a tunnel through the permafrost but soon realized they would never be able to find him without potentially cooking him first. Lyuda and I told them to stop. So there he stayed, buried under the wreckage.

I furtively dropped some pancakes in the snow to bribe the gods to protect my papa. Theo wiped his nose against the sleeve of my papa's old coat and stared at the plane with wet, freezing eyes. A child's love has its own vastness, bigger than its container. Like the ice inside our exploding trees, it must have somewhere to go.

It was time to turn back, but Theo didn't seem cold anymore.

"My parents died here," he finally said.

"I know." Then I said, "My father, too."

Theo followed my gaze to the snowy mass. My grandfather's boots were silent on his feet. He squinted, and then his eyes cleared. "The herder," he said. The crescent of his exposed face turned pale. "That was your father. Of course."

The scant sun was already beginning its descent under the fog.

"I don't want to talk to you anymore," I said.

Theo caught me with one mitten. He pulled me close and kissed me. His lips were freezing, coated with particles of ice, countering the white-hot current shocking my body, deep inside my furs. Something started to melt, and solidify, and melt again. *This is how we die*, I thought. *This is how body parts stick to metal, how dogs lose their tongues.* But I couldn't stop, and we didn't die. We stayed until it was dangerous, and I kissed the son of my father's murderers with all the heat I had stored inside myself for twenty-two years.

Theo went with me that evening to our town meeting. "Go," Lyuda said, waving him off. "Meet the village." We gather at the city hall twice a month. We arrange to fix broken pipes, talk about government back pay, schedule supply trucks, trade goods, plan *munkha* ice fishing and Yhyakh festivals, order vehicle parts for repairs, and decide punishments for crimes. Everyone welcomed Theo with pats, and I called the meeting to order in Russian. Theo watched while I heard a dispute between two brothers over a horse. We discussed permafrost sinkage from climate change and ideas for reinforcement.

A town elder said, "But who will do all this work? Our young people are leaving for Yakutsk or the mines. That's the real problem."

They were right. After graduation, we had no real way to advance beyond subsistence. Me either. No college, few jobs. The ceiling of achievement was directly above our heads, hovering in the low, cold clouds. "People come back sometimes," I said.

"Nobody ever comes back."

"Like your mother," someone muttered, and the room got very quiet for a few moments. I picked up my gavel, but didn't know who to throw it at.

"What just happened?" Theo whispered.

People had brought food to the meeting, and Yegor went home with a full sled: crucian carp and perch, half a gunnysack of flour, our reindeer fat, sacks of lingonberries, a block of frozen mare's milk, and plenty of *Russki chai*, vodka. Aytal gave him a carton of cigarettes and tried to give me a box of pasta, but I just thanked him and handed it to Yegor.

On the way home, Theo walked close beside me, and I was glad for the barrier of cold and coats between us. He asked, "What were you talking about?"

"Food, trucks, things like that."

"It seemed pretty important."

"It is."

"Someone else could arrange all that, though, right? If you weren't here."

"But I am here."

"But if you weren't."

Sometimes I imagine the town as a giant tangled ball of string and myself as the empty center. Theo said, "I thought being a mayor was mostly public relations."

"It's listening. My father was good at that. People need a lot of listening."

"Well," he said. "Anyone can do that." But he wasn't listening.

When we got home, I gave Theo the locket. I told him how his mother pressed the piece of metal into my mitten. How quickly she died, how his name was the last thing she said. How his father's arm had frozen to her torso within a minute, fused with blood.

When I finished, he sat at our table, opening and closing the locket with his thumb. I sat with him, resisting the urge to touch him, to try to soothe him. There was no soothing this. Time passed. A silent late dinner, *indigirka* with wild onion, frozen foal liver, and pickles. After dishes, Lyuda yawned and quietly played her mouth harp. She knows things without knowing them.

Theo finally spoke. "Tell your aunt I wish to marry you."

"Marry? You've been here two days." I got hot again, hotter than summer.

"Don't pretend you want to stay in this place."

"What did he say?" Lyuda asked.

"He wants a different towel. He doesn't like yellow," I told her.

"All our towels are yellow. Tell him," Lyuda told me.

"She said you can't marry me," I told Theo.

"She did not."

"You don't speak Sakha. You don't even speak Russian."

Theo leaned over and pounded the table with a pink fist. Lyuda's eyes flared. "*I*"—he stabbed a finger at his chest—"want to marry *Vera*." He mimed sliding a ring onto a finger. English, Sakha, or Russian, the meaning was clear.

Lyuda's chin rose. "*Da*," she said.

"*Da*? Yes?" Theo asked.

"Yes means no in my culture," I said.

"Vera, come on."

"You're upsetting her." I wanted Theo to stop, and I was also terrified he would. But this was not supposed to be the way I left home, with a marriage to this man. I was supposed to leave for a better life with my mother, one she had been building for me all along. If I left without her, how would she ever find me?

Theo sighed, exasperated. "You're so smart and beautiful." When I rolled my eyes, he said, "You must want more than this. In America, you can—see great art. Think about something besides survival. Go to college."

College.

I don't know what my face was doing, but Theo's eyes and voice softened. "A good college. Study Melville and Hemingway. Or anything you want. Libraries, with more books than you could read in your lifetime, all free."

I tried to imagine it. Houses filled with books in English. "Free?"

"Free books. Museums. Vegetables. Central heating. Summer."

"We have summer here."

"I've heard. It's when all the outhouses melt. You could live somewhere temperate."

"So where is this temperate town with the college and free books?"

Theo straightened and his voice deepened. "In New York. With me."

I envisioned a new city, made of metal. Buildings scraping the sky. No fresh air. People shooting each other with guns, people living in the same place who don't know each other and never will. Getting old and dying on the other side of the world from where I was born. I pictured my mother in a fur hat, walking on the pavement, clinging to the arm of a scientist. The idea stretched inside me like a crow testing its wings. I could go find her, not the other way around. I would recognize her somehow, touch her arm, and she would know who I was, that I took the same risk she did.

But maybe there are too many people in New York for anyone to recognize anybody.

Theo pressed a hand to his chest. "There's this ice inside my lungs all the time here. It hurts to breathe. Doesn't it get to you?"

Lyuda was watching us, her head swinging back and forth, and I felt suddenly ashamed to even think of leaving her. I said, "It costs over a hundred thousand rubles to go to America. And they wouldn't let me stay."

"I have lots of rubles. You'd get a green card if I married you."

I was used to proposals, but still I blushed. "You love me?"

He said, "I don't think there's a word for what I feel about you."

"*Da*," Lyuda said. She had already begun to cry.

I knelt before her and held her hands. "But, Lyuda. What would you do? What would happen to the town?"

"The town always survives. What will happen to *you*, Vera?" Tears wet her face, and she rocked back and forth. "You don't live. You only wait, watch, listen. Once I'm gone, you'll die alone."

"I won't die alone," I said. "We have men here."

She shook her head. She was right—they'd all proposed to me, even some of the married ones, and I'd said no, no, no. There was nobody left.

"Don't die alone, Vera," she said.

That night when everyone finally calmed down and went to sleep, I got out of bed and dressed in the dark. I slipped outside, quiet as a sable. I didn't know what time it was, but the frigid air staggered even me. A hint of a moon lurked under the fog. I trudged in darkness to Papa's plane as Theo's ice formed in my chest. I had never noticed it until he described it, but there it was, slowly freezing me from the inside.

I watched the mound in the scant moonlight. I didn't know how to say goodbye to a man under a plane under the snow. Would I miss this? The snow seemed to breathe back at me. Then I realized something was actually breathing and grunting. "Papa?" I stepped closer.

A brown bear walked out from behind the plane.

He was skinny. Hunger had woken him up. He winter-walked toward me, his empty belly swaying, claws puncturing the ice. He was made of two things—sleep and hunger. I'd brought no gun, no knife.

I realized I had now lived my whole life. I would die right there, beside my father. When the bear charged, I would run; he would catch me and eat me alive, entrails first.

The bear's big head rose to sniff the air.

"Stay where you are," my father's voice said in my ear.

I froze in place. The bear sniffed again, his head bobbing up with each breath. With one mighty paw, he scratched at the mound, the plane.

He swaggered closer. His paws crunched on the snow, erasing

my tracks. His breath hung next to mine and filled my own lungs. He smelled like deep musk and fat and the land beneath the snow. For a second he paused, wagging his head from side to side, sweeping so close the draft brushed my cheek.

Then he walked past me, as if I were already a ghost. I didn't move until the bear had retreated far into the distance, a dark exception to the infinite white.

We have spirits for everything here: animals, the sky, the forest, lakes and rivers, even the grass. But Cishan is our hometown god. Half-human and half-bull, he blows in from the Arctic Sea to bring us winter each year. He protects the cold, respects it, demands we do the same. Cold makes us who we are. It calls for a sign, forged in metal you cannot touch with a bare hand: OYMYAKON, THE POLE OF COLD.

My father loved the sign that brought the plane that killed him. I think he would have preferred that kind of death to one anywhere else. He never would have let himself become Ötzi, dying in some strange territory. Alone or together, here is where my people die. We claim a limitless, tough land, so cold it breaks our instruments to measure it, but it rules us all the same. Rules me, I realized.

And finally, I was honest about my life.

When the trucks arrived, the sun hadn't yet risen. Theo jogged in place, and Aytal nodded at me from the driver's seat. I nodded back. He is not a bad man. He said my name again, and it didn't irritate me as much this time.

So when Theo held out his hand for my bag, I dropped it in the snow. "I don't think I'm temperate," I said.

His eyes were unsurprised. "You haven't had practice yet. Just try it for a little while." But there could be no "little while." New York was ten thousand kilometers away.

"You should get in the truck. Get warm." I glanced at Lyuda's face, worried at the window.

"Please, Vera. Are you scared to go? Or is it me?"

"I need to stay where I am."

"Here?" Theo shouted, flailing. The fog ate his words. "This is the edge of existence!"

The scientists waited inside the trucks, but Theo didn't notice. I was trying not to cry so my eyes wouldn't freeze.

"I'm not scared," I said.

Theo was silent for so long, my fingers went numb.

"This cold is dangerous," I said. "Go."

He kicked some snow. "I'm not giving up. I'll come back for you."

"Keep the coat," I said in Sakha.

He squinted in a way that told me he was memorizing the words to look up later. In the truck, he would etch their outline onto a slip of paper with a frozen pen. He would try to translate them in New York with his libraries of free books, but there are no Sakha dictionaries.

Theo pulled down his scarf and kissed me. I felt an even stronger longing for him, now that he could no longer save me. I would be left alone again with the accident that shaped us both. And who in America would understand Theo, now that he had survived our cold? Survived knowing?

Theo removed a mitten to wipe my lips dry with his thumb, already freezing. He looked older, the truck's reflected light sinking into the thin lines beginning in his face. He walked backward to Aytal's truck. Aytal rolled down the window and told him in harsh English, "Go now." Theo tilted his head and paused, asking again without asking.

I shook my head.

Theo climbed into the truck.

The convoy belched smoke and drove away, the last truck with Theo inside. His face was pale against the glass, watching me, ready to tell them to stop if I moved. I watched, too, until my legs turned to pillars of ice beneath me, until the taillights blinked out and he was gone.

Sometimes I imagine this story moving like money from pocket to pocket, until it ultimately rests in my mother's. If enough people read it, every message finds its target sooner or later. So if you are finally reading this story, Starkova Tuyaara Zaharovna: Your husband is dead, your daughter is alive, and I am done waiting for you. I have grown up, and I'm nothing like you.

A winter in Oymyakon can change a person forever. You learn you can survive anything, just by standing in a deadly place and saying, "I live here."

I live here.

The Icewoman stayeth. Why? Because when a tree explodes in the forest, someone has to be there to hear it. I am the one who can stand in the taiga and listen. This is what I hear from the edge of existence: snow crystallizing in the fog, my father sleeping under the plane, Theo riding across the Road of Bones. My town beating back cold death. You. I hear you breathing on the other side of this page.

The Piano

The customer entered our piano gallery with her hands in her pockets, never a good sign. She was a few years younger than me, maybe forty, white, waiflike. She had combed her light brown hair neat and trim in the front, but her pale part zigzagged near the back where the mirror doesn't show. My wife used to have the same problem. My boss issued one sharp "Leron," so I headed toward the customer in a run-walk, but she had already sat down and started playing. It was February, the month after holiday credit card bills come due, hence our deadest month. My savings account was empty, and I hadn't made a commission in weeks.

Most piano buyers fall into two categories, and I was fairly good at diagnosing. With women, you can often tell from their fingernails—long or short. If long, they just want something cheap. Maybe they play "Für Elise," "Heart and Soul," or the beginning of *Moonlight Sonata*, or more often they can't play at all and just poke a few keys. They're buying for their children, who will quit after six months, or else they're looking for furniture to cover a blank space against a wall, so I direct them to the unrefurbished antiques. I'll point out the tigerwood grain or filigree, and they'll squint upward, imagining how it'll go with the coffee table. Or I'll unload a console with a black mirrored polyester finish ("Ebony, always a classic"), which they'll crowd with picture frames and knickknacks that would rattle and buzz if the piano were ever played.

Short fingernails are often the pianists, professional or amateur. They gravitate to the side of the store with the pricier pianos: the grands and baby grands, the Steinways, the Shigeru Kawais, the occasional Bösendorfer. They play scraps of the most elaborate, overwrought pieces in and out of their repertoire—Prokofiev, Liszt, Sorabji. They want to see how the piano performs under stress and strain, if it can handle complexity, if it's capable of sustaining the weight of its relationship with them over the years. A lifetime, perhaps, if bought when new.

This customer had short fingernails bitten to different lengths and an engagement ring that slid around until she slipped it off and buried it in the watch pocket of her jeans. Sales is matchmaking, and she puzzled me. She wasn't a terribly proficient player, but she played a shitty Kimball studio piano as if it were her reunion with music itself—stiff but sincere, discovering each note anew. I didn't recognize the music, and after eighteen years of piano sales and thirty-eight years of playing everything I can find, I usually recognize the composer, at least. Hers was a fraught tune in A-flat minor, melody-based, sensitive, not virtuosic but dramatic. She wasn't using her shoulder blades, and her fourth and fifth fingers weren't strong enough to play the difficult phrases, so the music stumbled and flew, stumbled and flew. She was rocking her body back and forth in her long-sleeve T-shirt, as if trying to eke every microtone from the flat-sounding piano. It's strange to play with such emotion in a piano store, on just a prospective piano, although some do. It feels wasteful to me. Some things are private.

When she stopped, I sidled up, bench-side. She was breathless, as if she had been exercising. I extracted her name—Dagny—and offered mine. "I think you found your piano," I said and stroked the lid like it was her dog.

"Not really," she said. She then picked up her parka and roamed from trade-in to trade-in on the cheaper side of the store, perhaps just to rid herself of me. She dismissed a Petrof in need of regulation and sat at a too-bright Young Chang. She tried a Yamaha from the '80s with a buzzing soundboard, and an untunable Baldwin. I pretended to be busy, far enough away so she could play but close enough so my boss wouldn't yell at me for ignoring the only customer in the store.

Finally, she stopped playing and looked up. I approached again and asked, "What exactly are you looking for, Dagny?" My boss took a sales seminar last month and told us to use the customer's name in every sentence.

"I want a piano that recaptures wasted time," Dagny said.

I paused, but she was actually serious, staring up at me from the bench. "If I had that, I wouldn't sell it. Dagny, is there a special occasion?"

"I'm getting married, *Leron*." I couldn't tell if she was mocking me. My boss frowned and opened a magazine. Dagny said, "Wedding's in three months; I just moved into his house. You married?"

"There's a wife," I said.

"Does she play the piano?"

"She sings. I play."

"Together?"

I nose-laughed politely.

"I bet that's nice. Anyway. These pianos aren't what I'm looking for. I mean, they're good," Dagny added. "Good enough quality."

Most people don't know that a piano's quality depends not on the brand but on the happiness of the people building it. Good manufacturers treat their piano-makers like master craftsmen. They appreciate their workers, pay good wages and benefits. Bad manufacturers take their employees' gifts for granted, and you can hear it in the piano. The unhappiness breaks it, makes it age out before its time. Often it's everyday things that wear a piano down. The price of rent near the factory can make the difference between a superior piano and a shoddy one. Or even bad weather. A piano built in spring is better than a piano built in winter. Right now, in this endless winter, factories everywhere were building terrible pianos.

I was about to begin pitching when Dagny said, "Whatever I buy has to be kind of cheap because I don't have much money. But I want something better than cheap."

"How long have you been playing?"

"It's been sporadic. Complicated." Her frown made the skin between her eyebrows pucker. Then she told me, "I used to be one of those prodigy kids."

"Musical prodigy?"

"I could play Prokofiev's Piano Sonata No. 7 in B-flat when I was ten. Like that. I wrote my first sonata a year later. I was just playing part of it."

She was probably lying. She had been playing a beautiful piece, too mature in its sadness to have come from the head of an eleven-year-old. Dagny's fingers trickled down the keys. "My father didn't want me to become a musician. 'It's too lonely a life,' he told me. Maybe he thought he was being kind. We couldn't afford lessons anyway. He sold our piano to our landlady, and it kept us from getting evicted. I could hear her playing my piano when I passed her house."

"That's horrible."

"Sometimes I played in churches and schools that left their doors unlocked. Nobody ever kicked me out once they heard me. But I couldn't afford lessons or sheet music. I got worse, not better. By sixteen, I stopped playing altogether. I was a coward."

"No, no," I said.

"Yeah. I was." Her greenish-blue eyes bore rust near the centers. "I tried to avoid music after that. But you know, you can't—it's everywhere." She raked her teeth over her lips, upper and lower.

Lately, every night after my boss left, I'd been staying behind to practice on the expensive gallery pianos. It's healthier for them to be played. I turned the lights off so customers wouldn't think the store was open, but the sound couldn't be helped. I didn't know how long it would last before my boss caught me and made me stop. My favorite time was when darkness overtook the store, I couldn't see my fingers or the keys, and the future held nothing and everything. I didn't know if I was awake or asleep, alive or dead, if anything existed beyond the sounds I was creating in the moment. I didn't even know if I was me.

"That's why music is my sadness," Dagny said.

"But now you want to buy a piano?" I tried not to emphasize the word "buy."

"Yeah. I can have one now. First time in my life! I'm in a real house, not some dive studio apartment with mold and gas leaks. I have a fiancé and a job teaching the internet to senior citizens. And all this empty floor space." Dagny laughed again, bright. "I just don't have much money for a piano—at least, not the kind of piano I want."

"It's always a compromise between price and quality—" I began, but she interrupted again.

"Yeah, but what I want is the kind of piano for the player I would have been by now, had I been practicing all those years. All that wasted time. The piano I would deserve if I had worked harder. I know I'm dreaming. But it's good to dream, if the dream is good." She softly played another melody I didn't recognize in Dorian mode, 7/8 time signature, fragmented and strange.

It was February, and beginning to snow, and almost five o'clock. My boss flipped a page in his magazine. Nobody else was buying a piano today.

"Come with me," I said and led her through the employee entrance to the back bay.

The warehouse was an echoey contrast to the cultivated posh of the showroom's low light, radiant heat, and lemon oil diffusers. Here, a skin of dust covered everything. The cement floor chilled the room by ten degrees. Severe fluorescent lights washed everything in cyan. The drywall had never been painted, and white patches of spackling paste covered the seams and nail holes. "This is where you bring your big spenders?" she asked, but I didn't answer. We passed the used Wurlitzers and Daewoos my boss accepts on trade against my advice, the scratched Yamahas needing finish work, the used Schimmels with drooping keys, all awaiting our beleaguered twice-weekly repair guy with gout and a sweating condition.

Then we reached the six-foot Steinway grand. "Brilliantly maintained," I said. "Until recently."

Dagny crossed her arms and raised one fist to her lips. She said, "I guess this is a good time to ask about the previous owners."

I HATE YOU had been patiently and repeatedly gouged into every surface of the piano with some kind of specific tool, the kind you'd have to buy specially. Maybe from a woodworker. The phrase was repeated about eighty times across every available space on the mahogany. The gouges measured between one and three millimeters thick. You could see the wood grain through the layers.

"On a *Steinway*," Dagny said.

"A six-foot-two-inch Model A grand. This is at least a seventy-thousand-dollar piano, even ten years old. I've played it. It's magnificent."

The nongouged wood was satin—old-style lacquer and a honeyed, cherried brown, polished to a luster. The case cradled an immaculate soundboard constructed from the straightest of trees, designed to absorb and reflect all the human experience has to offer.

"What the hell happened?" The overhead light tinted Dagny's skin blue, her mouth purple.

"The piano owner... his wife did this. Luckily, she left all the mechanics intact. She could have smashed the soundboard, or cut the strings, or poured water into it. She could have done anything."

"Yeah," Dagny said slowly. "She was real considerate."

I cleared warehouse dust from my throat and buried my fists in my pockets. "The previous owner saved up his entire life for this piano—bought it for himself on his birthday right here in this store a few years ago, slightly used. He played it three hours every day after work. He polished the case with paste wax and a microfiber cloth. He maintained and tuned it monthly himself. Not a speck of rust, not a hairline crack anywhere. He did everything right."

"He must have done *something* wrong."

"What do you mean?"

Dagny traced a gouged HATE with the tip of her finger. "Or maybe it's what he didn't do. Ignored her. Maybe he liked this poor piano more than his wife." She played a few notes, sustaining the middle C with her middle finger.

"Could you blame him?" I asked. "This is a Steinway grand."

Dagny frowned at me. Her middle C seemed to swell in volume before softening and dissipating. It's a special quality of this particular piano, defying acoustic science. Nobody even knows for sure if sound waves die, or maybe they just get smaller and smaller forever as they fling themselves from surface to surface. It's possible the first human note still haunts us now—a vibration of love, perhaps, or rage, too small to hear.

Dagny released the key abruptly. "So the owner has to give it up? Or wants to?"

"With his divorce and all, he had to move into a small apartment.

Not big enough for a piano, especially a grand. We took the piano on consignment but can't display it on the floor, of course."

"It might ruin the piano-buying mood," Dagny said.

"Anyone who buys this piano would likely be stuck with the, the cosmetics, to some degree. It's too complicated a fix. For a replacement case, you'd have to ship the entire piano to a Steinway factory."

"Could you replace it with a case here?"

"They call those kinds of pianos 'Stein-Was,' or 'Frankenstein-ways.' It would never sound like *this* again. And it would still cost a fortune." I pretended to think about it. "I can let it go for eight thousand dollars. Eleven percent of its worth. You could set up a payment plan."

Dagny's fingertips stuttered over the scars in the surface. "A person could fill in the words with wood glue and . . . paint it."

"You can't paint a Steinway," I said. Any other finish work would likely just highlight the ruptures in the grain, color, luster. There was nothing you could do to this piano without humiliating it further.

Dagny sat down at the bench anyway, pulled by the gravity of the instrument. In the zigzag part of her hair, one white wire shone amid thousands of strands of burnished brown. She lifted the fallboard and began to play.

The warehouse filled with big chords from a big, buttery instrument that forgives weaknesses rather than exposing them, that searches for intentions rather than actions, that soothes the ear, no matter how grave the technical error. The piano made Dagny better than she was, maybe even better than she could have been if, if, if. She played that piano.

This is what happens when you press a piano key: Each individual key has fifty-six unique parts. The key acts like a seesaw, engaging all those parts in eight consecutive mechanical actions so smooth they seem instantaneous. This design has not changed since 1880, through the microchip, the internet, the major inventions of the era. Because important things demand loyalty. Eight actions for every note, twelve thousand parts in a piano, thousands of notes in a piano piece. And we're not even counting the individual sound waves, each heard by an ear, or more than one, triggering the multitude of physical, electrical, and neurological responses that create what we understand as sound, music, beauty, art.

So if you look at it this way, "She played that piano" is not a sentence but the story of an infinite number of relationships forming, colliding, and dissolving in real time, before the last harmonics die and fade from the walls.

"Don't cry," Dagny said.

"No," I said. "I'm, I'm not—"

"She'll come back."

"You're making a—"

"And if she doesn't, fuck her. Really. I mean, look at this piano."

I removed my glasses and wiped my eyes. Dagny touched my arm, and I jumped, half-blind. She gazed at me, at my faded polo shirt tucked into my shiny khakis, my hair that always shows the comb's toothmarks no matter how recent the shampoo. Dagny said, "I'll take it. I'll buy the piano. I'll practice a lot, I promise."

"Sure. That's just fine."

"You're a good person," she said. "It's not over for you."

I played piano at my own wedding, accompanying my wife as she walked down the narrow aisle. She had asked me to hire a musician instead, but I was relieved to use my fidgety hands, to stare at the music instead of at her, a white blur in my periphery. I felt confident once at the bench with the score before me, the notes, the instructions to guide me through time. It was the only moment in my marriage when I was sure of myself.

Now all that's left at our old home is an empty carpet, scarred forever by the heavy feet of my piano, and the dirty, matted place to the right where my wife used to stand and shout in my ear, "Notice me, you selfish prick!" while I hunched my shoulders and banged hard at Rachmaninoff, but I didn't want her to leave me, just to leave me be, or rather to go back to the way she was in those early years when she used to stand in that very spot to sing as I played with abandon, back when her voice was in harmony with my fingers, back when she was the second-best place in the world to lay my hands.

Saturday, my piano will arrive at Dagny's house. Our moving guys will unpack it from its bubble wrap, attach the legs, and heave it into place like the object it is, noticing the scarification but not paid to care. They'll rub it down before leaving, the gouged splinters catching on their rags, the piano's flesh flashing HATE, HATE, HATE. Dagny's fiancé will look at the seven-hundred-pound wreckage in his suddenly crowded home and think, *What am I getting myself into?* He'll think, *Wood putty*. He'll think, *Ruined*, just as she's thinking, *Music*.

North of Dodge

My high school class voted me "Most Likely to Leave Dunfield," so a week after graduation I stole my uncle's station wagon and did just that. It's not a felony so long as you stay in Nebraska. I drove that trash can three hundred miles clear to Omaha, where I was supposed to begin college in the fall. I parked on Leavenworth Street for breakfast, forgetting the keys in the ignition. The car was stolen again by the time I left the diner, but I didn't care. My uncle was a dick, and all I needed was the ride out.

When you steal a car from a white supremacist, the safest place to stay is in a Black area of town. I asked a gas station attendant wearing a Confederate flag tank top what parts of Omaha he thought a white girl should avoid, and he said, "North of Dodge Street," so that's exactly where I went. As I marched north, my backpack chugged its own rhythm: *This is your chance, this is your chance.* I was walking backward in time, as far from Dunfield as I knew how to be. It was a Sunday morning; at that moment my uncle would be preaching his new sermon, "You Better Get Right or Get Left."

North of Dodge Street, I felt conspicuous: too white, too naive. Everything was dingy—buildings, cars, windows, puddles, clothes. Torn corrugated metal siding drooped from storefronts. Hand-scrawled flyers for lost dogs and children peeled from the sides of buildings. The occasional breeze smelled like cat sex and exhaust. Aluminum foil covered the insides of windows to deflect the sun, which was mostly a suggestion of brightness in an otherwise cloudy sky.

By afternoon I found a place to rent from a flyer staple-gunned to a telephone pole. The apartment manager said I looked sweet and honest, and that's what I am, except honest. I paid with the majority share of my uncle's cash—a bonus I had found in his glove box under an unopened box of expired condoms. I guess it was my cash, really, since my uncle had kept my dead parents' money for himself.

My new apartment was one musty room at basement level off Ames Avenue, next to a parking lot I later voted "Parking Lot Most Likely to Have One Abandoned Woman's Shoe on the Ground." First I found a white slingback sandal, then a gold flip-flop, and then a royal blue four-inch pump with glitter. I collected the shoes as they appeared, wondering if they were evidence. I lined them in my windowsill as a cautionary reminder not to leave the windows open while I slept.

The former tenant had left behind a mattress, a table, a lamp, and a vinyl chair I slid around in at night, greased by my sweat. No air-conditioning or fan. On the sticky table I found a notebook and three pens. In cursive handwriting, she (he?) had written an interrupted list:

Cookies
Lice shampoo
Vinegar
Toilet pa

By the time I read the notebook, it was too late to worry about lice—I had already slept on the mattress. I figured the notebook was a kind of gift, so I started writing in it every night before bed. I mostly wrote about how scared I was. I had never lived by myself before or worked a job besides detasseling corn, babysitting, or cleaning my uncle's church and the annex we lived in.

Shortly after I moved in, I was fretting in my notebook when two voices giggled at my window. My heart stuttered, but they were just children. The bigger one was African American, maybe eleven. In the dim light reflected from my lamp, he was chubby, his lower lip curling into the fat of his chin. His spectacular Afro reminded me of those disklike halos in medieval art. The white boy was skinny like an ermine, with curly, matted hair that might have been blond under the dirt. He looked much younger than the big one, who said, "We're going to rape you."

"Yeah," the smaller one said. "Rape you." They laughed. "Can we come in?"

"Hell no." I had babysat plenty, but never kids like this. "And don't say that."

"I'm going to break into your house and steal your TV, then," the big one said, but he could see I didn't have one. "You got a man?"

"He's in the bathroom."

"No way. We been watching you. You ain't got no man there."

I stood to shut the window and they scattered, afraid of me. So I left the sash open in hopes of a breeze and returned to my notebook, vowing to buy curtains. The boys crept back, like squirrels made brave from hunger.

"How old are you guys?" I asked. "What are your names?"

They grinned, flattered by my interest. The little blond one said, "I'm Kyle. I'm six, he's ten. He's Jarvis."

Jarvis said, "I'm twelve."

Kyle said, "He's ten. My dick's about from here to there."

"You're *six*?" I asked.

Jarvis told Kyle, "Aw, you got an itty-bitty dick."

Kyle told Jarvis, "Hey, that's your mom's shoe."

"That's *your* mom's shoe," Jarvis said, and they laughed, high-pitched, like jackals.

"She hates me," Kyle said.

I couldn't tell which mother Kyle was talking about—Jarvis's, or Kyle's own. It seemed inconceivable anyone could hate a six-year-old, that there would be anything to hate yet. Jarvis asked, "What're you writing?"

"I'm writing down the things you're saying," I said.

"Why?" Kyle asked.

"They're interesting."

"We're *interesting*!" They tried to high-five each other but missed and got embarrassed. Jarvis's nose looked like a fleshy arrow pointing downward. I wrote in my notebook, *Most Likely to Do Time for Someone Else's Crime.*

Little Kyle said, picking his nose, "I went to jail for drugs. I busted out." Then, "I'm gonna bust in here when you're asleep."

I began to get unrealistically nervous again until Jarvis said, "No way, not with these high windows. You'd knock into everything, it'd be dark, you wouldn't be able to find a lamp, you'd bump into a fan and cut your feet." Then he said to me, "Send him home. He should go to jail."

Kyle told me, "I love you."

Jarvis said, "I'm sorry for him. You're pretty. In the tits. Give me back my mama's shoe. Please, ma'am."

The "ma'am" shocked me even more than the rest of it. Jarvis had to point twice at the blue pump before I popped open the screen to hand it to him. He grasped it so gently, all I felt was a lightness in my hand as the shoe and the children returned to the streets.

Over the next week, I called every number in the want ads from a pay phone outside Bill's Bar-B-Que and Liquor, covering the mouthpiece whenever a truck passed. I needed a good job so I could save up enough for UNO in the fall. But nobody wanted me to come in for an interview, and I couldn't do half those jobs anyway: accounts receivable, security guard, hospice nurse, experienced meat packer. After a few calls, a ring of men started to form around the pay phone saying things, so I had to leave and return again later with a new pile of quarters.

There weren't any job openings nearby in my new neighborhood—hardly any businesses, even. When I walked down the street, men rolled down their windows to flap a wad of bills at me and ask, "How much?" I wondered it if might come to that. I didn't know how I would survive on my own, let alone make it to college in the fall or ever get the chance to be somebody to anyone. I considered calling one of the kids I had known in Dunfield, but I was mad at their nonproblems, mad at my parents for dying in Detroit and thereby indenturing me to my uncle in Dunfield, mad at the unrelenting bad luck that was starting to look like my due.

Still, there was something comforting about being in the worst place in America. At the bottom of the barrel, everyone floats. I voted myself "Most Likely to Have Nowhere to Go but Up" and called the number on a flyer I had chanced upon: "Icee Treats Is Hiring Drivers. Work Outside! No Background Check!! We Pay Cash!!!" The guy, Chip, hired me over the phone because I agreed to sell ice cream north of Dodge Street. He said, "I've had trouble filling routes up there, and you sound honest enough." "Honest," I was quickly learning, meant "white."

Icee Treats headquarters was a wooden shack containing about fifteen freezers, way down south by Q Street and next to a bunch of fields. I had to leave early to catch the right bus. At six forty-five in the morning, the white prostitute at the corner was still selling sex and also crank. My apartment was a half block from her corner. The moment I stepped onto the sidewalk she started chasing me, yelling and waving her fists. But I wore gym shoes and she wore stilettos, so I left her wheezing and screaming, "Get your own fucking corner!" I prayed she would never change her shoes, because her thigh muscles bulged in her fishnets. I probably couldn't outrun her if she came prepared.

The ice cream truck drivers all arrived at 8:00 a.m. to stock the ice cream trucks and fill them with gas. I didn't know why we had to check in so early, but we did or we were fired. So I sleepily counted out Peanut Buster Bars and Bomb Pops and Fudgsicles and Creamsicles and Choco Tacos and Strawberry Shortcakes and Chocolate Eclairs and Drumsticks and Push-Up Pops and Chipwiches, hauling them to my assigned truck. I filled my cold plate freezer but wasn't allowed to drive the truck away until ten, when people might start buying ice cream. So I sat on the grass to wait with the other drivers, mostly Mexican people and overweight white ladies who pulled out peanut butter sandwiches and burritos wrapped in wrinkled foil. Everyone was older than me, and nobody talked. At the windowless slaughterhouse across the field, food chain workers slumped across the yard toward the smell of manure.

Sometimes the cows screamed. Their throats sounded hoarse, as if it were their first time screaming, pleading for their version of God. There was nowhere to escape the sound—just fields and dead weeds and wide streets and the prairie beyond. After a few days, I bought a pair of spongy orange earplugs that dulled things until I could tear out of there at ten, driving ice cream back to my neighborhood north of Dodge.

North Omaha was even poorer than Dunfield or Detroit, but people found money for ice cream. It was cheaper than air-conditioning. Kids, adults, everyone was happy to see me. I was usually one of the only white people around, but people were nice about it, except they called me Vanilla. Lots of people begged for free ice cream. "Vanilla baby, I ain't eaten in three days," one old guy kept saying, but it was always "three days," even after I had given him a free Crunch Bar the day before.

Jarvis and Kyle showed up every day, calling me Vanilla Bonilla. They acted nice when they had money, but when they didn't, Jarvis would say something like, "Yo, Vanilla Bonilla, can I get my mouth on your ice cream?" and Kyle would laugh so hard he had to sit down.

Even the prostitute bought from me once, but I was safe inside the truck with the doors locked. She flipped the strands of her blue wig, slapped a five-dollar bill against the side of the truck, and said, "Bitch, get me a Bomb Pop." She peeled the wrapper and began to fellate the ice as she walked away, sliding it in and out of her purple mouth and smiling at the men in the street. Even I could get it—she was sexy in the way people are when boundaries evaporate, when the world is simply about the fucks and the fucked.

Customers kept asking, "I know you got ice cream, but do you got anything else?" That's how I found out over half the ice cream truck drivers dealt drugs. It was why they guarded their routes so fiercely. Mostly weed and meth, and one of the fat Icee Treats moms sold oxy; they'd ask for the "special sprinkles" and she'd layer it over the Cookie Dough Cup.

I had a crush on one regular customer, a serious guy in a do-rag and thick Buddy Holly glasses, a combination I found arresting. He always bought the Big Sundae Cup. He was older than me by six or eight years, maybe more. He held himself straight and tall, his lips parted, his skin satiny. He was often flanked by other men and boys who checked his expression after every joke they cracked. They were in love with him—everyone was. He and I rarely said anything but order-speak, but his glasses lent his gaze depth, and he often lingered at my truck like he was waiting for me to ask him something.

When the kids were around, Big Sundae Cup sometimes grabbed Jarvis's magnificent hair, shook it a little, and said, "Is this where you keep your extra brains?" He'd order "one each for the little soldiers," and I'd hand down Jarvis's and Kyle's favorites, tossing them smiles meant to boomerang back to Big Sundae Cup. He always sauntered away without asking for my number, which was fine, because I didn't have a phone. He made the wide street feel small, spooning fudge into that mouth, slapping hands of friends who then joined him, shoulder to shoulder.

I would be safe if he had my back. He was all the way inside this place, its hot heart; I was just a split end. I half hoped someday he'd brush aside the kids from my window, take me to Fuzzy's Lounge, and buy me enough drinks so I wouldn't care what I did. I was eighteen and already mourning the lost chances of my youth. The university sat just across town, but I had never even seen it. If I retreated to Dunfield, my uncle would probably have me arrested. I had no people, no money. My future was clearly visible every day in the women sitting on sunburned grass outside the slaughterhouse, chewing soggy burritos to the soundtrack of dying animals. I couldn't leave Omaha if I tried.

I didn't meet my uncle until I lived with him. After my parents died in a car crash when I was fourteen, I had to move from Detroit to the Dunfield church annex where my uncle lived. He was my mother's brother, and my father called him "a donkey's ass," which I always thought was funny because a donkey *is* an ass. But I had no one else to take me in besides the donkey's ass. As Dunfield's only pastor, it would look bad on my uncle if I went to a girls' home. He repeated this often, like it was my fault for threatening his standing with the community.

It's impossible to invite friends over to a church, so I mostly read alone in my room when I wasn't in school or doing chores. Kids in my classes treated me with an awkward distance, like I was just a visitor. They had all known one another since kindergarten or before, and I didn't understand their jokes. My uncle didn't approve of me going to parties, so I didn't do much besides study. I wanted to go to college to read literature and philosophy. I liked Nietzsche, especially this: "Everyone who has ever built anywhere a new heaven first found the power thereto in his own hell."

Nobody in my family had ever gone to college. My father had installed gaskets into Fords for eighteen years, and my mother had worked at a secondhand store. They got into a car accident on the way home from a church fundraiser. The last thing they did on this earth was eat hot dish and watch a polka band. I loved them and missed them until my teeth hurt, but I still wanted a bigger life than theirs. They wanted that for me, too. They used to call their change jar my "college fund," and a few times a year they emptied it into a bank account for me.

When I was accepted at the University of Nebraska Omaha, I didn't know how to tell my uncle. I finally brought it up one night at dinner, and the unexpected sound of my voice startled him so much a chunk of biscuit dropped from his mouth. I said, "I need the money from my parents' bank account." I didn't know how much it was—I hoped at least a thousand. I showed my uncle the UNO acceptance letter. "I'm late on the deposit. I think I can get a loan for most of the rest."

With a flick of his hairy wrist, my uncle tossed the letter back at me without reading it. "That money's spent," he said.

"Spent?"

He pointed at the kitchen I cleaned every day. "You think your food and shelter don't cost anything?"

Rage blurred the room. "The church gives you this place to live. And I work for my food. For you. For *free*."

"UNO's a shit school."

"I want my money."

"So you can go live with the wetbacks and Blacks?" he asked with more contempt than I had ever heard from him, even on the pulpit, for Satan. "Trust me. You wouldn't be able to stand the smell."

I didn't know what to say. I never did with him.

My uncle wedged a paw inside his belt buckle and leaned back in his chair, like he was waiting to see what I would do. He cleared gristle from his teeth with his tongue and took a swig from the red beer I had mixed for him. Then he forked a slice of sausage back toward his white molars, chewing slowly. The liquid sounds inside his mouth made me want to vomit.

That night after the engine of my uncle's snoring thrummed into gear, I stuffed clothes into my backpack and stole out the door. Fingers of wind clawed through my hair. My uncle's keys dangled from the ignition, where he always left them.

This wasn't how I always pictured going off to college. It was the first time I fully understood just how alone I was. No one, no one loved me. I wondered if this was how people felt during tornadoes, when the roof ripped off and there was nothing above but storm and sky. I shoved my uncle's car into neutral and pushed it half a mile down the road before starting the engine. Back then, I was afraid he would follow me. Now I know it's worse he didn't try.

Every night after work, Jarvis and Kyle sexually harassed me through my window, and every morning they shouted and ran to my ice cream truck as if they hadn't seen me in years. They presented me with all sorts of things that were supposed to stand in for money: bobby pins, hair bands, coupons, and once a torn and rained-on lottery ticket from two weeks ago. "It could be a winner," Jarvis said.

"Guys, I have to buy this ice cream myself. It's not free."

They looked so innocent, which made it hard when they paid me in meth. "It's my mom's," Jarvis said. "Don't tell her I took it." It looked like rock candy in a tiny ziplock, the kind beads come in. I accepted it so Jarvis wouldn't have it anymore. I thought there must be something to meth, if people were willing to give up their whole lives for it. I considered selling it, or smoking it. But I didn't want that kind of heaven. I emptied the drugs into a puddle.

One night at my living room window, the kids grew polite. "You're beautifuller tonight, Vanilla Bonilla," Jarvis said. "Did you comb your hair?"

Kyle said, "He wants to suck your—" But when I looked up from my notebook, he said, "I don't know. Your nose." He ducked away and then back into view. "I do, too."

Jarvis said, "Kyle had an abortion." Something dark and weapon-like lurked in his hands.

"Get rid of that stick," I said. Jarvis threw it away. Kyle bent over from a coughing fit deep in his chest. I felt crappy, too, probably from my diet of frozen food, which was all I could find to buy in the neighborhood convenience stores within walking distance. I hadn't seen a single fruit or vegetable since I left Dunfield.

I said, "Jarvis, it's late and your friend is sick. You should take him home."

"He's not my friend; he's my brother," Jarvis said in a habitual way that told me he wasn't using a metaphor. I must have looked surprised because he said, "It's okay. Nobody thinks we're brothers 'cause he's white and I'm Black. Are you drinking beer?"

"No, milk. You kids want some?"

Kyle stopped coughing and asked, "What does it taste like?"

"You've never had milk?" I asked. "Jesus, where are your parents? Why do they let you out so late?"

Jarvis said, "Milk tastes like Kyle Strobe."

Kyle said, "It tastes like Jarvis Wells. Can I marry you, Vanilla Bonilla?"

"Vanilla wants to marry someone else." Jarvis turned at a passing siren, so I didn't catch the name he muttered. "If I was the boss man, would you like me then?" Jarvis whipped a finger-gun from his pocket and shot me three times in the heart with *pew-pew-pew* sounds. Then he holstered the imaginary weapon in the front of his pants. "Would you give me a chance if I was in the game? Could I lay down by you and look at your butt?"

"Go home," I said, but nobody moved. I saw myself as they saw me—a white girl made of sugar and milk who drove a truck full of happiness and slept in a locked dungeon. They were chasing a flavor but not chasing it hard, because they couldn't pay.

Kyle said, "Whatcha writing now, Vanilla Bonilla? Are we still interesting?"

This and microwave burritos ruled my life. Sometimes I considered unlocking my door and letting them in. They could wrestle each other on my carpet or play hangman in my notebook. At their age, my parents would never have allowed me to run around at night and talk to strangers through windows. They would never have let me buy ice cream out of a truck. Or sell it.

My truck's loudspeaker was broken, so Icee Treats installed a cheap bell. My arm ached from clanging that bell all day, and the sound punctured my dreams. The boredom of Omaha was remarkably like the boredom of Dunfield, plus financial worries. Every evening, we checked in the ice creams we didn't move and reconciled our cash with Chip. I left there with bills and change that worried me. I kept my money in my gym shoes, under the insole. I still had hope that I could save enough for UNO somehow, but collecting dollars was so slow, and nobody else was hiring. I didn't know what I would do after the summer ended and people stopped buying ice cream. I imagined one of my own shoes abandoned in the parking lot.

One night I was out of tampons, so I started for the convenience store. My own invisibility distracted me, my body slipping through the dark, as if a woman had every right to walk on these night streets. I forgot about the prostitute—I was never out after dusk, so she mostly functioned as an early-morning hazard. We didn't notice each other until I had almost passed her. Then she whipped around so fast her blue wig tilted askew. "Bitch," she shrieked. "Get off my corner!" She grabbed me, and her press-on nails sliced my arm when I started to run.

She was fast behind me, her silver skirt flashing in the streetlight. I beat feet toward home, blood seeping from my arm and between my legs. From the crabgrass in front of my apartment building, two boys shouted, "Mama, *stop*! She's okay, Mama!" I wouldn't stop. I was nobody's mother. Then I realized the boys were shouting at the prostitute behind me, who had slowed and stopped to talk to them. And the boys were Jarvis and Kyle.

The kids stopped coming around. Where were they? I had nobody else to talk to. Maybe they were in summer school or gone on vacation. Ha! Vacation. It rained, and all my regulars deserted me.

Even Big Sundae Cup was off the streets. All day I drove around with a full truck. My voice atrophied into a raspy mumble.

On my day off that week, I stopped at Bill's Bar-B-Que and Liquor for a pork fritter sandwich drenched in Dorothy Lynch dressing. Attached to a gas station and liquor store, Bill's was the only restaurant in the neighborhood. They flavored their beans with beer, and the shy guy behind the counter always gave me free gizzards, which I didn't like, but it was food. I ate outside on the curb because Bill's didn't have tables. It had rained that morning, and the street was cleaner but soggy. Just after I bit into my sandwich, someone sat down on the curb next to me.

It took me a moment to recognize the prostitute without her makeup or the wig. Jarvis and Kyle's mom. She had showered. Irregular bald patches shone through her shorn light brown hair. Sores circumnavigated her mouth. I wondered if they interfered with her job. She was smoking, and I was afraid she'd burn holes in my arms.

"This is a public street," I said. "I can be here."

She exhaled smoke onto my plate, and it curled around my cooling sandwich like steam. "I ain't mad at you," she said.

The street was quiet except for the distant skin-like sound of old tires peeling off new asphalt.

The prostitute pressed her pop bottle to her forehead and asked, "You seen Jarvis? He been staying away a couple nights."

"No. Have you called the police?"

She snort-laughed and picked at what remained of her eyebrows.

"Is Kyle gone, too?" I asked.

She inspected her cigarette and drew on it. "*That* little shit ate all my waffles before he left." I saw—little Kyle didn't matter, but Jarvis did. She blew her nose in a tissue delicately. "So, Jarvis. Ask his daddy next time you see him. He knows everything north of Dodge. Everything in Omaha, Neblastya. Neblast-yo-ass."

"I don't know Jarvis's daddy."

"Sure, you do." Something nabbed her attention. "Did him," she said, chin-pointing at a large man with a beard across the street who slouched against a utility pole and lit a cigarette. "Bam!" She twisted to face me, her gaze lagging far behind. She was high. "Anyhow, it's on you. Go find my boy. Jarvis."

"Me?"

"You got that truck, bitch." She stabbed her cigarette into my sandwich to extinguish it. "Us crackers got to stick together." She stood and smiled, a pit where her left incisor should be.

"I'm not a cracker," I said, but she was already swanning down the street, waving at the man with the beard, who didn't wave back. Better to have a dead mom than that one. Maybe. I ripped out the part of my sandwich with the cigarette in it and ate the rest, but it wasn't enough.

All right, I thought. The next day, I crawled the ice cream truck down each street, peering into alleys and parking lots, looking for Jarvis and Kyle. It was a bad idea to drive slowly because kids liked to ride on the back bumper, and I couldn't see them from the driver's seat. So I had to accelerate and then stomp hard on the brakes so they'd bounce off. That's how I drove the neighborhood, in jerky bumps and rushes. Even I was carsick. I worried I'd crash the clumsy truck and lose my job. And for what? If I needed dirty children and bad treatment to mark my days, there was something wrong with me. But I missed them.

That afternoon, Big Sundae Cup was walking down the street, alone for once. I pulled over and opened the service window. He swiped the corners of his mouth and gave me a slow smile. "Vanilla," he said. He offered me the last few bites of his cheese frenchee. It tasted like my mother's used to before she died, with layers of fried batter, bread, cheese, and mayo, hard and soft and hard again.

"I like to watch you eat," Big Sundae Cup said.

I swallowed a few times and stammered, "Have you seen Jarvis and Kyle? Those kids you buy ice cream for sometimes?" I must have looked like a pervert in my ice cream truck, asking after young boys.

Dark drops of rain began to dot his button-down shirt, which was yellow and ironed. Big Sundae Cup stroked his chin with a smooth finger and thumb and watched my mouth as he asked, "You sure you're not looking for me?"

I thought about closing the window and driving away from this older man who made me so out of breath. But if there were answers in this neighborhood, Big Sundae Cup knew them. My vocal cords shrank and I almost squeaked, "Their mom's looking for them. They've been gone for days."

Big Sundae Cup's eyes melted behind his glasses. "Hey. Don't worry. Those kids are okay. They been crashing with my friend nights. I got Jarvis doing a few things for what I got going on. I'll tell his mom."

I was so relieved I forgot to feel nervous. "She scares me."

"She scares *me*!" Big Sundae Cup scratched the back of his head. "Those boys just need a chance. I gave Jarvis a Huskers jersey—you'd think the boy never saw a shirt before. He's pretty proud of himself, now he's got a little folding money. Little soldier's ready to work." He half smiled. "Maybe you could work for me, too."

"And leave all this?" I gestured around the truck. We laughed together, a bright sound in the gray day. I wanted to ask what Big Sundae Cup's work was, but he acted like I already knew.

Besides the guys at Bill's Bar-B-Que, I had the only legal job in the neighborhood. But whatever Big Sundae Cup did for money couldn't be that terrible. He was nothing like the gangbangers and drug addicts with bad teeth, shuffling over broken concrete in torn shoes. He was more like a superfly incarnation of the freshly showered men at my bus stop every morning, riding away to work in the cleaner parts of the city, or the young fathers pushing strollers home from day care each evening. Big Sundae Cup had something to live for.

"Jarvis said he wanted a job so he can kiss you, Vanilla. Can't blame a guy." Big Sundae Cup leaned on the frame of my service window. "Speaking of, you know I read books? One a week. Done that since I was wee. Jarvis says you write in a notebook. Just figured you'd want to know that about me."

"What do you read?"

"Anything, you know. Thriller, mystery, crime. Classics. I'm reading *The Autobiography of Malcolm X* from the library. Now, that's a good book. His daddy used to preach on this street a half mile thataway, before the KKK chased them out of town. Fuck the man, right?"

I nodded, although Big Sundae Cup was the only man there. He glanced left and right and then leaned in. "So what you got going on in your truck?"

Thunder rumbled in the rain. Big Sundae Cup was getting wet, but he didn't seem to mind. I was so lonely. I had only slept with one boy once ever, in a barn, and it was over so fast I wondered if I was still a virgin. I had come here to build my new, wild heaven, but all I had gotten so far was small change.

So I unlocked the truck and opened the door for Big Sundae Cup. He climbed in. We closed the service window.

Fingernail rain tapped the metal roof. He inspected the freezers. "Now, then. This all you're selling?"

Selling? Was he calling me a prostitute?

Before I could get mad or sad or anything but confused, Big Sundae Cup's brow smoothed. He stepped toward me. "Hey. My bad. I'm sorry. It's okay, right?" He was so close. Humidity built between us. I didn't know where to look until he kissed me and I could close my eyes in relief.

His skin radiated heat, and it was cool and damp inside from the ice cream freezers. Big Sundae Cup called me Vanilla, and it was too late for me to ask his name without feeling cheap. He folded his glasses, rested them on a freezer, and spread his good shirt on the cold floor of the truck before laying me down on top of it. He draped himself over me, warm and sleek. "It's nice in here," he said, and then stopped talking as he got to work. He smelled like bread, and I smelled like ice cream, and the summer wasn't so bad anymore. We used my uncle's condoms I had swiped from his truck, stashed in my backpack all this time as a good luck charm for my new life.

By eight the next morning, the sun had unhooded itself and it was 90 degrees with 90 percent humidity. My makeup slid off my face. Even my scalp sweated, greasing my hair. My cold plate freezer wasn't working and the ice cream had already started to melt, so I had to fight with Chip behind the counter for a replacement truck. I asked, "You think it doesn't matter if they get melted ice cream north of Dodge?" and he shrugged. I waited until he sauntered to the back before I swapped keys from the pegboard and stocked a different truck. My earplugs had fallen out of my pocket when I took off my shorts for Big Sundae Cup, so I had to plug my ears with my fingers when the cows started screaming again.

The temperature rose to 100 degrees by the afternoon. Everyone wanted ice cream, and nobody had money. Yesterday's rain steamed the air, and my skin felt poached. My crotch was raw and sore, and Big Sundae Cup was nowhere. I tried to decode the unhurried way he had rebuttoned his shirt yesterday, his goodbye kiss on my forehead instead of my lips. I voted myself "Most Likely to Get Dumped After One Ice-Cream Truck Fuck." I hated him a little, and hated myself more. Take a chance! Screw a stranger! Turned out being wild was pretty close to being pathetic.

I was selling a Bomb Pop just off Ames Avenue in the late afternoon when someone started setting off firecrackers one at a time.

People ducked, and it took me a second to understand it was gunfire. My customers fled, except for an old guy who seemed unable to move. I pancaked myself next to my service window, hugging metal. The only movement in my side mirror was two small figures running down the sidewalk, toward me.

Kyle and Jarvis. Kyle ran in front, so pale he blended in with the concrete. His arms flailed, swimming through the sodden air, heels tossed out at angles behind him. Jarvis chased him, wearing an oversize red Huskers jersey. He ran like a child unused to running. His belly jiggled with each short stride, his giant floppy hair bouncing into his eyes.

I leaned out the service window and shouted, "Kyle, Jarvis, get in here! Someone's firing a—"

But little Kyle darted past, glancing once as he hurled something through my service window. It was so fast I had to duck. The object bounced against the wall and skidded on the truck floor.

A gun.

I had shot old rifles at cans back in Dunfield like everyone else. But I had never seen a handgun before. It was new-looking, black-shiny. Kyle was six. Why was he throwing a gun? The shooting had stopped. Because the gun was on the floor of my ice cream truck.

I leaned out the window again. The tail of Jarvis's red shirt disappeared as he turned the corner at the end of the block. "Kyle! Jarvis!" I shouted again, but everyone had vanished, including that frozen old guy next to my truck. The street stared back at me. I didn't even know where those kids lived. All I had ever given them was ice cream.

I opened the door, kicked the gun out of the truck, and drove until I didn't recognize the street names.

But I kept orbiting the neighborhood, like a crow circles a carcass. I finally pulled over in the pocket parking lot behind the Church of the Living God and ate five ice cream sandwiches, letting the wrappers drift wherever. The streets reflected the sun and heat all around me.

Who had fired those shots, and why? Kyle was the one with the gun, but Jarvis was the one with the job, whatever it was. The "job." I should have kept the gun. The police were probably dusting it already, lifting Kyle's tiny fingerprints.

And Big Sundae Cup, oh God. I was the stupidest person on earth. Not stupid, oblivious—conveniently, desperately so. Nobody had fooled me but me, the fool. I stared at my fists in my lap. I wanted to close my eyes at the sight of myself, but I couldn't. It was better to be awake than to dream fake dreams. It was better to see this than nothing at all.

It was five o'clock. I didn't turn the truck south toward Icee Treats like I was supposed to. Instead, I drove back into the neighborhood. I inched down every street, every alley, avoiding the flashing lights on Ames. Those kids wouldn't be anywhere near the cops, anyway. I didn't know what to do if I found them. Could a six-year-old even fire a gun? Of course he could. It just takes the spasm of an index finger and all possibilities collapse into the shape of a bullet going in one direction, too fast. After that, what were his chances?

By some instinct, I ended up by the empty auto parts warehouses near the highway. When I turned a narrow corner, two familiar figures huddled in the late-afternoon shade of a warehouse bay.

My relief transformed into fresh worry. Kyle was crying and coughing. Jarvis rubbed his back, leaning in a posture of persuasion. *Go to a doctor,* maybe he was saying. *Go to the police. Go to Mom. Go to the boss man. He'll save us. No, he won't. No, she won't. No, they won't. No one, no one.*

When I pulled the emergency brake, both kids looked up. Their gazes caught mine across the abandoned street, but they didn't run away. Jarvis said something. Kyle made a gesture I couldn't decipher, fingers curling inward. The steering wheel blistered my palms, and I licked salt from my upper lip, unsure what to do now that I was most likely to do anything.

At the south end of town, Icee Treats was just beginning to wonder where I was. It was eight miles east to Iowa—one night's drive to Chicago or somewhere beyond, even on back roads. I had a truck full of ice cream, half a tank of free gas, and a hundred dollars in each shoe. All we needed was the ride out.

I stepped out of the truck. The children stood and slapped gravel from their pants, watching me. I pointed at the truck door I'd left open for them. You can correct the tilt of the earth if you just follow the signs. Two scared and dirty kids, an interstate pulsing beside you, and a reason to live. That's all the world gives you: chances.

Eat My Moose

Who knows what a euthanizer is supposed to look like, but judging from my clients' expressions when they answer their doors, they don't expect a sweat-sopped middle-aged guy in overalls, nauseated from a bumpy flight or a long truck drive on a chip seal highway. Sometimes I'm greeted by a lady with a walker or an entire family dressed in springtime colors to cheer their soon-to-be-departed loved one. Either way, they're always relieved to see me, even me. It doesn't matter what death looks like, acts like, smells like. It only matters that I'm there.

I've assisted 221 suicides all over Alaska. My job: I get a name and address, go there, help them die, and then travel to the next address. I source the materials, do most of the jobs, and maintain the planes, trucks, and helo. When Bonnie was working, she dealt with the clients and money, and she vetted every job to make sure it was euthanasia and not murder.

Bonnie loved subterfuge and never gave me an assignment by phone. Instead, she'd write the time, date, and GPS coordinates on a cigarette using a microtip felt pen, replace it in the pack, and hand it to me when we met for coffee and muffins. There were usually two or three cigarettes in the pack, and if she wanted to add a stop to a bump run, especially north of the Arctic Circle, she'd express mail another pack to me up there. After the job, I'd smoke the evidence under that pearly sky, thinking of her. We had both started smoking again, once we achieved temporary immortality.

There are three reasons people hire me:

1. Life insurance policies don't always pay out for suicide, even if you're terminal.
2. The stigma. Some sick clients fear judgment from loved ones who expect them to endure the pain. Pain is easy to understand until it's yours.
3. Some clients have nobody. They're afraid to die alone.

Although we usually serve families, a number of my clients are lonely subsistence hunters and fisher-folk, hours from roads or towns or other people, scared and sick enough to let a stranger help them die in the agreed-upon manner. Even after Bonnie quit, they still find my number somehow. Some of their cabins are so remote and clotted with forest that I doubt even a bush rat like me could fly back out without bending metal. Some of them change their minds once they see me, their first person in months or years. Then I just fly them back to Anchorage and drop them off at a bar or a hospital so they can live their remaining days surrounded by the humans they didn't know they needed.

But most folks go through with it and die. I provide the chamber, which is a plastic bag to fit over their head, like a turkey bag with elastic on it. Tubing to funnel the gas in there. A tank of helium, or nitrogen for an extra fee. They work the same and neither one shows up on your general autopsy, but some people like to be expensive.

I don't:

- Pull the bag over their head or snake the tube in there. They have to do that.
- Turn on the gas. I usually have to loosen the knob, however. These are very sick people.
- Hold them down.

- Tell them my name. They only know my alias, Clyde, a name Bonnie used to give them for humorous reasons. My real name is Colum.

I do:

- Give them instructions.
- Wear gloves.
- Hold their hands. Tightly, if they ask, if they're worried they'll struggle.

After they're dead, I check the carotid pulse with one hand and bag up the kit with the other. Bonnie always said, "Get out of there before the magnitude hits anyone, especially you." I keep my coat and boots on to make leaving faster. I'm gone inside a minute, lighting Bonnie's telltale cigarette inside my truck or plane or helo. It's like flying away from a bomb site. Which I've also done. But with these jobs, I feel ephemeral, anonymous once I'm in the air or down the road. Under the roof of clouds, I check my body for pain and find none. Oftentimes I forget the client's name within miles. Sometimes I sing.

It's common for clients to offer me whatever they're leaving behind: a truck, a plane, a shack, jewelry, food. In a poorly chinked ten-by-ten log cabin in the Interior, one man's last words were "Eat my moose." A whole winter's supply of butchered bull moose hung in his shed; he thought he'd live long enough to eat it all. It became a joke for Bonnie and me. "Eat my moose," she'd boom randomly, and I never knew what she meant, only that it made her laugh.

But I never take anything clients offer, not even the untraceables like meat or money, not even if they beg. I leave it there with the body, for someone else to find, or never find. I don't like to take my work home with me.

I first met Bonnie at an Anchorage VA cancer support group four years ago. The group was for terminal patients, stage 4 with months or weeks to live, many of them combat vets like me and Bonnie. Bonnie had deployed five times to Afghanistan, and I was Air Force with four deployments to Iraq and three to Afghanistan. We only talked about it via our cancers, the chemicals we had inhaled in theater, the nightmares we still got.

My cancer was pancreatic, Bonnie's a brain tumor, both inoperable. We were in our fifties, a couple of years apart, retired from service. Everyone in the VA support group had quit their civilian jobs. I had been flying for the Postal Service. Four out of five Alaska towns are only reachable by plane or boat, no roads in or out; bush carriers like me kept them connected to the world, even as we floated above it. Bonnie had been running a high-stakes underground poker game that sent her back and forth between Anchorage and Fairbanks until she began to forget where she was.

Bonnie's chemo did squat besides steal her hair. Without eyelashes or eyebrows, she always looked like she had been crying—almost funny given her hard-edged character. Her nose was continually pink, chafed from the oxygen tubes, with little red lines veining the thin skin under her nostrils. She sometimes forgot her last name but never forgot to bring doughnuts. Even with the monster eating her brain, she said things like, "The only reason we need this shitty support group is because no one else can love killers like us." It's like the things inside my head came out of her mouth. I had three ex-wives I had shredded on the altar of my viciousness, my drunkenness. Bonnie had the foresight never to marry at all. "And now I'm too bald to wear the veil," she said.

By the first time Bonnie asked me to kill someone, I, too, was on oxygen and Bonnie had stopped chemo. I was at home trying to decide between raisin bran and stale bread, knowing I'd barf up either one, when I heard her knock on my door. Bonnie wore a green dress soiled at the hem, pulling her own oxygen tank in a trolley behind her. She stood at the bottom of my steps and said, "Wyatt needs a spot of help."

Wyatt was the other vet I liked at the support group. Stubborn like me and Bonnie, an old sourdough. Leukemia. Bonnie said, "He wants to go out on his own terms." She explained the mechanics he wanted, the gas and the chamber and everything. "It's kind of cool, actually. No pain, just a little panic. *Et voilà*, it's over!"

I have to explain something special about Bonnie: she always gets her way. Bonnie ran that underground poker game for six years, and it was the most peaceable business in all of Alaska because you don't say no to Bonnie. You can only try, and grumble, and fail.

"Why doesn't he just take pills or something?" I asked, but I already knew the answer. Wyatt was twenty years sober. And why shouldn't a man die any old way he wanted? How you die should be up to you. How you live is usually up to everyone else.

It was not a small favor. Bonnie and I were likely weeks from death ourselves and could barely go a few hours without a nap. Bonnie used a cane because her balance was fucked, plus both of us had our oxygen trolleys. It was spring, with snow and rain and snow and rain, and it all added up to sticky, half-frozen mud. I still had the same damn pain that had alerted me to the cancer in the first place, a bizarre burning that radiated outward from my belly and back, like a pulsar inside me. It got so it made me stop drinking, and even three wives couldn't make me do that.

So we drove to Wyatt's cabin and hobbled inside. He showed us how he wanted it done, but he was too weak to turn on the gas himself, so Bonnie twisted the knob. It took a couple of tries, but we finally got him dead. It wasn't nearly as horrible as anything we'd seen or done in theater, me or Bonnie. The only really bad part was that we were next and we knew it and so had Wyatt, so fuck him for asking us. But who else did he have? He barely had us, and now he was just a bunch of cells beginning to decompose. Nevertheless, when we pulled the bag off his head, his face bore the smirk of a man who had gotten his way.

"He's lucky we're still around to help," Bonnie said, staring at Wyatt's empty body. "Maybe we should consider going out this way." But nobody would ever help us besides each other. Which meant one of us would be left alone in the end, literally and figuratively holding the bag.

We called 911 and left the phone off the hook by Wyatt's hand. Then we drove away in silence. At a Fred Meyer, we parked to wipe down the tank and plastics and dump them out back. Bonnie walked inside the store without even using her cane and emerged a few minutes later with a bag of apples and wrapped tuna sandwiches. We sat on the bumper of my truck and ate like healthy people.

"Are we going to hell now?" Bonnie asked.

I thought we were already in hell. Except at that particular moment, pain had hit a miraculous pause. That pulsar was gone from my abdomen, my head. None of the shakes, none of the nausea.

"You know what?" Bonnie said. "I feel kind of amazing. I'm not even dizzy right now."

"Adrenaline? Or maybe we're rallying," I said. They called it "terminal lucidity" in our support group, and I had come to anticipate it, that last flame before death, like the fire of maples in autumn.

"The end is nigh," Bonnie said.

We smoked and ate apples, and then I drove Bonnie to her place and went home alone to recover from what we'd done. But the sickness didn't return. I spent the next days and then weeks cleaning my house, scrubbing all the grime that had accumulated while I was feeling so damn awful, airing out the old man smell. I rose early each day, packed all my belongings into boxes, and scribbled FOR GOODWILL on the side. There was nothing worth sending to my ex-wives. They wanted to forget me. Anything that remained of my life was best left for strangers.

Three weeks after Bonnie and I helped Wyatt die, my place was all packed up and I still felt tremendous. I even started running five klicks a day like I used to. I waited for the crash, and waited. Finally, I visited my doctor, who gave me blood tests, ultrasounds, X-rays, and a CT scan and then said, "I'm sorry, but you still have stage four cancer. Your tumor is the same size. The good news is that it hasn't grown as of late."

"Aren't I supposed to be dead by now? Or at least sick?" The doctor began to explain terminal lucidity, but I cut him off. "All this time, though? A month ago I was on oxygen. And now?" Like a dork, I dropped to the floor and did some rapid push-ups. "Explain that, doc."

He said, "I'm glad you feel good. We'll schedule a follow-up in seven days." He hesitated, hand on the doorknob. "Just . . . what comes up must come down. Have a good week." The way he said it made it sound like a week was all I had.

I didn't think so, though. I knew what dying felt like. This sure wasn't it.

I only wanted to talk to one person. But how do you explain your good luck to someone with such bad luck? I had stopped going to the support group after Wyatt, and I waited to call Bonnie for so long, I worried there wasn't a Bonnie to call anymore. When

I finally dialed her number, she answered on the first ring. "You're still alive!" she exclaimed.

"I'm feeling better. Like . . . all the way better." At her silence, I said, "Ever since we did Wyatt. The cancer's there. I'm just not sick."

I winced, waiting. Bonnie's laugh bubbled. "What the everlasting fuck."

"I'm sorry. I feel funny—"

"No. I'm saying, I haven't been sick since Wyatt, either."

"You're better?"

"Well, I probably couldn't parachute out of a helo. But I did just clean my gutters and eat most of a pizza. Meat lover's."

I don't know why relief feels so damn funny, but we were like kids, laughing until tears, until urine threatened, until we felt empty and scared. I stared at the Goodwill boxes stacked against the wall. "How much longer do you think it'll last?"

Her voice was dryly amused. "A day, forever, who knows? You and I defy science."

"But we're assholes. People like us don't deserve miracles."

"This is no miracle. We still have cancer," Bonnie said. "It's a stay of execution. We're on loan to God."

"Okay, so what does God want us to do with the rest of our lives?"

"It's obvious, dummy," Bonnie said. "He wants us to kill people."

I said no, and then no again the next time Bonnie asked, and no the next two times we saw each other. I couldn't make a vocation out of killing again, not after Iraq and Afghanistan. That was a hole with no bottom, and this time, I had a choice. At a coffee shop, Bonnie pressed my sleeve, and it had been so long since anyone but a doctor had touched me, my skin quaked under my flannel shirt.

"Not real killing," she said. "This time, we'd just help them kill themselves."

"How would we even find people for that?"

"Aw, dying people are everywhere. We could start with our support group. Word of mouth is everything."

"Corpses don't make referrals."

"Don't worry about advertising. I'm a wizard." Bonnie's underground poker games ran for years, and everyone knew about them except the cops. "We could make some real money. My doctor's bills are fucked. Anyway, you've got to do it; it's too much for me alone. We're already trained for it. It's like we were talent-scouted by God."

"Or the Devil," I mumbled.

"The Devil is us. Come on, Colum. How else do you explain you and me getting better at the same time, directly after we offed Wyatt?"

I admitted it did feel like a sign.

"Not a sign," she said. "A signing bonus. Our lives."

Someone once told me Buddhists don't believe in suicide, not even assisted suicide. Karma comes in the form of pain, and if you avoid that pain, you lose your big chance for enlightenment. So what Bonnie was proposing wasn't just illegal; it might be cosmically cruel, setting humanity back lifetimes. But I'm no Buddhist. I couldn't imagine it was good karma to watch someone suffer without trying to end it.

"We both know the suffering side, too," Bonnie said, reading my thoughts the way she did. "We could actually help people."

When I was about eight, I found a rat whose spine had been broken, probably by a dog or cat or hawk. Its back legs were paralyzed, and it could only drag itself forward in an army crawl. I offered it water and my bologna sandwich, which it refused. I sat with it in the forest, sang to it to make it feel better. My parents didn't allow rats in the house, so I had to leave it alone at dinnertime. When I came back, the rat had slipped into the underbrush and I couldn't find it. I felt like I had failed that rat, that instead of singing to it, maybe I should have run it over with my bike or stomped it to death. In my mind even now, over fifty years later, it's dead but also alive and still suffering, Schrödinger's rat.

When you do bad things in the military, you tell yourself they're not bad because you're on the good side. You might kill people, but only to protect the innocent. That's how you lose your own innocence.

And suddenly, *you're* not worth saving anymore.

"Okay," I told Bonnie. "What the hell. Let's kill some sick people."

Those four years Bonnie and I worked together, our tumors slept. No pain and no gain. Bonnie's hair grew back, dull silver and yellow, and she put on some muscle in her shoulders, legs, arms. She was always beautiful to me, but now she was pretty, too. The doctors finally conceded we were in remission. They wanted to do tests to find out why, chemo, radiation, but we said no, no. We knew how to survive. Just do the work. And I had Bonnie, who became my reason to survive at all.

Bonnie and I were both used to teams from the military, but this was different because nobody else told us what to do. Bonnie worked locally, and I took most of the distant jobs so she could rest her leg, which throbbed on flights ever since her chemo. Even alone in the air or in the forest or on the tundra, I felt like she was there with me.

But I never asked her on a date, not once. What would we even do? Everything we had in common was death. Bonnie moved back to Juneau, where she was from. "No roads in, no roads out," she said. It was an odd choice to headquarter an illegal death operation, but she liked the rainforest and marine lifestyle. "Don't get comfortable," I told her, but she did anyway, outfitting a cabin out the road that she had inherited from an uncle. My Goodwill boxes were long gone, and I lived in hotels.

Bonnie never invited me to stay over, instead shooing me off at night after dinner. Sometimes she touched my hand or booped my nose. She was a booper. But I wasn't allowed to boop back, or touch her at all.

About a year into our new arrangement, it was raining that misty kind of rain that envelops, and we sat on her porch swing. Bonnie started talking about her high school cannery jobs as a fish sorter and a gutter. It wasn't romantic talk, but I went silent, sitting next to her. I leaned closer in gradual millimeters, like I was trying to catch a fly, and Bonnie talked faster and faster. When I reached out to turn her chin toward my face, she jumped in her seat and smacked my cheek hard.

My eyes watered. Because of the slap. It's just a physical response. She looked shocked, distressed, like she couldn't understand what had just happened.

"Bonnie, I'm so sorry," I said. "I don't—"

"I'll get you another iced tea," she said and walked inside.

I almost cried for real then, from relief that she wasn't kicking me off her porch and out of her life. So stupid, I was. It was a while before Bonnie came back with two bottles of iced tea and two scratchy wool blankets. She handed me one of each and said, "It's cold," tossing me an even, fake smile.

"Thank you, Bonnie. I really didn't—"

"Stop." Her face zipped up. And then, an uncustomary "Please."

I shut up. Bonnie never talked about Afghanistan. I didn't try to touch her again.

I'm proud to say that there's scarcely a town in Alaska that I, like Death itself, haven't visited. Most of our work was in Juneau, Fairbanks, or Anchorage, but I also did jobs everywhere from Utqiagvik to Chicken to Adak Island. I've flown the wetlands and deltas of the Alaskan Southwest and landed on tiny volcanic islands to wait out gale-force winds. I've skimmed the rainforest and fjords of Southcentral and Southeast Alaska, the boreal forest and extreme tundra of the Interior, the barren ice fields of the Far North. I've watched the glaciers get smaller each year I continue to live.

Seems like every other person in Alaska is a pilot, so our comings and goings never drew suspicion. Bonnie used a sliding scale, and some jobs we did for free, but somehow we still made fistfuls of money, too much to spend it all. I outfitted my high-wing Piper for the bush with a taildragger, skis, and tundra tires, and bought a dinged Cessna with floats. I was used to landing on empty roads, seawater, sandbars, fields, and even glaciers sometimes, but I needed more agility, so I also bought a used helicopter and rebuilt it myself. With that eggbeater, I could drop down anywhere and then vanish again in minutes.

I'm originally from Missouri and only moved to Alaska after I served twenty-five years in the Air Force. It's what some soldiers do after service, disappear into the biggest state in America, capacious enough to absorb even our contamination. I can float over ranges upon ranges of mountains, like wrinkles on the face of a weathered god. I land, end a human life, and soon I'm back in the air again, flying for hours to make it on time for a sunset cigarette with Bonnie.

Times when things go wrong: When a client tears off the chamber but is too loopy to give consent to continue. When a client is too far gone to consent in the first place and I discover it's all a murder ploy for the relatives to get their inheritances. When the relatives accuse *me* of murder and I have to abort. When I arrive but the client changed their mind or checked into a hospital. When I finish the job and their healthy bereaved spouse begs me to kill them, too. Winter weeks when the cloud ceiling is too low to fly, while sick people wait in pain, sometimes for a month or more. When all I find is an open door to the tundra and I have to search for the body. When the client got tired of waiting for me and botched the job themselves, lying prone on the floor for who knows how long, in pain but paralyzed, dead but still alive.

The job that made Bonnie quit was this: A few months ago, her aunt Elspeth asked us to do her. Elspeth was the only relative who still spoke to Bonnie, and that was only because she had dementia. But the Alzheimer's killed Elspeth's resolve to die. She had waited too long before taking action against herself. It wasn't Bonnie's fault. Impossible to tell if Elspeth's body was rebelling against the gas or if she changed her mind, but Elspeth fought like a murder victim. Bonnie pushed me away and ripped off the chamber, but it was too late. She started CPR, pushing on that old chest until the ribs cracked, breathing into the dead mouth. Crying. Me pulling at her hands, but she wouldn't stop. "She didn't want to go!" Bonnie cried. "She changed her mind!" I had to half carry Bonnie back to the Piper, fly her home, and tuck her into bed cradling a fifth of rye.

She called the next day, slurring, still drunk from the night before. "Colum, I'm settling my debt with God. You can keep going. But I *quit*." She drawled the word into two syllables, stoppered by a hard *T*.

I didn't know if she was quitting the work, quitting me, or quitting life. I was scared. "But what if you get sick again?"

"Fuck you, buster," she said.

Bonnie didn't take my calls for days. I apologized on her voicemail over and over. I didn't really know what I was apologizing for.

Then, the next week, she finally picked up the phone, said, "Hey, sugar booger," and started talking about her geraniums. It was as if she had never hated me, as if there was nothing at all to hate in me.

Life is precious here, and also not. We kill a lot, right? Even vegans own guns up north, but sick folks don't always like shooting themselves. With zero repeat clients or advertising, people nevertheless find my number and I always have work.

But it was hard for me after Bonnie quit. I missed her. I spent as much time as I could on her red plaid Barcalounger circa 1975. We'd eat and smoke until she yawned and I left, aflame. I could love her just as well from afar, maybe better that way. There's such a thing as a safe distance.

About a month into my solo career, Bonnie's voice started sounding thin and distant, like dry twigs more likely to splinter than bend. She stopped inviting me over. I didn't see her for a few weeks, and then a few more. When I asked if she was all right, she snapped, "Get over yourself. I just don't want you around." Increasingly forlorn, I took more jobs in the Interior, sulking in Fairbanks hotels when I didn't have work, feeling stupid now that Bonnie didn't need me anymore. It was fair. She had told me back in our VA cancer support group: Nobody can love a killer. Not even another killer.

One day I called to ask how she was, expecting her to blow me off again. But this time she said, "Party's over."

The world emptied. "Oh no."

"The tumor's begun to impair my facilities. Is that the word?

No. My faculties." This time, her voice was clean and clear. "Colum, I need a favor. One I won't be able to return, I'm afraid."

I was crying already.

Bonnie said, "You have to eat the fucking moose now."

"I never know what that joke means. It's not a funny joke."

"Tomorrow's good. I'll make Danish. Put on your hero goggles and stop blubbering, you gigantic pussy." But she sounded like she was going to cry herself. Then she hung up on me.

I don't know if there's such a thing as redemption for someone like me. Not after what I did in-country in Iraq and Afghanistan, who I was there and how I came back from each deployment someone else entirely, drowning my marriages one bottle at a time. Even now, in my late middle age, I don't know who I am. I only know how to smoke cigarettes with Bonnie.

Next morning, I flew to Juneau and drove straight to Bonnie's house. She lived out the road, past the Shrine of St. Thérèse with its whole island dedicated to murdered souls of the unborn. *My Bonnie lies over the ocean*, I sang on the way. When clients ask me to sing to them, this is the song I sing. When they hear my terrible voice, sometimes they interrupt their dying to tell me to stop, that I'm ruining it for them.

It was the kind of day I would've hated to leave behind: fog shrouding the mountains that stretch up from the water, petrichor surging from the earth, the whisper mist that coated my skin and XTRATUFs. Western and mountain hemlock, Sitka spruce, Alaska yellow cedar. Bonnie's place is remote, and animals roam everywhere: bears, porcupines, beavers, marmots, deer. You can smell them when you can't see them. Rain pushes life from every crevice.

So when Bonnie answered the door, her hair curled, wearing a freshly ironed yellow dress, I wanted to ask, *Are you sure you want to leave all this today*? But she was starting to look the way she had when I first met her. Blue-tinted fingers, shadows on her skin. I shoved my hands into my pockets and eyed her oxygen tank, the battalion of pills lined up on her coffee table, her pocketbook open and upside down on the floor. She had kept the suffering to herself. Soldiers don't complain.

Bonnie handed me coffee, asked, "How are you?" and then laughed. "It's weird, isn't it? I want to catch up, but why? Pretty soon there will be no me to remember anything you tell me. Might as well start."

We organized the plastic, the tank, the tubing. It all felt different this time, everything too slippery, cheap, and disposable for the permanent thing we were about to do. Bonnie said, "You're going to have to restrain me, just in case I fight the gas. I don't want to risk turning into a vegetable if I pull the bag off. I refuse to be a rutabaga for the rest of my short, painful life. Boop!" She booped me on the nose and I smiled, and then felt like an idiot. "Seriously, Colum. Finish the job."

"I promise."

"I know you're in love with me. But don't fuck this up."

My lips turned numb. I had never told her that. But of course she knew.

Bonnie said, "Oh, and once I'm gonzo, will you do the dishes? I didn't feel like it on my last day. But they shouldn't just sit in the sink like that."

"I'll do the dishes."

For the first time since I met her, I saw doubt skid across her hard-lined face. "We thought we were following the signs. We thought we did good. Helped people."

"We did help people, Bonnie."

"Did we, though? Were we right or wrong, how we used our

second chances?" I struggled to say something comforting, but I had no answer for her. Finally, she shrugged. "Well, I guess I'll find out in a few minutes."

I offered the chamber, but she turned her face away.

"Don't mess up my hair," she said.

I fitted the chamber over her head, snaked the tube inside, cinched it all up around her neck, and turned on the gas.

Bonnie held out her arms. "It's hug time!" When I hesitated, Bonnie barked, "Fuck's sake, Colum! Hug me! Hug me!"

She had never let me touch her before. I reached for her timidly, like she was made of butterfly scales, and then just grabbed her and held her tight.

She flinched, tensed, and then relaxed. She still had residual strength from the military, gardening, snowmachine driving, shoveling, splitting wood, and firing her shotgun at the lynxes who tried to invade her henhouse. She was assigned to multiple Special Forces combat teams in Afghanistan, even though Army women weren't technically allowed to fight yet, even with her bum leg from a childhood chain saw injury. She did a hundred push-ups a day until she got sick. She once faced down a bear who broke into her home and tried to eat her cupcakes. Now the tense cables of Bonnie's muscles vibrated under my hands, in her mostly healthy body, run by her diseased brain. We were closer than we had ever been, close enough for me to kiss her if we weren't also separated by a dome of plastic, fogged by her breath, which quickened into panting. It had been everything I wanted, to hold her, but not like this, too late. The gas hissed, and Bonnie caught my gaze with her green eyes that deepened into midnight at the edges. Her skin smelled of lavender. She gave me a shy smile. I already missed her.

"I love you, too," she said through the clear plastic.

Then she began to gasp. Then she began to struggle.

I had watched this fight hundreds of times. But I had never felt it in my arms. Bonnie twisted away from me, her stubby nails reaching up to scrabble at my chest and face. I didn't know if she was already half-rutabaga like she feared and if she still wanted me to kill her, or save her, and what was which. It was a miniature revolution inside her—short gasps, tiny pants, growing softer. Her expression split in half, top and bottom. Her mouth pursed into an O, sucking air, but each breath only delivered more death. Above her mouth, Bonnie's face was perfectly blank, the way it never looked in life, forehead smooth, free of thought, despite the chaos of her mouth, her hands. It felt like her body stayed behind to fight, but her mind had already made its decision and left the room. But maybe I was just telling myself that.

How would I know if she changed her mind? I thought. I searched her closed eyes for answers. *I have to stop this. Ask her if she still wants to die.*

But instead, I tightened my grip as she tried to escape me. I couldn't fuck up. Bonnie needed me to be strong enough to kill her when her body still wanted to live. It's like we had been training for this all along, Bonnie and me, and I knew this was my last time. Now, instead of stopping, I wanted to hurry her death so she wasn't fighting herself any longer. *Should I break her neck?* I thought wretchedly.

Time bent itself into unrecognizable shapes, but it probably took only one more actual minute for Bonnie to sag heavily in my arms. She drew her last breaths, and then she wasn't breathing at all. I lowered her body to the ground and checked for a pulse, my fingers on her neck and my ear to her chest. Nothing moved inside her. Her body was the stillest thing I had ever known, like outer space must be, like the point between exhale and inhale, like the eternal present moment. Like love.

I didn't leave within the minute like I usually did. This time, I stayed. I lifted Bonnie from the floor, tucked her into bed, and waited with her as her body cooled. I smoked her cigarettes. I talked to her. I sang in my terrible voice. I did her dishes, scrubbing them long after they were the cleanest dishes in all of dish-doing. I vacuumed, dusted. When there was nothing left to do, I dialed 911, kissed Bonnie's cold face, and left the receiver tucked in her right hand. I didn't want her to rot there. Even though she wasn't my Bonnie anymore.

Driving away from the house, I grabbed my side. There it was, that fucker, just like I expected. As punctual as the military. Only one thing ever felt like that, a pulsar radiating pain from inside, shocking my nerves, thudding through my organs. My tumor had waited me out, and now my Bonnie lies over the ocean, it's come back for me.

It's a relief, you know. Nothing but. I've traveled alone over a frozen white world longer than my due. I've eaten your fucking moose, Bonnie; now I've got to eat my own. And after, maybe I'll be able to answer your question: if we hurt more than we helped, if we bet on the wrong side of God, or if all our killing managed to save us enough so I can see you again.

—For Kim Hayashi

Save Me, Stranger

I was buying my daughter hot cocoa at a convenience store when the three of them walked in, pulling guns from their winter coats. At first it looked like nothing, like, hey, look at those almost-men over there with guns and pantyhose on their faces, and some part of my brain told me they were just carrying the guns, not that they'd use them on us, although I still couldn't justify the pantyhose. I held two empty cups. Mina pointed, her ten-year-old lips parting to say, "Look, Mama—" But she didn't make it through "Mama" before the young men started shooting the plaster ceiling.

We all dropped as if we'd done this before, except Mina, and I pulled her down hard. "Wallets, phones, jewelry, now," one of them said.

The cash register pinged above my ear. I threw my purse like it was a grenade and lay on top of Mina. She tried to wriggle free, to see. I pulled her head down by her braids. "Close your eyes," I whispered. She squeezed them shut, and only then did I see fear shiver through her jaw.

An untied sneaker stopped on the linoleum. It kicked my head, hard enough to hurt but not hard enough to knock me out.

"You," he said. "Up."

I didn't move. He kicked my head harder, and everything turned white for a second. "Up, bitch." His knee popped as he crouched down next to me. The cool circle of his gun pressed against my temple. I smelled his socks under his wet shoes.

Mina opened her eyes.

"No," I said. At least, I thought I said it, but I was already getting up, palms against the gritty linoleum, watching my daughter's face constrict as she realized I was leaving her behind. I tried not to look at the big man's face, blurred by pantyhose so his nose was pushed flat. His eyes looked sleepier than he could possibly feel, panting through the mesh. The hand on the gun had a Band-Aid with Spider-Man on it. This young man with the gun and mean friends had a child or baby brother, or else why Spider-Man? So he wouldn't kill me, the obvious mother of my obvious daughter, who was starting to cry.

But that's not what the gun said as the man jammed it into my ear. He told his friends, "I got someone. Let's move." His friends had already collected all the cashier money and handbags and wallets in a black trash bag and were now stealing candy. The gunman wrenched me toward the door by my arm. I wasn't sure what they wanted me for, only that it would be bad. I thought, *But I promised Mina that hot cocoa*. Mina started to scramble up, and I snapped "Down" in a voice that horrified us both. She looked around for someone to help us, but the other customers looked away. There was no one. "Save me," I whispered to the God I didn't believe in anymore.

But another person was standing now. It was a teenage boy who must have been lying near us because he was next to me so fast, the gunman jumped a little at the boy's clear voice: "Take me instead."

All I saw were the boy's eyes, black against white. He said, "Not this lady. Me."

The man with the gun said, "Sure, asshole," pushed me back onto the ground, and pointed the gun at the kid's hooded temple. "Hands on your head."

The kid obeyed. Mina and I grabbed each other on the floor. The boy leaned over me. His jacket smelled like dirt and cold. His perfect lips said, "Olivia." He held my gaze to make sure I heard him.

I was about to ask who Olivia was when the gunman shoved the boy. The boy stumbled, recovered, and they were all out the door.

Everyone lay still on the ground, like they were asleep.

I could see the men and the boy through the glass double doors, and I blocked Mina's view with my shoulder. The men packed themselves into the back seat of a navy-blue car, snow dusting its hood, no license plate.

The boy stood in the street, staring at the men in the car. His fingers were still laced behind his head, elbows wide, as if he were lying on a beach with a magazine on his lap, except it was snowing and he was standing up before bad men on a bad street in a bad neighborhood. He seemed dizzy, glancing around and swaying a little, like he didn't know how he had gotten there.

A hand stuck a gun out the window. The boy who said Olivia lowered his arms. He was just a teenager, maybe younger than I first thought, all gangly muscle and knobby bones. He shook his head over and over, his hands now pushing the air, as if he could spirit the car away, his mouth moving fast, his words against their gun, nobody to help him now. *No no no no*, I said, as if it would change anything.

Snow swirled around the gun, the bare hand. Then that hand shot that gun, shot the boy three times into his collapsing chest before the car drove away. By the time I made it out the door, the boy had already fallen, alone in a vacant street, eyes rolled back; he was already dead.

Aurora means "dawn," which is supposed to mean "hope," but I'd never seen either thing in my neighborhood. A couple of times a week I cleaned the G Spot in north Aurora while my daughter was in school. An unlisted place in a frigid basement, they had those peep show booths where a lady danced on the other side of the glass and the men did whatever. I don't know. They don't make rubber gloves thick enough. I always had to lay down a layer of hot water and then come back a half hour later when the dried semen loosened into a paste I could scrape up. When the peep door closed, people fumbled their money and nobody picked it up, so I got that, too. I would dump the coins into a jar filled with bleach and empty it out when I had enough for a pizza, but even that was hard to eat.

I also cleaned the plasma donation center after Mina went to bed. They needed me more often because their customers stank and sometimes vomited. I worried about Mina the entire time, from when my key left my doorknob until I slipped under the blanket next to her after I returned, relieved and still smelling like bathroom cleaner.

Plasma is your blood with all the good stuff sucked out. The same drunks and drug addicts wandered in every week and said they were clean now, so could they donate so they could buy more drugs and alcohol? The answer was yes. Yellow payday loan signs surrounded the center, and every store wanted to buy my gold. I'd lived in Aurora since I was seventeen but never crossed the city to the Rocky Mountains on the other side. My daughter didn't pay attention to their outline through the smog and dust, but she knew they were there.

After the shooting, Mina and I didn't leave our monthly rate motel room for over two days. I called in sick at the plasma center. We clutched each other, stared at the TV, and forgot to eat. "Why would he do that?" I asked over and over. Mina cried into my shoulder. She had nobody but me on this cold earth, and "what if" ricocheted through my head. The gun still pressed against my ear.

Between our two memories, we patched together a picture of the boy who said Olivia. Maybe seventeen ("Eighteen," Mina insisted), tallish, wearing a hoodie over his thin chest. He looked like an older boy version of Mina, plus a few pimples. His dark eyes were what made him beautiful, gaze flashing back and forth between Mina and me and the gun.

While Mina pretended to sleep at night, I wandered around our room with a rag in my hand, unable to clean or cry. I tried to make sense of everything that happened—the boy, the killers, the scared silence of the other people in the convenience store, even after the police finally arrived. The police's professional disinterest in this Olivia. In us, the spared.

By the third morning, I knew I wasn't returning to work at the sex shop, that day or ever, just like I knew Mina wasn't going back to that elementary school where the kids said racist things and tried to sneak butterfly knives past the metal detector at the front door. "Enough," I said. "We can't hide in here like rats."

Mina said, "I think we need to find Olivia and—" But she didn't know what we were supposed to do, either. Still, we dressed and left. In the motel lobby, we found yesterday's paper in the bin and flapped it open on the counter.

The boy who said Olivia had his own name: Vance Farris. He was seventeen, a junior at the high school near our neighborhood. His next of kin was his aunt, Darcy Farris. The story was on page 2, upstaged by a dead politician. Vance's article didn't say anything about Olivia or that he took my place. A police officer mentioned gang initiation shootings, but because Vance "left with his killers voluntarily," they said they couldn't rule out possible gang connections between them.

Voluntarily? Is that what they called leaving at gunpoint? I didn't realize I was shaking until Mina tugged my finger away from the boy's picture and the sound of rattling paper stopped. "We're going to find this boy's aunt," I said. "And apologize."

"For what?" Mina asked.

"I don't know. I don't know. I said, 'Save me.' Maybe he thought I was talking to him, like, 'Hey, save me, stranger,' and that's why he did it. He was a child."

"*I'm* a child," Mina said. "He was saving *me*, Mama."

I stared at my daughter, who still had me and wasn't alone in the world or at Child Protective Services or dead in a convenience store or dead later with a needle in her arm because she watched her mother get shot in the street. She was safe. But I didn't feel any better about myself.

Mina searched for Darcy Farris's address on the hotel lobby's ancient computer, and I wrote it on a receipt. We walked to save bus fare, and we were numb by the time we arrived at the apartment building. Graffiti covered the brick, no artsy tags, just FUCK YOU WHORE in white paint, pissed out large from a fire extinguisher.

We knocked on a hollow door until it opened. A middle-aged white lady said, "Yeah?" Her bottle-blond hair stuck up from where she had been lying on it. Her eyes were so dilated, they were more black than blue. She looked nothing like Vance.

"Mrs. Farris?" I asked. Mina shrank from the open door, which burped canned peas and microwave bean burritos. "It's about Vance."

"He's dead," she said.

"He saved my life," I said. Mrs. Farris stepped aside. We edged past her into the heat.

The room was uncleanable. An ammonia ring scarred the carpet around the litter box, which sat in front of the TV. A cat squatted on the kitchen counter, licking its way down the assembly line of dishes. Mina was already sneezing into her elbow. The bathroom had hinges but no door. A sofa sagged against one wall, but I shot Mina a look forbidding her to sit on it.

Mrs. Farris sat. One of her eyes leaked tears, while the other stayed dry. She dabbed at the wet eye with a hardened tissue but never looked at me once. "So what's this?" she asked.

I told her the story. Throughout, Mrs. Farris's one eye kept crying, crying. I wanted to sit next to her and take her hand, but I was afraid of that couch. "I'm just so sorry," I said.

Mrs. Farris jutted her chin at Mina and said, "No, you're not."

I stuttered a little, then got out, "When is the funeral? Can we come?"

"I don't know. I can't pay for that." Mrs. Farris flapped a hand in the air. "We weren't well acquainted, Vance and me. I just knew him the couple of months he lived here. Vance is my ex-husband's nephew, not mine. I only let him stay here out of my goodness." She was so high, her head swayed every few words. Mina kept glancing at the door as if it might lock by itself and trap us.

"Where was Vance living before he came here?"

"No idea. We don't really talk. Didn't. His mother was somewhere in Denver." Mrs. Farris said "Denver" like it was another country instead of two miles west. "But then she left town with some man, somewhere. Vance's father could be anyone."

Mina began squirming, so I touched her shoulder. *That's not you*, I told her silently, and she settled down. "Was his mother's name Olivia?"

"No."

"I think I'm supposed to find someone named Olivia."

"Well, I don't know no Olivia."

I groped in my pocket for a clean tissue to blow the cat dander out of my nose. "Did he have friends? A girlfriend?"

"Boy didn't say much. I didn't even know he was in a gang until the newspaper. I wouldn't have let him stay here if I knew. Bringing that into my home. He should've told me."

I said, "I don't think he was in a gang—" But Mrs. Farris had already shifted her wet-dry gaze to the TV and started coughing, a deep, phlegmy cough that shook the couch.

Mina said, "Mama, let's go," but I couldn't move from that disgusting carpet. I asked, "Can I see his room?"

Vance's room was half of a bedroom, divided in two by Mexican blankets hung from the low ceiling, Mrs. Farris's sunken bed on the other side. Wool covers lay rumpled on Vance's bed, with a naked pillow, no sheets. His clothes rested in two wrinkled piles next to a stack of textbooks on a blue plastic crate. A notebook sprawled open on the brown carpet.

Looking over Vance's room, two emotions seemed to battle within Mrs. Farris before whatever drug she was on quelled them both. "This is it," she said.

We nodded. She led us out.

In the hallway, I almost knocked Mina over, scurrying back to Vance's room to pick up the notebook. I shoved it into the back of my pants, under my coat. "Where you at?" Mrs. Farris yelled from the living room, and I hurried out to her.

"I just wanted one more look at his room," I said.

For the first time, Mrs. Farris's eyes focused on me. "It's for rent," she said. "Let me know."

I was late for work that night because I couldn't stop looking through Vance Farris's notebook. HISTORY, the cover said, but inside I didn't find one date, one fact, one name except for "Olivia."

It was on every page, in all kinds of handwriting. Olivia on the side of an intricate airplane, on the heart of a robot, inside a rose. In one drawing, the O was the back of a girl's head, hair in a ponytail. In another, the O was the front wheel of a race car.

One sketch featured the silhouette of a boy with a cape, or maybe just a flapping shirt. He was catching a girl in his arms, both of them shaded opaque with blue ink. From their postures, it was hard to tell if the two people were flying or falling.

My head hurt, and I turned pages, looking for relief. On the back of one Olivia-covered page, Vance abandoned her for a moment to draw a blossom with triangular leaves, spiky at the tips, layered inside one another. It looked almost like a machine. The bloom grew out of some scribbled liquid. Underneath, Vance had written, "In a pool of shit grows a lotus."

I didn't even like it, but I couldn't stop looking at the page.

I tore it from the notebook, folded it, and slipped it into my coat pocket. Even there, it bothered me. I checked on Mina, her eyes blinking in the dark. I kissed her and hurried to my job at the plasma donation center.

The sign by the bank said 1 degree Fahrenheit. Every winter night I worked, there were two men loitering by the air vent. I guess it was warmer there. The men always pretended they didn't see me, and I pretended they didn't frighten me, and we all did our jobs—me cleaning an empty building, them staying alive one more night.

But this time, after I unlocked the door, I didn't walk through it. That's the funny thing about being alive when you're supposed to be dead. Nothing is a habit anymore. Instead, I felt the ghost of the gun in my ear, alongside this sensation I couldn't name. It was so powerful, it shook my whole body inside my coat.

"Get inside," I said to the men.

They didn't hear me at first, so I said it louder, meaner. "Come on. You're letting the cold air in."

They looked at each other and then shuffled over. There was ice on the beard of one, around his mouth. I held the door open, and they walked in as if they had been waiting all winter for me to come to my senses.

Inside, the men faltered from the heat. Or maybe they were drunk. But drunk people need to rest, too, especially drunk people. "Stay in the lobby," I said and waved them toward the chairs, but they flopped onto the floor. They were asleep on their sides before I took off my coat.

"Well," I said, and started my cleaning.

It was much nicer to clean a building with people inside it, even if they did smell. Cleaning isn't a bad job. The plasma place had Berber carpet, an easy glide for the vacuum, and they always paid me on time. I had even begun to come out a little ahead, before I lost my money in the robbery.

In between scrubbing toilets and mirrors and counters, I pulled out the piece of paper Vance had drawn on—the lotus on one side, the Olivias on the other. Was Vance's heroism for us or for this Olivia person? I had to know so I could sleep at night like these men on the floor, snoring even as I vacuumed around them. "Olivia," I wrote in foam on the front desk counter and scrubbed it clean. I folded Vance's drawings back into my coat pocket.

Then, "Time to go," I said in a not-scared voice, and the men woke right up. They stood, straightened their dirty coats, and walked out the door into the freezing air. I sprayed Lysol behind them, closed the door, and locked it.

The cops said over the phone they couldn't discuss the case, so Mina and I went to Vance's high school to look for Olivia. The kids were already leaving for the day. Hormones thickened the air and gave me a headache. I gripped Mina's arm too hard as the teenagers bumped past us to get away from the building.

Inside, a security guard ran a gloved hand through my purse and scanned my ten-year-old daughter in a way I didn't like before pointing us to the office. We passed a small white sign that said: IF YOU NEED TO TALK TO SOMEONE ABOUT THE MOST RECENT SHOOTING, PLEASE SEE THE GUIDANCE COUNSELOR.

We waited at the counter for the secretary to notice us. I've always been invisible to most people, but I was suddenly tired of waiting. "Hello," I said. "I need to talk to a girl who goes here. I don't know how to find her."

"Is she expecting you?" the secretary asked her stack of paper as she attached a paperclip.

"No. Her name is Olivia. I don't know her last name. But I think she takes history."

"All the kids take history." The lady finally peered at me over her thin glasses. "And there are over two thousand kids here. And school is out for the day."

"She was in Vance Farris's class."

"Oh. The boy—" The secretary squinted. "Are you related to either of them?" she asked Mina, for some reason. We shook our heads, and she said, "Then I can't give out confidential student information."

As I struggled for a way to explain the situation, Mina said, "Where's the bathroom here?" Once we were out of the office with a visitor's pass, Mina grabbed my hand and pulled me in the opposite direction from the restroom sign. She said, "The school library will have yearbooks."

I stared at her. "You brilliant girl. How did you figure that out?"

Mina pressed down a smile and her shoulders rose half an inch.

We passed miles of lockers until we hit one decorated with wilting carnations and paper crosses. VANCE FARRIS was printed in uncertain block letters next to R.I.P. Mina stared. I nudged her past, into the library. Nobody was in there, not even a librarian.

It took Mina three minutes to find the section with the yearbooks. She grabbed the one from last year and said, "Ta-da!" I slipped it under my coat. I was getting good at this.

"Mama, that's stealing," Mina said.

"Not if I bring it back."

Out in the hallway, a teacher turned a corner and headed for us. That book was under my coat and the bathrooms weren't anywhere near us, so I pulled Mina through the first open door into a little room.

"Hello?" A young man with a silver earring squinted up from a messy desk. "Are you here to sign up for GED classes?"

I read the paper sign taped to the door. "Yes," I said. "GED." The teacher in the hallway passed by, glancing once.

The young man handed me a sheaf of stapled papers, and I had to pinch the yearbook against my ribs with my elbow to take them and shake his hand. "We offer childcare on Monday and Thursday nights." He tapped a brochure. "We have translations for this in four languages."

"Her English is perfect," Mina said. "Better than yours."

"Mina. Okay. Thank you, sir." I pushed Mina out the door and toward the building's exit. The security guard at the door didn't glance at us. They don't care about you if you're on your way out.

At the taqueria down the street from the high school, Mina and I sat side by side in a booth. Mina sipped horchata and asked, "What's GED?"

"Like a diploma. If you didn't graduate high school, you can get a job with it. Or go to college."

"Am I going back to school?"

"Do you want to?"

"I want to be a lawyer," Mina said.

"Why a lawyer? You want to send those men to jail?"

She shrugged. "I don't want to clean buildings."

This stung so hard, I had to look away for a second to organize my face. Then Mina asked, "Lawyers send bad guys to jail?"

"Right to jail, where they can't hurt anyone. You could do that."

But Mina eyed the calluses on my hands, the nurse's scrubs I always wore because they were easy to clean in. Did she think she'd end up like me because she was my daughter? Maybe a GED wasn't the worst idea.

We thumbed through the pages of the yearbook together, Mina's foot on mine under the table. We scrutinized every page. The kids looked older than high schoolers. *Don't grow up*, I silently entreated Mina, pressing my nose into her clean hair. She elbowed me away but gave me a smile, a secret one between us, before looking around to make sure nobody noticed.

We found Vance's picture first. It shocked us a little to see him in there among all those other kids. It was the same photo they used in the newspaper. His eyes were dark and flashing, like they were that day in the convenience store, like Mina's eyes. Her finger traced his head, as small as a nickel.

We turned almost every page in the book before Mina pointed to a face and said, "Here. Olivia Zarate."

Even then, I kept looking for a different girl. "Maybe there's another one," I said. But this was the last page before the ads, and by some trick of fate, there was only one Olivia in Vance's grade.

She was as dull as a paper towel. Olivia had uneven bangs, plastic glasses, blotchy skin. Hers was any face you'd see on the other side of a fast-food window. She smiled like she was happy to have her picture taken, flattered at this common courtesy given to her, too. It was the face of someone who would be forever praised as a great worker before she was replaced with someone younger. Who would smile when people shortchanged her, and thank them.

"Hey, Mama," Mina said. "She looks like you."

Where there had been two homeless men, the next time there were five, then nine, and that night, seventeen men waited for me to unlock the door at the plasma donation center. They staggered in as I took off my coat and draped it over the counter. The last one smelled like aftershave and looked just like the man who hired me for this job a year ago, which it turns out he was. He said, "Everyone out of the pool."

The way the men shuffled back out told me they had been expecting this development far more than I had. The air was so cold it felt aggressive, and the men didn't look at me before disappearing into the night.

"What the hell?" my boss asked once they were gone, his eyelid twitching. "You had half of Aurora waiting outside like we're the fucking Rescue Mission."

"It's cold out." I could see him thinking, *Bleeding heart*. But I wasn't bleeding. I didn't even feel bad.

"You're fired, of course," he said.

And I couldn't help it; I was laughing.

For a moment, my boss laughed with me. His was a bewildered, barking laugh, which made me laugh even harder. The sounds swirled in the gray space between us.

Then he frowned and shoved me out the door.

The next afternoon, I finally found out where "the disposal of the remains" would take place for Vance, after calling almost every cemetery and mortuary in town. "Does his school know?" I asked the voice at the other end. "Did you tell people?"

"I don't know," he said. "I just work here."

Mina and I walked to the high school again to watch the kids leave. I stood at the foot of the stairs. Mina ran to the top, right at the edge of the doors, so she could see faces better.

The kids erupted from the building as before. We scanned the crowd for anyone recognizable, but there were too many kids swarming around me. They smelled like sweat and baby powder and supermarket cologne and sex and pot and meth, and I couldn't find Olivia. Soon, there was only a slow dribble of students left. Another day lost. I let my arms collapse to my sides and called up the stairs to Mina, "We'll try again tomorrow."

But Mina had already slipped next to a half-asleep girl in a light brown coat. If Mina hadn't noticed this girl, I wouldn't have. Mina plugged her mitten into the girl's gloved hand; the girl didn't jerk her hand away, so she probably had little siblings of her own. Mina pointed at me, and I waved.

They approached, my butterfly of a daughter dragging this lumpy girl dressed entirely in brown. I asked, "You're Olivia?"

She nodded.

"Is there another one at your school?" I asked. "Some other Olivia?"

"I don't know. Maybe." It was cold, and Olivia clenched her teeth to halt a chatter. "I'm sorry. What's this about?"

"Vance Farris," Mina said. "We're friends of his."

I asked, "Can I buy you a hot chocolate or something? Coffee?"

She regarded my face and shoes, a film of distrust on her face.

"I have to get home to watch my sisters. You mean that kid who got shot, Vance?" We nodded, and she said, "I'm sorry, I didn't really know him. Maybe someone else—" She glanced backward at the school, already deserted. "I might've sat in front of him in one of my classes."

"History class," I said. "He drew you in his notebook." I pulled the picture of the lotus from my pocket and showed her the doodles on the back—Olivia the rose, Olivia the car, the *O* head with the ponytail. "That's the back of your head, right there."

Her eyes widened at her own name in his handwriting. "I'm sorry. He never talked to me. I think he borrowed a pen once." She tried to hand the paper back, but I wouldn't take it.

I told her, "Your name was the last thing he ever said."

So much steam escaped Olivia's lips, it looked like her mouth was on fire.

I explained about the robbery, about *Take me* and *Olivia*. I don't know what I expected from her, but it wasn't stillness. My feet grew numb in my sneakers.

When Olivia finally spoke, she stammered. "There must be another—"

"That"—I flicked the piece of paper in her hand—"is from his history notebook. He sat behind *you* in his history class. It's you."

"I'm not that important." She wasn't looking for a compliment; she was stating a fact. She said rapidly, her eyes reddening, "I'm sorry. I didn't know him, he borrowed a pen, I just, I'm sorry—"

"Stop saying you're sorry," I snapped, and Olivia teared up. I didn't care, stupid girl. I grabbed her coat sleeve and shook it. "Why did he say your name? What are we supposed to *do*? Think, girl."

Olivia twisted herself from my grasp. "I have to go. I'm sor—I have to go." She scooted down the sidewalk, moving fast without looking back.

"The funeral is tomorrow morning at Mount Nebo Cemetery," I yelled after her. "Ten a.m." The wind had picked up, and it was hard to tell if she heard me. She was making time in her polyester coat, chin tucked, hands weighing down her pockets like stones.

"She wasn't even his girlfriend. I don't get it." I turned to Mina. "What did Vance want us to do?"

Mina frowned after the girl. "Did he love her?"

I remembered the sketch of the boy catching the girl in his arms. "He must have."

Mina stared into the distance. "Maybe . . . he wanted her to know who he was. Why he took your place with those men, and did that for us."

"That's just it," I said. "Why did he?"

"Mama." There was pity in Mina's eyes, and I realized I was already eating her dust. "Because he was a hero."

The cemetery was a sunken square near Olivia's high school. We waited next to the Jesus like I had been instructed over the phone, our feet numb. The statue stretched out its concrete arms, and I don't believe in that stuff anymore, but I made the sign of the cross anyway.

The outline of mountains was faint through the smog that morning, like a torn piece of pale paper. Traffic noises bounced off the gravestones. With no buildings on that whole block, the sky was enormous. Mina stared at a pinwheel stuck in a small grave. Pink plastic flowers rested in an empty whiskey bottle.

"I want to go back to school tomorrow," Mina said.

"Maybe a different school," I said. "One without bullies."

"It doesn't matter where I go," she said. "I'm not afraid anymore. Of anybody." Mina's face was smooth as a petal, one eyebrow slightly raised in challenge. I didn't know if it was good or bad for a little girl to feel that way. There was plenty to fear. But being scared wouldn't keep us safe.

At ten past ten, a young man in a narrow tie and overcoat approached us, carrying an urn. He wore a badge with the name of the funeral home on it. "You're here for Vance Farris?" He glanced around, but we were the only ones there.

"Maybe we should wait for his aunt," I said. The young man shook his head. He must have already spoken with her.

He led us across the grounds and stopped at a small hole in the ground, about a foot wide, with upturned dirt beside it. He pulled the lid off the urn and tugged on a plastic bag full of ash, whitish-silver and powdery. The young man asked, "Would you care to do the honors?"

I shook my head, then nodded. I reached for the urn, but he pulled out the plastic bag and handed it to me instead. It was heavier than I had expected. I didn't know how to feel about this, a child in a bag. This child, Vance.

I placed the plastic bag in the hole.

The young man opened a small book with a ribbon marking a page. He began reading, "Return unto the ground, for out of it wast thou taken, for dust thou art, and unto dust shalt thou return." He asked me, "Is there anything you want to add?"

My throat was full of tears. Mina stepped forward. She shouted down the hole, as if the plastic bag were very far away, "Thank you for saving my mother!"

With a collapsible spade, the young man nudged the dirt back into the hole, and then he stepped on a metal plug to fasten it onto the grave. There were similar metal plugs everywhere, dotting the flat lot, shining their dull faces to the sky. Vance's said 947 on it, no name.

"What's that?" Mina asked.

"It's his marker. So we know where he is. May he rest in peace," the young man added. He bid us goodbye and walked out of the cemetery, cradling the empty urn.

We were alone in this place. Mina studied her cheap shoes. She didn't question my judgment, didn't complain about the cold or ask to leave. So I just told her: "I lost my job."

"Which one?"

"Both."

She squinted up at me. "That's okay, Mama," she said.

It wasn't, though. We had the cheapest hotel room in town because a lady got murdered there before we moved in. But after we were robbed, even there, I wouldn't have enough to cover the monthly rent.

"Do we have to go to the shelter again?" Mina asked. "Or can you get a new cleaning job?"

"I don't know." The paying world will always need cleaning. I just wasn't sure I was the one to do it anymore.

I reached into my pocket for the picture of Vance I had cut from the newspaper and gotten laminated. I had planned to prop it against his gravestone, but he had none. "Dust thou art," I murmured, and I wanted to believe it. But I couldn't, because it wasn't true, and that strong thing rose up in me again, past my frozen feet and cold legs to my heart, which was wild with beating.

Someone panted behind us, and we turned around.

It was Olivia. She had run here, it looked like. She was almost pretty in stockings and a black skirt under a borrowed-looking black coat, and high heels that teetered on the bumpy ground. She caught her breath and stared at the vast cemetery. "Nobody's here."

"We didn't know who to call," I said.

"I really didn't know him," she said. "I swear to you."

"It's okay," I said.

"He never talked to me except that once, asking for this." She pulled something from her coat pocket, a pen. "It's the one Vance borrowed," she said. "It has his toothmarks on it." She gave it to me. I traced the ridges and dents his teeth had made on the white plastic. I handed her the laminated photograph of Vance. Traffic noise surrounded us.

Olivia closed her eyes and pressed her lips to Vance's picture.

When she looked up, her face was a mess of tears. Mina was already reaching for her, but Olivia flailed her arms, choking out, "What do I do now?"

I wanted to tell her to go back to her crappy life. That she'd forget this dead boy, the way she'd forget live boys who would forget her after they were done. But no, because Vance Farris had blocked every direction but up, daring us to find a way to rise, too. Here.

—For Cort McMeel

When in Bangkok

The morning after we landed in Bangkok, my father tossed some baht onto the hotel restaurant table without counting it. Enough eating, he said. My sister and I stood immediately, still chewing.

But the girls haven't finished their breakfast, my mother said, and stood up to file out of the restaurant behind us.

Our hotel was only two blocks from the Patpong district in Bangkok. Patpong was a different place during the day. All the bar girls slouched around, hungover, in flip-flops and dirty tank tops. Crumpled cigarette butts clutched one another in the gutters, and the garbage from the night before brewed in puddles of rainwater and aging urine. We walked looking down to sidestep the vomit and sticky spit that would never dry here.

My father hustled down the street, his arms swinging, knocking into other people and swearing. He darted in and out of stalls, peering at the signs written in Thai and English. He strode so quickly, it was hard for the rest of us to keep up.

What are you looking for? my mother asked. A bathroom?

Not a bathroom, he snapped, and then he stopped.

A little girl sat alone on a stool outside a place with a sign that read OIL AND THAI MASSAGE. She wasn't more than four. Couldn't have been. She licked a red Popsicle that was melting rapidly in the heat. Red syrup dripped down her hand and arm and onto the pavement. Her lips were stained red, by the Popsicle, maybe. She licked until it was gone, until she was forced to insert the stick into her mouth for the last nub. Black kohl lined her eyes, and a small snarl rooted in the back of her hair. She wore an orange tank top and hot-pink shorts, and she kicked her brown legs in the air, toenails painted in coral.

A woman walked out of the store. She was middle-aged, with short hair dyed black. She looked almost like my mother, except she was Thai and wearing a stretchy yellow tube top. She poked my father and asked, You like? You buy her another?

The little girl gazed up at my father, juice on her chin. He shoved his hands into his pockets. One of his shoulders was up by his ear.

The woman said, You come back. I member you.

My father shook his head no and shuffled away. My mother and sister lingered by a vendor grilling food on a stick. My father pulled his wallet from his back pocket and selected some charred meat for them. His gaze caught on the window of a nearby tailor. He jutted his chin at a white female mannequin wearing a boxy black suit and a skinny piano tie.

I need a new suit, he said.

It's a *family* vacation, my mother said, swallowing her meat.

But my father stared into a horizon he couldn't see, obscured as it was by building after building full of more people than the world could possibly hold. He looked relaxed now, and even smiled.

I've earned myself a new suit every now and again, he said.

This was our fourth trip to Bangkok, for our fourth family vacation since we moved abroad to Singapore four years ago. The company paid airfare for the vacations, one per year, anywhere on the globe. My mother had asked for Paris this year, but my father said no, we were going to Bangkok again. I was twelve, and my sister was fourteen. She didn't ask why, so I didn't, either.

My father was a genius, my mother said. He had graduated college when he was nineteen, some kind of prodigy in engineering.

He was bossing people twice his age by the time he was twenty-one. Now he was forty-five, he wasn't a prodigy anymore, and he just bossed people his own age. He had worked for the same oil and gas company his whole adult life. I never saw him wear anything but suits and white dress shirts, short sleeved on weekends. His eyes had started to pucker at the edges. When he first got assigned to Singapore, it was like he had won some kind of big company prize. Now he often growled to our mother, They've all but forgotten me Stateside. Nobody cares about me; they're just saving themselves. Fucking Reaganomics.

To make him feel better about himself, sometimes my mother told me to ask my father for help with my homework, but nothing is for free. There was a space in the hallway of our Singapore apartment where we were meant to store bicycles, and for a while I slept there at night until the building superintendent caught me and that was over.

Last year, on our third trip to Bangkok, we traveled with another family. My father was the other father's boss. We stayed in a different district, one with normal hotels that had other kids and families and no neon go-go bars in the street outside. The two men went off shopping for menswear while we all lay around a muggy hotel pool. The other family's kids ignored us, and my sister and I ignored them. The sky was that particular kind of gray I've only ever seen in Asia, like a felt muffler above us. The two mothers drank bitter Tiger beers, tried and failed to suck in their chubby stomachs, and said, Now, isn't this pleasant! Isn't this pleasant!

When the other family's father—why can't I remember his name?—returned from his outing with my father, he walked straight over to the pool and toppled in with his clothes still on. He still wore his dress shirt tucked in, belt, shoes, pants, everything.

He stayed underwater for a long time while everyone pretended it was normal to do that.

He finally surfaced after I thought he had already drowned. I realized with shame I hadn't even considered jumping in after him, or calling for help. He gasped, face tinted yellow, his glasses wet and flashing. The front of his white shirt darkened from the shadow of his chest hair beneath.

Are you okay? I asked.

He shook his head and coughed for a long time, tethered to the pool by one elbow. Then he pulled himself out of the water and sat a few feet from me at the side of the pool. Water streamed from his body and pooled under me, wetting my suit. His wife laughed at something my mother said, a laugh like a scream.

I asked, Where's my dad?

The sun slid out from behind a cloud. The man said, Your dad, he's, he's, he's. We got lost.

Should we go look for him? I asked.

But the man just shuddered at the water casting its reflection onto his face and said, When in Bangkok, man. When in Bangkok.

The next morning, that family was gone, checked out of the hotel in the middle of the night. That man quit his job before we returned to Singapore, and the family flew home to the States. We never saw them again.

On each trip to Bangkok, we visited one tourist hot spot, alternating each year between the Grand Palace and Wat Arun (*What a Ruin*, my father liked to joke). This year it was the Grand Palace again, which was the shinier of the two and didn't have puns. My father said he wanted to get it done our first day there, so we taxied straight there from the massage parlor and suit store in Patpong. Our bodies were slick and slippery under the long skirts we had to rent to wear on temple grounds. The air felt breathy. Our clothes chafed our armpits, soaked up the humidity, and stayed wet.

The buildings in the Grand Palace scintillated with gold, and gems encrusted the ceramic statues. Or maybe it was steel that was painted gold, and chips of glass made to look like gems. The place was so shiny, it hurt. Monkey demons propped up buildings on their shoulders. Massive currents of people pushed us around the grounds, and my father poked our spines with two stiff fingers. Move it, he said, his gaze pointed past us at the distant exit. When we didn't walk fast enough, he kicked at our legs.

But he lingered at the Inner Court, where the king had kept his harem. My mother read aloud from her four-year-old guidebook: *King Rama V kept a consort who was his half sister, and he is single-handedly responsible for reviving the Thai national sport of muay thai, a.k.a. Thai boxing. The Inner Court was governed by a series of laws known as the Palace Laws, punishable by death.*

My mother's thin eyebrows lifted. She said, Girls, aren't you glad we don't live under Palace Laws?

I'm sure glad, my sister said.

My sister reminded me of a doe, or a cow. A mythological doe-cow creature. Her eyes were far apart behind her glasses, and she was a stranger to outrage. Her grades were perfect. She was here but not here. Nothing affected her, as if she were constructed from some nonstick surface. She never cried or hurt herself, even when we were little. I had already broken an arm, a foot, a collarbone, and three toes, but she had never even sprained a finger. I couldn't understand what she was made of.

Behind me, my father said, The king was a lucky man. He had a whole harem to choose from.

Words seemed to shudder up from my feet, and then they

were in the air before I could stop them: And some of them young enough to be his daughters, I bet.

Shut up, Elsa, my sister said, and then she kicked me herself.

I blocked my head with my arm as I turned to face my father. But he had stopped to count his money a few meters behind us. He hadn't heard me.

Don't screw this vacation up. He's in a good mood, my sister said. She pulled me close, and her clammy fingers dug through my shirt, her voice hot in my ear: Don't you know when you're lucky?

The next morning, we put on our bathing suits and rode the elevator to the rooftop pool, where we would spend the next four days in Bangkok while my father shopped for suits he never bought. As my father was leaving the hotel, my mother said to his back, Bring up some sunscreen before you go. We're all out, and the girls will burn.

Even though it was hotter and wetter up on the roof, the air lightened with my father gone. My mother and sister wore matching purple bathing suits. I felt embarrassed by my mother's cowering shoulders and the clots of fat on her thighs. She never did anything, said anything. When other pool guests passed by, she grimaced behind her orange lipstick, as if they might hit her.

My mother and sister opened their books, lying on the plastic lounge chairs. I swam in the pool until I was dizzy. When I pulled myself out of the water, my mother and sister had fallen suddenly asleep in the sun, like matching princesses in a fairy tale.

Standing on a plastic table, I hoisted myself up onto the high wall, kicking my feet over the side. I steadied myself and dangled my legs. Twenty stories below, people dotted the ground, streaming in two opposing directions in the street. There was nothing beneath my feet but thick, wet air, with my father somewhere down there.

I could just let go and fall, I thought.

I scooted closer to the edge, my bathing suit catching on the concrete. I felt strangely human at that moment, with the street below, the crumbly concrete beneath my thighs, the smell in my armpits. I wondered what I'd have to do to land on my father's head.

I leaned forward and willed myself to find that edge between here and there. Just to see if I would. The street rippled below me. Gravity pulled at my ankles. A rare breeze lifted my hair from my neck. A siren sang.

I lost my balance but caught myself with my hands before I realized it, saving myself. It surprised me that I did that, that I would want to.

I climbed back down to the terrace. My mother and sister didn't wake up, so I slipped my T-shirt and shorts over my bathing suit and rode the elevator down to the lobby. Bottles of sunscreen cluttered the gift shop, a battalion of sunscreen, but my father was the only person in our family allowed to carry money. I walked out of the hotel instead, retracing the path we had followed after breakfast.

The suit shop hadn't changed its piano tie mannequin, so I recognized the storefront right away. Inside, a family sat in chairs, sewing. They peered up at me and then back down at their dull fabrics. A man stacking rolls of cloth asked, You want suit?

I'm looking for my father, I said. He's buying a suit.

Not here, he said. Your first time in Thailand?

We've been to Thailand four times. Just Bangkok, though.

Bangkok is not Thailand, darling, he said. Why you sad? It's okay. *Mai pen rai.* He plucked an orchid from a bowl of water and handed it to me. It was wet. I thanked him. Back in the street, I dropped the flower on the ground and stepped on it.

A few doors down was the same massage parlor as before, its sign a geriatric pink beneath the sleeping neon. The same woman loitered in front, pinching a cigarette.

Is my father in there? I asked her.

The woman said, No English.

I said, You knew English yesterday.

The woman assessed my sandals, my damp T-shirt and shorts. You go home, she said.

A little girl was here this morning. Eating a Popsicle? I said.

The woman slipped easily into her anger, shoving me backward, yelling at me in Thai as people wove past us. I retreated, hands up, until she broke off and walked away, her butt wiggling from long habit.

She said over her shoulder, Girl gone. Just like you, *farang*.

That evening, my mother and sister were nearly magenta. They hadn't woken from their naps until the sun finally hid behind a cloud that promised afternoon rain but never delivered. Their burned skin shone, too tight for their bodies. My father drew a scalding bath and made my sister get in, her bright legs turning purple. It draws the heat out, he said. He was so happy, his voice went singsong.

When my sister emerged from the bathroom in a towel, stepping gingerly, my father shot her the disgusted look reserved for her and me now we're not little kids anymore, as if we have taken something from him.

We need food, my mother said. We haven't eaten since breakfast.

Then get your clothes on, my father said. It's terrific out.

You forgot our sunscreen, my mother said. We're lobsters. Why can't we stay in the nice parts of Bangkok? Those hotels have room service.

Too expensive, my father said. Takeout tastes better, anyway.

He took me along to help him carry the food and set off down the street. I followed like his dog, staring at the dark hairs on the back of his head, the square outline of the wallet in his back pocket. I always avoided his eyes. You could get sucked into that blackness and never come out. I didn't want to see anything as he saw it. It was good he never looked at me anymore.

Patpong had woken up. American music played everywhere, and we walked from Duran Duran to Genesis to Tears for Fears, winding through vendors. I didn't know where we were anymore or how to get back. The night market was doing good business under its plastic canopy, and we passed stalls of T-shirts and cassette tapes until my father veered into one of the bright side streets where the go-go bars were.

Neon illuminated the street in primary colors and pastels, and everyone was friendly-mean. Doors opened to women swinging around poles. Some places displayed menus, but instead of food, there were photographs of girls next to the prices. Men stood outside and chanted to us, Ping pong show. Watch pussy open bottle, shoot banana, put fish in, write letter, blow out candle. Hey, you bring your daughter.

A grown-up woman in hot pants sidled up to my father, hips first. She stroked his arm and rubbed her big breasts against his chest, saying, I love you, you love me? Why you no love me?

My father shoved her away with a snarl. That's just disgusting, he said.

Dad, I said. Can we go somewhere new tomorrow?

Where? he asked, flailing his arms, as if there were no other place on earth.

Mom's guidebook says there's a Tiger Temple. Or maybe an elephant ride?

I have a fitting at eleven, he said.

What about the bridge over the River Kwai? Or a jungle tour. It's supposed to be a beautiful country, a big, nice country. Dad? Wouldn't that be funner than . . . another fitting?

But he strode ahead before I finished talking, and city noise strangled my voice.

I let my pace slow. My chest loosened as my father's body grew smaller and smaller. Soon, I couldn't distinguish the back of my father's head from anyone else's in the street.

I took an abrupt right, down a road with a sign I couldn't read.

The new street swallowed me. It looked the same as the one I was just on with my father. The same pink and orange neon blinked above my head: Pussy, 69, Cocktease, and words in Japanese as well as Thai. Same girls in bikinis, same men chanting. I walked as quickly as I could, trying to avoid the barkers standing outside the go-go bars.

He'll have to miss his fitting now, I thought. He'll be at the police station, trying to find his daughter in a city of millions, saying we just arrived two nights ago and he already lost me. He'll make a deal with himself. He'll say, If Elsa comes back, I promise never to do a bad thing again.

But the farther I walked away from my father, the more my body buzzed with audacity. Maybe I'd never come back. I didn't know where I was, or even the name of our hotel, except that the word had a kind of backward *N* in it. I might walk out of this district into the real city, with normal people pushing strollers or holding their kids' hands or going to work. I could get a job as a babysitter with a new family. Or find the American Embassy and ask them to fly me back to the States. Maybe California, or Hawaii.

But I didn't have my passport. My father had it.

"American girl," a barker yelled at me. "You want job?"

No money, either. Not one baht. No money, I realized in a panicked rush, no money no money no money no money no money and I would rather shred my flesh from my face than return to him but *no money*, so I turned around and bolted back down the street the way I had come, cutting right at the corner, scanning for my father's dark head against all the dark heads in the night street as I ran and ran and I was lost until there he was, hands in pockets, still ambling ahead, neon reflecting off his hair in ice-cream colors. He hadn't noticed I had left.

I'll never escape, I thought. I'll never even try.

I caught up to him just as he stopped at a stand for food, the kind that grabs you by the shirt, with chili and soy sauce and ginger and garlic attacking the air. The cook unstacked Styrofoam boxes while my father pointed at things: beef and pepper stir-fry, roasted duck, green papaya salad, black sticky rice. I couldn't catch my breath from running. Please, I said as he filled my hands with plastic bags and sauntered back in the direction of the hotel, whistling through his teeth. I stumbled behind, bags rustling.

Back in the hotel room, everyone but me sat on the floor and ate the corrugated strips of meat and vegetables gleaming with cornstarch. Then they lay on the beds, drowsing toward sleep until my father turned off the lights. I couldn't stop trembling. My sister was also sleepless, flopping around on our double bed, trying to find a cool place for her burned skin. It was like there was too much blood inside me. I strained to force air into my stony lungs. I think I'm dying, I whispered. I need to get out of here.

My sister whispered back, You just don't know how to be happy.

I once read a Thai myth of a family with several daughters. The youngest daughter talked too much. Her family hated her. Her town hated her, because wherever she opened her mouth, she told malicious stories that turned into rumors. They just couldn't be true. So the family and the town put her and her sisters on a raft and sent them down the river to the sea.

After a few hours adrift, a pirate brought the daughters aboard and made them his wives. But after listening to the youngest daughter talk, the pirate couldn't stand it anymore and threw her into the ocean. A pair of sea eagles picked her up and flew her high into the sky. The girl kept talking. She couldn't stop herself. So the sea eagles dropped her back into the sea, where she drowned and fed the fishes.

Years later, a monk found her bare skull washed onto the beach. He made it into an incense holder for the chanting room at the monastery. But the monks began squabbling among themselves. The girl's skull had poisoned their minds, you see. So they turned her skull into a water scoop for the toilet. Then the monks got rashes all over their bodies.

Finally, they burned the girl's skull and scattered the ashes to the four winds. Pieces of her flew in every direction. She was everywhere. And she was silent.

Our third morning in Bangkok, I started vomiting at 10:40, when my father began to tie his tennis shoes. My mother came over and gawked at the bilious wisps floating in the toilet bowl. Stench swirled in the dead space there.

Elsa is sick, she called to my father, and bent over me. What did you eat for breakfast?

I ate what you ate, I said. I leaned into the toilet bowl and retched again. I said, I need medicine or something.

My mother said, Stan, could you please go to a pharmacy?

She's a kid, he said. Kids throw up. I have my fitting.

She snapped, You can be late for your tailor. Your daughter needs Pepto-Bismol.

You go get it.

How can I when you won't let me have money?

You always do this, he said. I work hard, while you sleep on the couch all day and make spaghetti for dinner. I ask so little of you.

Yet you take everything, my mother whispered.

While they argued in the hallway, I stuck my finger down my throat again.

My father left anyway, of course. I crawled back into bed and told my mother, I just need rest. You two go ahead.

My father had relinquished a short stack of baht to my mother, so she left with my sister, promising Pepto-Bismol upon their return. As soon as the door shut, I jumped out of bed and jammed on my shoes. I took the next elevator down, slid into the crowd outside, and walk-ran until I reached the massage parlor.

This time, I didn't ask. I waited behind a pay phone and watched in the heat. The same lady paced in front, in a green tube top this time. When she drifted to the bar next door with a plastic cup in her hand, I held my breath and ran into that lady's stinky hole of a building, down the narrow hallway with its line of cheap doors.

I threw open each door. Inside the first room was a bucking white man and a Thai woman, both naked. Naked people in the next one, and the next one. Before I touched the knob of the last door, it opened on its own.

It was my father. He was still dressed but with a wad of baht in his hand, one finger hooked around the neck of an empty beer bottle. He stared at me as if he had never met me before. He stood just behind the door, his mouth open.

I pushed at the door, but he stayed it with his hand. Inside the room was a bed. On it, I could only see one wiggling foot with coral toenail polish, impossibly small.

Daddy, I said.

Then I lunged at him, clutching his shirt and clawing at his face. He shoved me away with one palm on my forehead and slammed the door. I tried to turn the doorknob with both hands, but he held it tight from the other side.

Now the woman in the green tube top ran down the hall and tried to pull me away, joined by women streaming half-clothed from the other rooms. They yelled at me in sharp voices. I was shouting, Come out, you asshole, you coward, I hate you, I hate you. I held on to the slippery doorknob, but the women seized my limbs and counted in Thai. On *saam*, they gave me a good yank, and I lost my grip and fell backward to the floor. I scrambled back up, grabbing everything within reach and flinging it at the door—a lamp, a door mat, a credit card imprinter, a magazine.

As the women reached for me again, I took one final run at my father's door. At the last second, I dropped my arms and rammed myself into it, headfirst.

I woke up on my back. The air was orange and pink. Everything hurt. Someone was running fingernails through my hair, probing my scalp. Women surrounded me, sitting on the floor.

Did he come out? I asked.

The women leaned in, murmuring soothing vowels: *Mai pen rai, mai pen rai*. They patted me all over. My father's door was still shut.

I turned and pressed my face into the women's laps, crying into their perfume and sweat and that rubbed-off smell from their crotches, that underwater stench of everything I had tried to forget and now knew I never would.

There's a dead place where no one else goes. It's warm and still and deep and perfectly quiet, like the bottom of a full well. Up above, you can be walking, talking, smiling, but all the while, you're asleep down there in the silent, dead place where no one else goes, and no one is your enemy. That is your enemy.

That evening, my father came back late, bounding around the hotel room, elated by his own nerve. He smacked a fist into a palm and sang, Girls, get your shoes on. He wouldn't look in my direction.

I was having trouble walking. My head hurt, and a big lump swelled just past my hairline. I threw up again in the bathroom, this time without trying to. I asked for an aspirin, but they all just stared at me as if I were speaking another language.

My sister said, Those aren't even words.

Elsa, my mother said. Try.

Something tickled the inside of my ear, and when I wiped it, a pink smear of blood streaked my finger.

Outside the hotel, the sky had lowered like a brow. My father shuttled us into a *tuk tuk*, loud and metal and dark, and it carried us down the road. My father rested his arm around my mother's burned shoulders. My sister tried to keep her balance in the *tuk tuk* without her skin touching anything. I was sweating on the outside and the inside.

Then the rain began. It was so much rain, like a giant hose was trying to wash away all the human dirt and spit and urine and shit from three thousand years, wash it all into the Gulf of Thailand, where the waves would beat it and sterilize it with salt until it had disintegrated into nothing but seawater again.

But that didn't happen, and instead the city just smelled like wet garbage. The *tuk tuk* sloshed through roads like shallow rivers.

By the time we made it to the muay thai stadium, it was raining so loudly we didn't even try to speak. Our waterlogged tickets broke apart when the man tried to rip them.

This is Thailand's national sport, my mother said. It has ancient royal traditions, girls. We're fortunate to experience it.

My father grunted at her.

Inside, the stadium was hot. We sat on red plastic seats. We were only about fifteen rows from the ring. The ropes were so old the red paint had worn off in spots, revealing brown underneath. Rain pelted the tin roof. It was hard for me to focus on anything. I tried to rest my head on my sister's shoulder, but she shrugged me off.

The children fought first. The adults headlining the posters would come later. The matches began slowly with traditional dances and ended quickly with fights between scared seven- and eight-year-old boys. I wondered what happened to the losers and the winners. There were no proud parents I could see, just trainers and some kind of handler who shouted things to make the kids kick and punch harder and grab each other in clinches.

People were already betting. A special group of gamblers sat at one end and waved baht in the air, although the betting was all around us, too, people pointing fingers in combinations that made no sense to me. They were loud, yelling and scoffing at the little boys, who held their wrapped fists by their faces and fired off stunning leg kicks. The crowd cheered whenever flesh struck flesh. My head felt like it might crack at every sound.

Isn't this pleasant, my mother said.

The next match was a couple of boys a little younger than me. The first boy was much bigger than the second—fattish, with a head so big the wreath they placed on it looked like a miniature crown. His muscles shone, and his eyebrows looked like two thumbprints.

The second boy looked strange. His features were delicate, and he was much thinner yet fleshier than his opponent, with narrow shoulders. He wore red trunks, his skin dark and smooth. His hair was shaved so short he was almost bald.

Of course, the big boy was beating up the little boy. Each time the bigger boy landed a heavy kick, the crowd cheered in unison. He smirked around his mouth guard. The blades of the ceiling fans pivoted around and around. Gamblers touched pieces of paper to their lips.

The small boy's arms dangled from exhaustion. The music circled him, whining and wheedling—a sireny horn, bells in relentless beat. He was losing, but he kept going. The small boy's handler slapped the floor of the ring and pointed at him, shouting, face sweating. The small boy then came alive, punching and kicking with everything he had left, as if this were his last chance to live.

The small boy was a girl.

She looked the same as any boy fighter—skinny chest, polyester trunks, hips tight. But I knew she was a girl, as surely as I knew I was a girl.

I suddenly realized: Everyone knew. The betting men knew, piling on the odds. The announcer knew, the referee knew, everyone knew except for my fat, dumb mother who cheered when the fat, dumb boy delivered a sharp knee to the center of the girl's stomach. Or maybe my mother knew, too, like everyone else.

But no one would say it.

My father's gaze flicked once in my direction and then away. He cupped his hand around his mouth and shouted, Get 'er. His eyes were half-moons. I deserve this, I heard from inside his head. I deserve happiness.

Now I was standing in front of him, trying to catch his gaze.

You, I said. Hey. He veered his head to glimpse the fight behind me. I weaved to block his view.

I told him, You can't do this anymore.

Startled, my father looked at me, maybe for the first time ever. Fear rose from him like an odor.

And I'm telling everyone, I said.

The crowd ignited around me, loud with surprise. Something had happened. Men exchanged large stacks of baht, waving hands by their noses. My father didn't speak, didn't cheer. I turned around.

The girl was on her feet, and the fat boy was on the ground.

The sour referee grabbed the girl by the elbow and raised her arm in the air. Her wet cheeks shuddered with exhaustion. Still surrounding us was the eternal siren music and the men trading on her pain. But this girl knew it was over now. She was going to make it out of this place alive.

—For Lucia Berlin

The Standing Man

I recognized the Standing Man the second he stepped into our ramen shop, although I had never seen his eyes open before. I'd often passed him on my dawn walk to work. Each time, the Standing Man leaned upright against a pole near the entrance to Komae Station, profoundly asleep.

The Standing Man wasn't the first person to sleep standing up outside a train station on the outskirts of Tokyo, but he was the only foreigner I'd ever seen like that. His suit was always left over from the day before, American cut with flap pockets like I wore as a young man. His tie was still cinched tight, shirt wrinkled. I had to brush his suit jacket aside to press the WALK button. Even snoring, he managed to hold up his head. A thread of drool sometimes stretched from the corner of his mouth to his phone, which he cradled in one hand. His wire-rimmed glasses had curved arms where the ears go, and they swung above the pavement, looped around his thumb.

Ojiichan and I felt sorry for the Standing Man, so after preheating the oven early one morning, I warmed up a bowl of ramen broth from the day before. I carried it to the station with some dumplings and left the takeout tray on the ground near his feet, resting a pair of chopsticks across the top. But before I got five steps away, the Standing Man woke up and stepped in the bowl, spilling the scalding soup, swearing, still caught in the gauze of his dream. Since then, I've let him sleep undisturbed.

The Standing Man was awake now, fidgeting in his seat in our ramen shop, across the counter from me. I positioned his order before him and wiped an oily drop that had spilled. He began eating without noise, but that's the American version of manners. The Standing Man appeared different in the yellow light of the shop, rather than the gray glare of dawn. He was tall but delicate, maybe thirty, with orange hair and thin, chapped lips. He smelled like butter. Behind his glasses, his pale eyes were the color and depth of a skating rink, and freckles marred his nose and cheeks. He looked exhausted; everyone in Tokyo Prefecture looks exhausted. Komae is a bedroom community outside Tokyo on the Odakyu Line, fifteen lumbering local stops from the hub terminal at Shinjuku Station. From there, everyone disperses to their corners of the city, like marbles dropped on a wooden floor.

After a few minutes, the Standing Man said "Excuse me" in lousy Japanese. His voice unnerved me; I preferred him as a silent cautionary tale. He asked the same question everyone does: "You never write your orders down. How?"

"We remember them," I said in English. "Pardon me." I hastened to the line of customers standing against the west wall, waiting to sit. I took seven new orders, memorizing and queuing them in my mind behind the six orders I had already taken and those of the twelve people slurping ramen at the counter.

When I returned, the Standing Man said, as if I hadn't left, "But there are too many people to remember them all." His gaze roamed as he struggled to extract the Japanese words from somewhere inside his head.

"You can just speak English," I said in English.

He relaxed and leaned forward. "Thanks, man. I moved here from Chicago six months ago, but I still don't have the language. I'm Abraham."

I said, "I'm Satō. This is Ojiichan—he doesn't tell people his real name. Ramen Komae is his shop." Ojiichan ignored us, working the toppings with shocking speed for a seventy-five-year-old.

"Satō," Abraham repeated, forgetting to add the -san honorific. "And Ojiichan means 'grandpa,' right?" He looked skeptically at Ojiichan, who was only twenty years older than me.

"Very good Japanese," I said. "Wonderful." I set down an order of tonkotsu ramen two seats away, for a man with a cold.

"It's nothing compared to your English. You sound like me." Abraham pointed at me and then himself with his chopsticks, oblivious of his bad manners. "My neighbor told me about you guys. And I've been watching you. You take about twelve orders along the wall without writing them down. You remember each one when they sit down a half hour later, right?"

"That's Grandpa's system."

"And when we're done eating, you still remember what to charge us," Abraham continued. "All while taking new orders on top of the ones you've memorized. That's . . . about twenty-four changing meals in different stages at any given time, supposedly kept in order in your head. All while you're helping cook, too."

"Grandpa doesn't like a paper trail." I lowered my voice. "Taxes. Very high here."

"Yeah, but, like, how do you remember them all? It's incredible. Unbelievable, actually. You shouldn't waste that kind of memory cooking ramen in Komae."

We were low on bamboo shoots and I had nearly overcooked the noodles. The corners of Ojiichan's mouth pinched upward, his version of frowning. "I'm sorry," I said to Abraham. "It's the rush. I need to pay attention to what I'm doing now."

Abraham wiped his mouth with his hand and said, "It's just that I've been here for almost an hour, and you haven't made one mistake."

"Is there a problem with your ramen?" I asked.

"No. That's the miracle. It's exactly what I ordered. Salt ramen with pork and leeks, butter on top."

"And egg. Nine hundred and fifty yen, please."

Abraham stood to reach deep into his pants pocket and paid for his meal in fifty- and hundred-yen coins. I felt sorry for him again. "I've seen you before," I said. "At five a.m., sleeping outside the station."

"Oh. Well, the hours for my job are pretty long, and then sometimes it doesn't even seem worth it to go home, so I stay out. Hostess clubs, you know." He blushed. "I can't miss my morning train and show up late to work. I'd get fired and lose my visa. My bosses were clear on that—'*Okureru na!*'" He shook one stiff index finger in imitation. "If I go home, I forget to set my alarm. Or forget to wake up at all, for that matter."

"Is that why you came to our shop? To see how we remember things?"

"No," Abraham said. "I came here to taste the ramen I stepped in last month. But I'll keep coming until I discover the secret to your memory."

"You're after the wrong secret," I said.

It begins with the bones. At five in the morning, I preheat the old oven that was left here when Ojiichan's father bought this shop. Other ramen shops don't have them. I rinse and crack chicken backs, necks, wings, and feet with my hands to release the flavor, spread spareribs and neck bones in a big pan, and take a hammer to pork femurs and joints covered in ragged tendons. It's the bone connections that nourish our own bones, washing them in collagen and amino acids. Since I started procuring pig joints, Ojiichan's arthritis has gone away and his back has straightened a bit. Since I began spreading roasted marrow on toast for his lunch every day, he's gained fifteen pounds and says he has lead in his pencil again.

After I roast the bones and pick out Ojiichan's hot marrow, I deglaze the roasting pan and pour the whole mess into the iron cauldron Ojiichan's father used long ago. I pull the marinated pork shoulder from the refrigerator, roll and bind it with twine. Then I fry it in oil with mounds of scallions, carrots, garlic, and ginger Ojiichan chops up with the chipped cleaver he refuses to upgrade. Everything melts, becomes heavy, until I ladle in dashi stock and dump everything into the cauldron with the bones.

Foam carries impurities to the top. We simmer and skim, simmer and skim, and continue simmering. The restaurant fills with steam, invading the pores of the pine paneling, expanding the small rectangle of the shop outward. The walls gleam with oil and ramen sweat. Ojiichan works his rag, squirting from a spray bottle filled with vinegar water.

Finally, I sieve the broth. I scoop out the pork shoulder and wrap it in plastic wrap, which Ojiichan says is humankind's greatest invention. I chill the meat for slicing later, and together we prepare the rest of the toppings: marinated eggs, fermented bamboo shoots, sliced leek, mizuna, and nori, which Ojiichan insists on toasting himself over a hot flame. If he's bored, we prepare more: cabbage, corn, bean sprouts, spinach, fish cake, bacon, mushrooms, kimchi. Spices: dried chili, fermented yuzu peel, toasted sesame seed, curry and garlic pastes, pickled ginger, chili oil, and black garlic oil. And two kinds of tare: shio and shoyu, Ojiichan's secret recipes he won't even share with me, who calls him Grandpa.

Last are the noodles, made with flour, water, and alkali salt. I knead yesterday's dough and use the hand-crank pasta maker that Ojiichan says is a lie. But before I came along, he bought the noodles from his friend Nobu and pretended he made them himself, so who's the liar?

Ramen was Chinese before it was Japanese. In the early twentieth century, Japan conquered parts of China and most of our surrounding countries. This was how we treated our neighbors. Eating Chinese noodles—ramen—was a way to symbolically gobble up our enemy. A minister of war was born right here in Komae, perhaps where this ramen shop stands. He was indicted for Class A war crimes against China when Ojiichan was six.

But we have changed. Now ramen is a humble meal, made with compassion. It is food to alleviate pain. It is our shame and despair, transformed into nourishment.

The next week, Abraham the American brought friends to the shop—five office workers, all Japanese but Abraham. I took their orders and returned to my noodle pot.

When I turned back around, everyone had switched places in line. Abraham, who had been first in the group, was now last. The two women had traded spots; the same with the three men. Ojiichan shook his head.

"I can't believe it," Abraham said when he finally sat and I placed the correct bowl before him. He had ordered bone broth ramen with mizuna and enoki mushrooms. As if we'd forget. He gestured to his five friends, scattered along the counter because we don't have party seating. "They're all on me." The others didn't talk to him, not even the salaryman sitting next to him. It was clear they'd just come for the free dinner, except for an Office Lady with short hair who stared down the counter at Abraham with the same expression others directed at the ramen.

"Who are your friends, Abraham-san?" I asked during a lull.

"My students. I brought them to watch you and tell me your

secret. But they couldn't figure you out, either." Abraham rubbed his nose. He wore funeral attire—white shirt, black suit, black tie. Maybe nobody had ever told him what not to wear. "Do you use a mnemonic?" he asked.

"I just pay attention."

"And your English is so good, like, scary good. Did you live in America? Were you a spy?"

It was hard not to like him. "I was once a salaryman for a big company. I traveled a lot. What's your job, Abraham-san?"

He straightened on the stool. "I work for a large educational organization, teaching business English to executives and middle management, also international sales departments. They subcontracted me out to Tōgō Electronics Corporation for the year."

"Oh! Tōgō," I said. "Very good company."

"It's boring. I teach the same stock phrases over and over. 'I look forward to continued good business relations.' 'Together, we can hit a home run.'" Abraham mimed swinging a bat. His eyes were pink and blurry at the edges. Tōgō Corporation is huge, and they often work their English teachers twelve hours a day, with only Sundays off. Or at least they used to, back when I took English classes there myself.

Abraham said, "I'm fascinated by your memory. Ever since I got here, I can't seem to remember anything at all. Like, I even forgot where I lived once. I'm not getting much sleep." His gaze drifted to the slit in the *noren* and the dark night beyond it. "My students are tired from their own jobs and don't remember anything I teach them. I doubt they even understand what I'm saying to you right now." The student-salaryman next to him pulled out a manga and began to read. "Or maybe they just don't give a shit," Abraham said.

I wanted to talk more about Tōgō Corporation, but eight people entered the shop at once. You have to rest while moving. I hurried to the west wall to take more orders. When I got back behind the counter, Abraham's coworkers were bowing out the door except for the Office Lady, who lingered.

I told Abraham the total price and he laid the bills on the counter. I rang up his change, but he and the O.L. were gone when I turned back around. I used the money to give extra pork to the next person in his seat, a skinny young woman with purple hair and a silver bone inserted through her septum.

Abraham began bringing the short-haired Office Lady to the restaurant and stopped sleeping against the pole near the station. The Office Lady taught him how to slurp noodles properly, with noise. They usually arrived drunk, his hand wading up her skirt at some point. "*Baka gaijin*," she shrieked, and he smiled ravenously. Since we don't have a liquor license, they would sober up over the meal, until the O.L. gaped at Abraham in bewilderment, like he was an orangutan she had found in her bed.

Despite his new relationship, Abraham hadn't given up on me, and the drunk O.L. played along. They tried disguises. They'd be wearing work clothes in line. Then after they ordered, they'd shed and put on different sweaters, jackets, hats when I wasn't looking. The O.L. rearranged her hair and applied lipstick. Abraham slipped on sunglasses and once even pasted on a false mustache.

"Really," I said. "The mustache. It's ridiculous. You're the only foreigner here. Do you really think I'll mistake you for someone else?"

Sometimes after ordering, they'd let customers ahead of them in line to try to confuse me. The two of them ordered bizarre combinations with ten or more toppings, reading from a scribbled list so they wouldn't forget the orders themselves. When I grumbled, Abraham said, "This would all be much simpler if you'd just tell me how you do it."

"I already did," I said. "You're not paying attention."

Abraham said he and the O.L. had a running bet on who would catch me in a mistake first. "If she wins, I renew my teaching contract for another year."

"And if you win?"

The way he gazed at the O.L.'s engagement ring made me understand he hadn't given it to her. Then he roused himself and said, "Seriously, man. You've got a fucking gift. Why aren't you working at NASA?"

I, too, have slept standing up. Ten years ago, I left my job at Tōgō Corporation at noon and sleepwalked from train to random train. I got off at Komae, for no reason except that it was a local stop and express trains shot right through it.

I had never been here before. That week, I had disgraced myself at work. I lost over five hundred million yen of Tōgō Corporation's money by forgetting to send a fax. I had been in charge of acquiring and destroying a competing American business. Everything took place on New York time, and the bidding process spanned my days and nights for a week. I could barely remember my own name. I fell asleep at my desk on the last night of the bidding war, and the deal unexpectedly closed in America without our final offer. My error cost the company a year's worth of research from my entire team, not to mention who knows how much future market share. At the annual rotation, they were going to move me to the window, to sit out my life with the other failures.

I was forty-five. All I had ever wanted was to belong to the kind of extended family you get in a gigantic company with thousands of other people who dress and think like you do, share the same slang, eat and work and exercise together. Outside my job, I had no connections except for occasional sex in love hotels with an O.L.

who was already looking for my replacement. I couldn't blame her, nor any of the higher-ups I had drunk with and hired prostitutes for and who now pretended they didn't know me. In one day, I had been deported from the inner circle. I was nothing I thought I was.

My apartment held only kitchen appliances I didn't know how to use and a closet full of suits in different shades of the same colors. That day I wore my best black suit, black tie, white shirt. Nestled in the inside pocket of the jacket was a neatly typed resignation letter next to my suicide note, both held together by the crease of the fold. My parents had retired and moved to Maui the year before, so they'd be able to avoid the hefty railway fine when I caused a train delay with my *jinshin jiko*, my human accident.

The usual station recording recited, *Abunai desu kara, kiiroi sen no uchigawa made osagari kudasai. Because the train is dangerous, please stay behind the yellow line.* I crept over the yellow tactile paving toward the edge. Everyone else on the platform was engrossed in their newspapers, novels, and pornographic manga concealed by privacy wraps. I had just shined my shoes. A waste to wear them, I thought. They were expensive.

The express train rushed toward me on a schedule maintained to the second. It would not slow down here. My toes dipped over the edge, and I teetered, but I didn't want to jump too soon and risk survival. It would be like falling asleep, I told myself. The air was drenched in fumes. My vision flattened. Silhouettes circled me. The train was the only thing that looked real. It bore down fast, all metal and oil and sweat and suits and time. The conductor blared the horn, outlined in horror. I leaned forward into my final step, off the platform and into whatever was next.

Then I was strangling, falling backward as the express train blasted past me. Someone had hooked the back of my collar and pulled me hard. My tie was too tight, and I landed on the raised yellow bumps of the platform, choking, thinking what an idiot I was. I didn't even loosen my tie for suicide.

Abunai desu kara, kiiroi sen no uchigawa made osagari kudasai.

The express train was gone. A man stooped over me, but he was stooped over anyway, his spine curved from scoliosis. He seemed old even then, although he was just sixty-five. His hair was cropped short under a black cap. He sweated silently, his cheeks quivering with emotion.

People flowed around me as I sat on the platform, legs splayed in my best black suit. I said, "Please, Grandpa. This has nothing to do with you."

The old man hooked me under my arms from behind and wrenched me to standing. He was stronger than he looked. His nose was crooked, ears jiggling, eyes black. He led me to the exit and paid the ticket master, since I had dropped my ticket somewhere on the ground. I hadn't thought I would need it.

The old man pulled me out of the station and onto the drab, drippy streets of Komae. His hand sweated against mine, and I stumbled behind him like a drunk. Exhaust from trucks and taxis hovered close to our faces. When we reached a small ramen shop near the station, the old man released me to unlock the gate and raise it. Then he brushed aside the *noren* and pushed me inside.

The restaurant was shabbier then and empty except for us. It was just one room in a dim old house constructed of warped, unfinished wood. It smelled of grease and old dust. Still wordless, he sat me down on a stool at the middle of the counter like I was a child.

The old man went to work behind the counter, chopping and boiling. I don't know how long I sat there. Hours, certainly. I could have left. But time passed, time borrowed from my death. The air filled. Steam dampened our clothes. The HELP WANTED sign had long ago curled into a roll from the humidity. The old man scrubbed the floor around me on his hands and knees with a rag, squirting from a bottle of vinegar water every now and then. "Grandpa," I said. "Let me help you." But I couldn't move from my stool. Actually, I don't think I even said those words aloud.

Before turning the sign to OPEN, Ojiichan placed a hot bowl before me. It was ramen, of course: noodles in an unctuous shoyu broth, a pink-swirled slice of fish cake, scallions, thin slices of pork shoulder, and a rich, jammy egg split in half, just like the ones we serve today. When I leaned over the bowl, steam seeped into every pore of my face. The soup smelled meaty, animalistic. I hadn't eaten in days. Out of habit, I split apart the chopsticks, but it didn't feel like habit anymore. I still remember that ordinary sound, the thin snap of wood releasing.

I folded a cascade of chewy noodles into my mouth. They tasted like fields of grain, earth, sun. Each individual noodle curled against my tongue. The old man stood behind the counter, arms crossed while I ate. I picked up the bowl and drank deeply from it. The soup was salty, magnifying my thirst even while slaking it. Liquid slid down my throat and warmed my core, then spread outward in radiant bursts. I couldn't contain it, the heat. It grew inside me, magnificent, terrifying, until it seemed to explode outward all at once in silent red shock waves that bounced against the walls of the small shop and back into me. And I woke up.

This is the secret Ojiichan taught me: we add the broth from yesterday. Each day's broth is different, according to our moods. Adding flavor from yesterday makes today's soup richer, more complex. Using this method, it's conceivable that today's bowl of ramen would bear memories of the first broth ever made in this cauldron, from when Ojiichan's father opened this shop after the War. Except sometimes we run out.

Abraham didn't return to the shop for a long time, months, although he had occasionally resumed sleeping at his early-morning pole next to the station. Then one day he showed up at closing time, when the restaurant was empty. His shirt was rumpled, his funeral suit shabby. The skin under his pale eyes looked bruised again, like he hadn't slept in weeks. I set his order before him: bone broth ramen, nothing extra. He didn't even look at it.

"Is there a problem with the ramen?" I asked. Then, "Where's your Office Lady friend?"

Abraham closed his eyes. Ojiichan glanced at him and signaled me not to charge.

"Abraham-san, are you all right?"

"No," he whispered without opening his eyes.

The express train's horn sounded through the door Abraham had left ajar. He was so still. I started to feel afraid for the pale foreigner on my stool, living so far from home, if he even had a home. His eyeballs shifted under the thin membrane of his closed eyelids, and one hand twitched. I wondered if he was asleep until he murmured, "Tell me your trick."

He was asking out of habit, according to the unsigned contract between us—he would ask, I would deny. But his voice was vacant. Abraham was worked nearly to death in a subsistence job, dumped, alone. Tokyo is a city of concentric circles, with millions of exhausted people crowded inside each rigid layer like atoms, bumping against each other and quickly away. At the center of all the rings is loneliness—dirty socks in a laundry basket, a bed you roll up each morning, shoes placed in the entryway facing outward, for leaving.

"Shoes," I said.

Abraham opened one eye. "What?"

"The secret. That's how we remember," I said. Ojiichan knew enough English to turn and stare at me.

Abraham's eyes were open now. I babbled, half in relief, half disgusted with myself, "Ojiichan and I, we imagine the soup spilling and splashing on the customer's shoes. All the toppings. Like, scallions and such on a black lace-up. We make a picture of it in our minds. It's why you couldn't fool us with your costumes. It's a memory trick."

"The shoes," Abraham said in wonder. "I never thought to change my shoes."

Ojiichan mumbled, "*Kawaisō*," but I wasn't sure who he was calling pathetic, me or Abraham. He was right; this was the wrong way to help anyone. I just wanted to give this sad man whatever he needed. And even after a decade of feeding people, I still didn't know what that was.

I stammered, "Actually, it's—I'm not being entirely—"

But Abraham wasn't listening to me anymore. "The shoes," he said. He was inside a dream. He would spend the next week looking down at people's feet, but he was doing that already. "Shoes," he repeated and stumbled into Komae's street-lit night.

Another train moaned past. Ojiichan patted my shoulder and said, "*Shōganai*." It can't be helped. He shuffled off to the refrigerator, this man to whom I owed everything. I didn't know what to do with the empty space the Standing Man had left behind. I'd never see him awake again. I had lost him.

Ojiichan once told me a story about a young monk who climbed a mountain, searching for enlightenment. On his way up, the monk met an old man who was on his way down. He seemed wise, so the young monk asked, "Old man, do you know what enlightenment feels like?"

The old man beamed and released his heavy bundle from his back. It thudded to the ground. The young monk marveled at the lightness of the old man's shoulders before asking, "And what happens after enlightenment?" The old man lifted his load again, hefted it onto his back with a grunt, and resumed his long walk down the mountain.

That night after Ojiichan went home, I prepped the noodle dough and dashi stock and marinated the raw pork and a dozen medium-boiled eggs for the next day. For the shells to come off without a mess, the eggs have to be old—"opposite from people," Ojiichan says. Tonight, I was an old egg, my shell peeled cleanly off. I tried to preserve that feeling as I cleaned the restaurant, locked up, and began the walk homeward. At a convenience store, I bought a salmon onigiri and a Choco Monaka ice cream bar. I ate them as I walked.

The air bit at me. Dry leaves huddled around a half-dead tree planted in the sidewalk. Salarymen sleepwalked past me, their suits stinking of nervous sweat and indoor air and spilled beer and sake and soy sauce and women. Office Ladies dreamed of their futures as they wobbled home on high-heeled pumps well on their thirteenth hour of use. Children slept in dark apartment buildings, with or without their parents home. In a bar somewhere, Abraham drank himself into a standing sleep. My mouth tasted like ice cream, and the city smelled like old fire, burning out slowly before the next day would fan the embers into new light.

I wished I could tell Abraham: It was a lie. There is no trick, no secret. We don't look at anyone's shoes. What he called "memory" is just taking care: paying attention now, and now, and now. Practicing it all day as if your life depends on it, because it does. It's work, and as Ojiichan says, work will save us. Or kill us.

I'm philosophizing. Some would call me pathetic, *kawaisō*. Poor man.

Once back home, I removed my shoes in the entryway of my six-tatami apartment. The place is narrow and dark, two by six meters, but what do I need now? A week's worth of clothes and aprons, my hot plate, saucepan, bowl, cup, chopsticks. That's enough for happiness. I skipped my bath and lay down on my futon, still smelling of the day: work and sweat and food and the pain of others. I thought of my mistake with Abraham, my human accident. I won't forget it—can't. That's the price of waking up. I closed my eyes and rested my legs for the next day of standing.

Jude

1. *Judenfrage*

We visited my grandmother Roberta once per season while growing up, always in her crowded Brooklyn apartment. It was almost a three-hour drive from our house outside Philly, but my grandmother never smiled when we arrived. Even when her husband was still alive, she sat away from everyone at a side table, photographs trapped under the heavy glass tabletop, her bitter iced tea sweating above. The blurry, pastel photos depicted every living Jew in the family, including my cousins, aunt, uncle, and father—no pictures of my gentile mother or her children, us two girls and my brother. But when I was born, my grandmother had baffled everyone by insisting on giving me my name, Basia, so she must have felt something special for me.

Even in the hottest, swampiest Brooklyn summer, my grandmother wore long sleeves to hide her number. That's what we all called it, even though it was actually five numbers in a row. I only saw them by accident when her sleeve hiked up. My grandfather also had a number. Kids in my neighborhood sometimes had grandparents with numbers, old people we never saw because they never left their rooms. "He has a number," we whispered, and then played again. I don't remember learning that my grandparents' numbers were their tattoos from Auschwitz, or what happened in the camps. It seemed like I always knew.

My grandmother paid attention to me and not my siblings, and I didn't know why. "Basia, sit here," she'd call out, patting the empty seat next to her at mealtimes. She would show me how to eat, pressing astringent onion slices into the cream cheese and lox on my bagel, ruining it. Or she would show me how she matched that day, flipping up her gray wool skirt to reveal a burgundy satin slip and then the burgundy bows on her gray pumps, snagging a shiny burgundy bra strap from beneath her suit with the burgundy piping, with a fingernail polished in burgundy. She had sewn all of it, even the bra. She was elegant but tired, her neck and ankles bloated with edema from her endless sitting, arthritis hooking each finger inward at its smallest joint. Nobody else sat with her, maybe because she didn't smile or look us in the eye, or because of the gloom that leaked from her like a noxious gas. But for me she held the same kind of fascination as Halloween or the heavy balls of mercury we used to roll around our palms when a thermometer broke.

By the time I was ten, my grandfather had been dead for five years and my father was sick of paying for my grandmother's expenses: her rent-controlled apartment, food allowance, prescription medication for her high blood pressure, bottles and bottles of single-lettered vitamins, discount acrylic paint for the thick pictures she painted of places she had never seen, and every now and then a five-dollar ticket for a local klezmer band that played at the Jewish senior center in nearby Crown Heights. My father added up her expenditures in a ledger, using a pocket calculator he brought on visits. I didn't know if money was as scarce as he said, but I did know our dinners were skimpy, and he complained that his overcoat was twenty years old. He wanted her to begin collecting her Social Security money, but the paperwork he filed for her kept getting denied. "They say no such person exists under that name and number," he said. "I need your Social Security card."

"Tell them to check my elementary school records in Poland," my grandmother said.

"Get real. I can't use Polish records to claim Social Security. And you can't get your money without the card."

My grandmother was poor, always had been. In younger days, she'd walk an extra twenty city blocks to save a dime on kosher meat, but she made sure both of her kids went to college. She now ate a quarter of a can of creamed corn for dinner each day, accompanied by a slice of stale Wonder bread she had saved from her lunch at the Jewish senior center. She also saved the paper plates and plastic forks they gave her, brought them home and nestled them in brown paper bags in her closet "in case there's an emergency." But now she said, "I'd rather not have the money in that case."

"Do you know how much money you're talking about? Enough to live on," my father said. "Do you have a card? Are you even legal?"

"Okay," my aunt said, resting a hand on her brother's wrist, and she asked in a nicer voice, "Mom, are you even legal?"

"I'm legal," she said. A minuscule sneer.

My father shook off his sister's hand. "I've been paying for you since Dad died. If you don't give me your card, I'm just going to stop paying."

"May God free you from your burden," my grandmother said dryly, and then muttered in Yiddish, "*Gey strashe di gens.*" Go threaten the geese.

My father squirmed in my grandmother's creaky chair. My aunt was an elementary school teacher and her husband was on disability, so they couldn't pay for my grandmother. People would know what my father did to his own mother.

Finally, my grandmother said, "Part of the problem might be you don't know my real name."

The table stilled. Even my sleepy uncle turned his gaze to my grandmother, who sat straight in her boiled wool suit, her red underwear hidden beneath.

"What are you talking about?" my father asked.

In drips, she told the story. When she was a little girl in Poland, the other kids made fun of her because she looked "like a little grandmother," so they called her Bubbe, and she began calling herself the same. Her goyish teacher was from England and thought they were calling her "Bobbie," so she marked my grandmother in the school records as "Roberta" Frydman. Eventually, even her own parents forgot about her real name. The only people who used her birth name were the Auschwitz administrators who seized her papers.

My father's nostrils were in full flare. "Two grown children and one dead husband, and nobody ever knew who you really are?"

My grandmother said, "I'm Roberta. Not that other name."

My father asked, voice rising, "Are you going to tell me your real name now so you can get the goddamn money?" My siblings and I glanced at each other and then away. We knew that tone. Suddenly, my father slammed both hands on my grandmother's table, shivering the ice cubes in the glasses, rattling the mismatched bone china. "Well, will you?" he shouted.

"No," she said.

2. *Untermensch*

In contrast to my ever-present grandmother, my grandfather was rarely around. He died when I was five. I remember him ripping sticks of chewing gum in half for us kids whenever he was at the apartment, though he usually slipped away soon after we arrived, having only so much tolerance for crowded rooms, for my grandmother's family lottery where we had to guess the number of matzo balls in the soup. Only she knew the answer, and the one who came closest won nothing besides the same bowl of soup the rest of us got.

One summer day, shortly before my grandfather died, four

friends from his shuffleboard club came over to visit while we were there. The five men gathered over a circular frosted-glass table in the apartment building courtyard, telling stories in Yiddish and gesturing, drinking my grandmother's foggy lemonade. Whenever she approached to refill their drinks, they stopped talking and then resumed their conversation as soon as she left with her pitcher. It was 90 degrees and humid, and the old men wore wilting short-sleeve button-down shirts. I couldn't stop staring at their arms as I decapitated my Barbie doll over and over. Five forearms tattooed with blurry iron-colored numbers rested on the frosted-glass table, like spokes in a wheel with no hub.

Even the tattooed numbers looked like they had accents, weird sixes and fives, the sevens with lines running through the middle like a cross-out. I wanted to touch their numbers, feel if they were smooth like my grandfather's or if their ink made bumps, if their skin tried to reject it. But even at five, before I knew what respect was, I knew to leave them alone.

My grandmother said my grandfather's first job in the camps was to plunder bodies for valuables, after gassing and before cremation. *Kapos* had already confiscated the victims' clothes, ordering them to strip for the "showers," then shoving them into the gas chambers as they clutched bars of soap rumored to be made from the rendered fat of dead Jews. The *Kapos* spread those rumors themselves to cultivate more despair. On their orders, my grandfather ransacked what was left after the gassing, pulling eyeglasses from faces, cutting women's hair for German ladies to wear as wigs. With pliers, he extracted gold fillings from warm mouths. Many of the gassed people were still alive. Many were children. Some, he knew.

My grandfather's work unit was called the *Sonderkommando*, dealing with the problem of the new-dead and still-dying. Trying to hold his breath, he pulled Jewish bodies from the chambers with meat hooks after the Zyklon B cyanide gas had aired. He shoveled coke into the crematorium furnaces and dumped prisoner ashes into the Soła River. He ate food scavenged from the bags of the people he had helped murder while the prisoner orchestra was ordered to play rousing marches to mute the screams, or maybe to mock them. All day, every day, my grandfather considered disappearing into the gas chambers with the other prisoners, but what saved him from early suicide was a drive for revenge.

We don't know how he got reassigned, but he was the only one from his work unit who didn't end his life in the gas chambers, too. "The *Sonderkommando* were also called *Geheimnisträger*, bearers of secrets, so it's a miracle they didn't kill your grandfather with the others to keep him quiet. He surely did something to survive that, Basia," my grandmother wrote me in a letter once I was grown. She didn't tell me what she had to do to survive.

My grandfather's new work unit was easier, packing the gassed victims' clothing for shipment to Lévitan, a department store in German-occupied Paris that resold the seized Jewish possessions. Prisoners called the warehouse where he worked *Kanada*, the German spelling of "Canada," because Canada was a rich country and this warehouse held the stolen wealth of Auschwitz. My grandfather had somehow hidden away the dull scissors he had used to cut the corpses' hair when he was in the *Sonderkommando*. So before he packed the coats and clothing for Paris, he secretly snipped every seam he could, so the clothing might disintegrate on the wearers' bodies and they would be suddenly cold in the street with no coat, dress, pants. He eventually carried those scissors from Poland to Germany to Switzerland to New York, where he used them every day at his greengrocer job on the Lower East Side. Those scissors now lived in my grandmother's drawer, and she sharpened them by cutting folded tinfoil, used them to cleave wrapping paper or nip the string binding a brisket.

My grandfather could never tell his lemonade friends he had been a *Sonderkommando*. They might have rejected him or tried to kill him. His children didn't know, nor his grandchildren. Only my grandmother knew, and she kept his secret alongside the secret of her name, until she wrote me a letter about him thirteen years after his death.

When I read Viktor E. Frankl's *Man's Search for Meaning* in college, I underlined, "We who have come back . . . we know: the best of us did not return."

There was no revenge for my grandfather. He never faced Nazis or *Kapos* at the greengrocer, nor anywhere in New York. He stabbed nobody with his scissors.

His only revenge was against himself when he jumped off the Brooklyn Bridge into the freezing East River, twenty-nine years to the day after liberation. When he washed ashore a week later at Buono Beach on Staten Island, his corpse was bloated beyond recognition. My grandmother identified his body from the number on his left arm.

3. *Daseinskampf*

As I grew, I studied my grandmother during every visit. I watched her clip pictures from *National Geographic* to paint later, or file her nails to a point, or stifle a burp into a hand-tatted handkerchief soaked in Estée Lauder. I learned details but not the part of her as unfathomable as the inside of a lightning bolt or a volcano or a swarm of bees.

This is what my parents told me about her: Her red puffy parlor chairs weren't for sitting; you can only sit on furniture colored brown. Don't ask her questions. The framed ladies' gown designs in the hallway had come from her pen. She painted Asia and Africa but never went on vacation. She would not move away from Brownsville, one of the few neighborhoods willing to rent to Jews after the war, but she was the only Jew left in her building now.

I have never been Jewish. My Jewish side is the "wrong" one, my father's and not my mother's, useless in a matrilineal religion. In Conservative and Orthodox Judaism, you are all-Jewish or zero-Jewish, so I was zero according to my grandmother. Hitler wouldn't have agreed. I was Jewish enough to be murdered but not Jewish enough for a bat mitzvah. Still, something strong stirred in the Ashkenazi molecules of my blood, only amplified once I left for college and began thinking for myself. I wanted accountability; I wanted evil to answer to *me*. I thought the secret of human existence must be hiding where nobody can stand to look—inside its total annihilation.

With every bowl of soup I ate, I wanted to know what it felt like for the prisoners to eat thin, half-rotten gruel in the cold, how every calorie was the thinnest armor against death. I wanted to know if Jews died from keeping kosher rather than eating the rationed slop, or how many died in the uprisings, how many of them wrested away a gun and shot and shot until they were shot. I wanted to know how I would fare in my grandmother's place, and if I had inherited her strength, and where that strength now lived in her. When I went away to college, I began writing her letters, and I always received a response in my college mailbox exactly two weeks later, beginning with "Basia" at the top, skipping the "Dear."

I asked my grandmother how she survived three years in an extermination camp. She wrote back that they had run low on white paint at the Jewish senior center, so she was mostly painting night scenes. I asked her which work unit she was assigned to in Auschwitz and what her duties were. She wrote back about her job at the garment factory after the war, with a sketch of a plum-colored dress she had designed but could not afford to make, so she had to wear a *schmatta* to work. I asked her what liberation had felt like, to be rescued by strangers. She wrote back about a day cruise she took once after the war; Mary Pickford was on board, but my grandmother didn't see her.

I asked her about my grandfather, and she told me everything.

Right before Germany invaded Poland, my grandfather miraculously managed to secure passage for my father and my aunt to a Swiss orphanage through Kindertransport. My father was three, his sister, five. Miraculously, the SS arrested my grandparents together, identifying them in the street from their yellow Star of David badges with JUDE inscribed in the center, and they helped each other survive the cattle cars. Miraculously, both of my grandparents separately survived three years in Auschwitz. Miraculously, my grandmother and grandfather found each other after liberation, even after the Nazis force-evacuated my grandfather to Buchenwald during the last months of the war. Miraculously, my grandparents found their kids six years after they parted. So many miracles for my lost grandfather. The reunited family traveled to America by boat, third class. The last time my grandmother gave her real first name was at Ellis Island, but the officials could spell neither their Polish surname nor their original Russian one, so they gave the whole family a German last name. "Your name," she said.

I asked my grandmother about the rest of our family lost to the Holocaust. In tight script, she wrote a long list of names and ages ranging from eighteen months to eighty-two years. It was two pages long. I wrote away for their records, but they didn't exist, part of the masses of Auschwitz evidence the Nazis destroyed during the last months of the war.

I asked my grandmother about her son, my father, what had happened to him in that orphanage in Switzerland. She wrote back she had never liked processed cheese, but the government still sent it to her every month in big blocks. "Tell me, Basia. How can one widow possibly use all that cheese?" she asked.

I still couldn't understand the costs of surviving, which pieces of yourself you have to trade. I abandoned the dorm cafeteria and stopped eating and drinking for two days to see what starvation felt like. I lingered on balconies, imagining my grandfather's last steps off the cold bridge, a free man. Nothing made sense, and I was missing my life. In the bleak Iowa autumn, I felt abstracted, like I was dissolving into my grandparents' legacy until I was nearly imaginary. Nobody I knew at college was interested in genocide or torture; everyone was living now. My father used to ask, "Basia, if everyone in the world jumped off the Brooklyn Bridge, would you?" I would. I didn't want to be the only person in the world.

When I switched dorms, I somehow lost the letters my grandmother had written me, even the one with the names of the dead. I couldn't ask her for them again. Maybe I lost them on purpose. My Holocaust obsession was over, and it felt pointless now. I had my own teenage devastations—boys, unfair professors, exhausting jobs. Next to my grandmother's story, how could mine mean anything?

How could *her* story mean anything?

4. *Endlösung*

I still sat with my grandmother every time I came home from college. The rest of the family clustered around the television, but my grandmother and I drank weak sugared coffee at the side table with the trapped photographs, still none of me. After the Crown Heights riot, she stopped visiting the Jewish senior center and refused to leave her apartment; she was lonely, and our letters had unlocked something between us. As family noises throbbed in the background, my grandmother shed her secrets.

She said our ancestors had been exiled beyond the Pale of Settlement before the emperor recalled them to the Pale's shtetls and pogroms a generation or two later. "Our family lived in eastern Siberia, north of Mongolia," she said. "That's why I look Chinese." I didn't know if she looked Chinese, but her eyes had a partial epicanthal fold. I didn't realize then that only men were exiled that far beyond the Pale; that they would have intermarried with Mongolic Indigenous women and not Jews who would carry the matrilineal line; that she was telling me she was not a true daughter of Abraham. That she was like me.

She said, "I hated being a mother."

She said, "I don't believe in God anymore."

She said, "Everything I know, I don't want to know."

She said, "They made us buy tickets to ride the cattle cars to Auschwitz."

She said, "I'm glad my husband is dead. He was dead when he was alive."

I hoped her confessions meant we were getting closer, but when I told her my secrets, she pressed her lips together, as if I had said something rude. Her gaze always drilled through my face to someplace else, some facet of herself reflecting a shard of light back to her eye. So when my grandmother told me her secrets, this was as special as I got. I felt wonderful but single-use, like a cigarette, treasured until it burned all the way through.

I was twenty when my grandmother said, "I'm going to tell you my biggest secret, Basia." She didn't look at me, so I wasn't ready when she said, "I'm going to tell you my name."

I thought, *No.*

I didn't want it. My grandmother had assigned her pain a name, her name, and made it as secret and unutterable as YHWH, the forbidden name of God. Even after a decade of asking, my father didn't know her Auschwitz name; he still wrote her a check every month and badgered her about Social Security. Priceless, his ignorance, his existence, when total extermination was just on the other side of his mother's name, a name worth everything and nothing, the price of stolen life. Hair, teeth, coats, eyeglasses, shoes, rendered human fat, złotych paid for a one-way cattle car ticket to Auschwitz. No. I felt myself disintegrating again against the pressure of her pain. I wanted to be happy.

I considered walking away, never talking to my grandmother again. It would be easy to slip her—by then, she rarely left her chair. I could try to chat up my cousins. I could bully my brother into a game of cards. I could see if I was needed in the kitchen. But nobody ever needed me besides her.

She had already grabbed my forearm with her red-tipped fingers, and I flinched at her touch. I felt sick, but I couldn't pull away. The kitchen hummed with family who didn't notice us, would not rescue me. My grandmother's silk blouse was cinched tight at the wrists as always, but her shoulder tugged the fabric up, exposing the corner of one blurry number. *Please*, I thought. *I'm barely here as it is. Don't erase me.*

My grandmother's chair creaked as she drew close. Her Aqua Net globe of hair shimmered in the lamplight. Black dots clogged the pores of her nose, and every tooth was chipped, as if she had eaten stones. Her hot breath lodged itself in the folds of my ear. It was too late. She was in me already. I was made of her.

"My real name is your name," she whispered. "Basia, you have my name."

She leaned back in her chair, and I joined her in the land of the missing.

Fear Me as You Fear God

I drove all night on tangled mountain back roads to escape my husband in Idaho, panicking at every set of headlights in my rearview. Dimly lit signs welcomed me to Wyoming and then Colorado, and come morning, I had no idea where I was. There was nothing familiar about this divot in the canyon or the roadside creek crashing against its banks. I had never dared make an actual escape plan, for fear Timor would somehow read my mind and stop me. Regardless, two states later, in this secluded parking lot off a craggy mountain byway, I felt like I had reached my destination. Away.

The parking lot was clogged with corroded junk cars that seemed too modern for what looked, even in the slanted morning light, like a haunted house. The blade sign read PYRENEES MOUNTAIN B&B, though we were in the Rockies. I was so bleary from driving, for a second I wondered if I *was* in the Pyrenees, if reality had twisted and folded so I was in Spain instead of in Colorado, twenty-six, broke, homeless, and husbandless. But of course, nothing is easy like that.

The manor was made of limestone and warped wood, faceted in stained glass, encrusted into the mountainside like a parasite that had colonized its host. I decided to ask the owners for room rates as a pretext for using their bathroom. I could never afford to stay there, of course. My car had burned through most of my money on the drive, and what I had left wouldn't cover one night in a place like that.

The morning air was wet and slightly gelled, like the inside of a refrigerator. Walking on the crunchy ground felt shivery, strange, and the burn on my forearm had dulled from fire to a numbness that almost scared me. Planters of red petunias brightened the porch, but when I touched the petals, they were plastic. I rang the doorbell before remembering to check my appearance, so I was still pawing at my hair when a middle-aged man opened the door. He said, "Check-in isn't until three."

By then I had spotted the HELP WANTED sign in the window, with NIGHT MANAGER FREE ROOM written in the white space with a black marker. "Actually, I'm interested in the job?" It was my first full sentence in days, and I was surprised at how normal my voice sounded. The man waved me inside without smiling.

Nearly every surface was oak or pine. An antique revolver hung above me from forged iron hooks hammered into the stone fireplace. It was a Colt Single Action Army, a Peacemaker, much fancier than the fourth-generation version my father had given me. I'm good at guns, always have been. I shot cans off fences every day with my father's Colt, using footsteps to measure out twenty-five or fifty yards. I even won an NRA teen competition once. I only stopped shooting when Timor confiscated my gun a month ago. The one in front of me looked like it was first generation, a few rust spots on the nickel plating, with pearl grips and leaf engraving on the cylinder, probably from the 1870s. They couldn't know how valuable it was, just hanging there in the lobby for anyone to steal. I wanted to flip it around like I was in a spaghetti western, *pow pow pow*.

The walls of the inn were burnished logs with seams of white chinking. Flagstone floors in the entry gave way to quartersawn oak in the lobby, and every step squeaked. Double-sided stone fireplace, antique player piano, giant animal heads mounted on walls, high-beam ceilings. I had never seen a hotel so flammable. I worried about the lit candles on the mantel, the fire in the fireplace, the possibility I myself might spontaneously combust from exhaustion.

The man, Jamison, smelled like some kind of medicinal hair treatment, but only a few wisps crept across his head. He had a habitual cough that sounded like *uh-HAH-huh*, too verbal and despairing to actually do any good. He said, "The job—uh-HAH-huh—is for a night manager, in exchange for the room. You would be here twelve hours a night to check in any late guests and answer their needs. Winter is slow; people lodge in town once it starts snowing. You could sleep most of your shift. We have a little room off the lobby."

"So I could live in the room full-time?" I asked.

"Just don't bother the innkeeper," he said with another cough. "Her name is Bigi. She works day shift every day—her choice—but you'd sub when she's sick. You'd also be responsible for my wife's dog when I'm away. Ex-wife. You'd feed and walk it. Her."

"I like dogs," I said, but he didn't seem to care. "How often are you away?"

"Whenever I can be. I don't want to be here anymore."

"Why not?"

Jamison waved his hand before his face as if there were gnats. Then he frowned. "You'd have to comb your hair."

I shoved strands behind my ears. The lobby smelled like something bready with cinnamon in it. The building was an insulated kind of warm, decorated to please pretend cowboys and old western bigots. Hope, that stranger, poked me in the chest. *Hire me*, I prayed to the God my husband said hated me. Finding this place seemed like a lucky fluke until Jamison said, "No days off, no salary, no benefits. Just the room."

"No salary?"

"This is more like a position than a job. You hardly have to do any actual work, only the occasional late check-in once or twice a week at most. You're free to sleep and work your day job the rest of the time." Doubt crossed his face at "day job," but he soon shifted back to indifference. "You just have to be on your shift here from eight p.m. to eight a.m. By law, the inn must always be attended by a responsible and sober party. Uh-HAH-huh." He leaned forward, that liniment scent drifting from his scalp. "That means no partying. No drugs."

"Not me," I said.

"You're young," Jamison said. "You should be out making bad choices."

"I'm twenty-six. And I don't make bad choices," I lied. I pulled at the cuffs of my shirt, covering my bandage.

"Well, you'd be the first," he said. He showed me the dog food and pointed at a black border collie mix cowering in a corner, her tail tucked so far between her legs it looked like a hairy penis. I crouched down to pet her, but she skittered away. "That's Zena. A rescue. My wife brought it home two weeks before she left. Her dog, not mine." Jamison said "her" like it hurt his mouth. "I'll show you the room."

He unlocked what looked like a closet door off the grand lobby, and we entered a small room with indigo carpeting. It brimmed with leavings, mostly cardboard boxes labeled with Sharpied question marks, and a paisley armchair with a broken front leg, sloped left. Junk obscured the windows outside: stacked terra-cotta flowerpots, lawn mower parts, a Weedwacker, a metal detector, stray broken boards, torn and bent screens, shovels. I couldn't see out.

Jamison seemed nose-blind to the clutter. "There's the phone for the inn, and there's the bed." He waved at what was actually a gorgeous bed, with a mahogany headboard and footboard. I salivated when I saw it. There wasn't a clear path to the mattress, but I didn't care.

"I can start right away," I said.

Jamison said, "I should probably check your references."

If I gave references, Timor might learn my location. Everyone I knew, he knew. I panicked about what to do. Cry? Act stupid? Take off my clothes? Instead, I stood very still, as if there were a bee on me.

After another halfhearted cough, Jamison said, "Oh, screw it. This place is my wife's. Ex-wife's. I don't give a ratfuck if it burns down again." Again? Jamison handed me a gold key. "Introduce yourself to Bigi when she's done with breakfast. I'm leaving town."

"Where are you going?"

"I don't know. Anywhere. This place . . ." He looked around and gave one last cough he didn't seem to notice. "It's someone else's dream."

After a soapless shower, bandage change, and long nap in that mammoth bed, I left my cardboard maze to meet Bigi, the innkeeper. She leaned on her elbows at the front desk, which was a long counter made of varnished rough-hewn logs. Bigi was in her fifties and looked like she had been squished, so everything about her—face, torso, ass—seemed much wider than it was long. Her hair and clothes were purple, and her voice could sand a deck. She offered me a muffin from the glass cake dome. I tried to eat it slowly so she wouldn't see how hungry I was. I couldn't quite believe I was there, two whole states away from Timor, who had probably just woken up and realized I was gone.

Bigi gave me the door codes and showed me how to use the credit card machine. "If you have questions, don't bug Jamison; I know more about this place than he does."

"Have you worked here long?"

"This is the only job I've ever had. I started when I was about your age." I couldn't imagine doing any job for multiple decades, but there wasn't a puff of restlessness in her. Bigi said, "This place is home."

"You live here, too?"

"No. In town with my husband. Are you married?"

I said, "I worked at an elementary school with a lady named Bigi once, a really nice German lady." From Bigi's grimace, I wondered if I had said something inappropriate. "This building is incredible."

"It was originally a whorehouse for cowboys and miners in the 1870s, before it burned down. Years later, some Texan rebuilt it into an Elks Lodge, and then Jamison and his wife renovated it after he died. Over there is an old newspaper clipping about the fire," she said, pointing at some framed pictures. "What brings you to this job?"

I said, "Hey, where should I put the boxes in my room?"

"Oh. Jamison's ex-wife's stuff. She's supposed to come back for it, and also for Zena." Bigi waved at the dog, who cowered. "You can move that crap to the third floor; Jamison just left for parts unknown. I hope you got your tetanus shot. Their apartment's a hoarder's den."

"Do you know why his wife left?"

Bigi said, "Spoiled. A fantasist. Maybe having everything wasn't enough for her. Maybe what she needed was a whole lot of nothing."

I spent the rest of the afternoon lugging boxes from my room to Jamison's apartment. I broke one box open first to see if I could use anything inside, but it was just drawer clutter and Chinese and Mexican knickknacks so racist I lost the urge to open any more boxes. I huffed them one by one up to the second floor and unlocked the door to the private third-floor stairway with Bigi's key.

Jamison's place was worse than my room. It looked like they had saved everything from their entire marriage—everything except for the marriage itself. The *New Yorker* and the *Economist* rose in stacks up to my chest. Costco remains smothered every surface: giant bags of dog food, new sets of multicolored knives, three boxed space heaters and four fans, an old-fashioned ice cream maker. A couch loomed, stacked on end and shoved into a corner. Propane tanks, clutches of power tools, and a lidless baby grand piano, a chain saw depressing the keys around middle C. They had left only a narrow path for walking. Dust furred everything. The room was alive with the spin of motes, sawdust, and the heavy stench of sadness, which in actuality smells like old dog.

Did Jamison not see it? Did he not smell the dried mold, the silica in the air, the human dander, the contrast to the pristine guest quarters just one floor below? But I still wanted to stay in his apartment, not leave. This had been some kind of a marriage, a bad one, and here was the evidence. I touched the burn on my forearm from when Timor held it over our gas stove until I smelled my own meat cooking. I huddled in a corner for a while, thinking, *I'm here. Not there.* I wanted to hide in this heap forever, soaking in despair, or maybe smash everything in this smashed-in place because who would notice? But Bigi might hear me.

After hauling the last of the boxes upstairs, I almost tripped on a steaming cup of tea left for me at the bottom step of the third-floor staircase. Freshly made, the water was still clear, the bag only beginning to seep its brown silt. I hadn't even heard Bigi unlock the door and enter the stairway.

I found her downstairs at the front desk and thanked her for the tea, sliding over her key. "What tea?" she asked.

I raised the steaming paper cup. "At the bottom of the stairs to the third floor."

"That wasn't me," she said. "Maybe a guest? Wait, no. They all checked out this morning."

Was Bigi messing with me? But she was giving me the same are-you-messing-with-me look.

Then her face cleared. "Aw," Bigi said. "She's back."

"Who?"

"The ghost," she said.

She was a thirsty ghost, Bigi told me. "Always doing things with cups, glasses, liquids. She's been around since before I began working here. I think she died in this building in the old days. Jamison's ex-wife hired an exorcist two years ago. You must have brought her back inside when you arrived."

"You think . . . I brought a ghost here?"

"Maybe she likes you," Bigi said. "She's harmless, just messy."

"Uh-huh." I grabbed the vacuum and retreated to my decluttered room so Bigi wouldn't see me roll my eyes. I didn't believe in ghosts, not even the Holy Ghost, which was part of why I wasn't in Idaho with my husband anymore.

Spirituality has always wicked right off me, so I didn't notice a single warning sign before a month ago, when my husband said God had revealed that he, Timor, was a prophet. "Fear me as you fear God," Timor said, "and you will be saved." He then began quoting what must have been Scripture but mostly sounded like words we don't say anymore.

This was new for Timor, although he had always been God-adjacent and full of himself after a few beers. I married him straight out of high school; I had known him since kindergarten, all his caprices and smells and changes. I usually just waited for one phase to wane and the next to wax: vegan, flat-earther, Communist, Republican, beekeeper, ghost hunter, Illuminati researcher. We had been married for eight years and recently moved from Oregon to Rexburg, Idaho, for his job in sales at a multilevel marketing company. Their lunch break was Bible study, which he said was "optionally mandatory."

The prophet revelation wasn't like Timor's other crazes. He turned into a stranger. Suddenly gone was the man who asked about my day or picked spinach out of my teeth. The one who crooned out-of-tune love songs to me, who had helped me study for my teacher's aide certificate. Who used to bring boxes of pasta and produce to my house every week in high school, after my parents bailed and I was alone.

Now he inspected our house and my body every day for cleanliness, next to Godliness. When I got my period, he locked me in the bathroom so I wouldn't sit on any chairs, and I had to sleep on the bath mat with a cold towel as a blanket. He emptied our joint account, allotted a food allowance, and demanded receipts. He took away my laptop and credit cards and shorts and tank tops.

I never knew what would ignite his fury. Once it was uttering the name of a false god (I had called my sneakers "Nikes"). Once it was picking up a grape I had dropped on the floor. He recited as he slapped me, *Neither shalt thou gather every grape of thy vineyard; thou shalt leave them for the poor and stranger: I am the Lord your God.* How did he know this stuff? He hadn't even memorized our wedding vows, and he was always forgetting his passwords. Whenever he bent to tie his shoe, I froze; if I hadn't shined it well enough, he threw it at me.

I had no idea what to do. I had known Timor for so long he felt like an absolute, like he was me, even, and that I was doing this to myself those past weeks. I didn't know who I was without him. So

I had to be the one to love him back to safety, to alchemize his pain into my own until it was gone.

But after he burned my arm on the stove, I wasn't that girl anymore. I wasn't anyone.

The very next night, I mixed six crushed Benadryls into Timor's beef stew. He fell asleep on the sofa he now made me clean every day with Lysol.

Once he was helpless before me, I was shocked by my old attraction to him. After everything he had done to me, how could I feel this way? But his was the only other body I had known, the only meaningful connection I had ever had. He was still handsome, but blurry, pliable, like gelatin that had melted in the heat. I could do anything to him. I could seduce him, or break his knees, smash his head with a hammer. If I had been holding a knife at that moment, I might have stabbed him. Or I might have stabbed myself and bled to death on top of him while he slept, so he would feel just a fraction of my horror when he woke.

Instead, I rolled him onto his side, pulled his wallet from his back pocket, grabbed the trash bag of clothes I had stashed in the dryer, and snatched his car keys from the candy dish. Then I stepped outside and started Timor's car. I drove out of there like I had every right to grab hold of my life and live it.

Over the next few weeks at the inn, I got in the habit of stealing food after Bigi left for the night. I scrambled eggs in the inn's pans, which I washed and dried and replaced on the shelf, handles facing the same direction as before. From the way Bigi's greetings became more and more frosty each morning, I realized she was counting the eggs, so I then restricted myself to cereal. After I found Sharpie lines on the milk cartons, I gobbled the cereal dry, and sometimes I stole a slice of bread or swallowed butter by the tablespoon. On nights we had guests, I raided the lemon bars left in the cake dome for them. Bigi no longer offered me muffins.

I also stole towels, bedding, soap, toilet paper, and trial-size shampoo bottles and conditioner. Zena followed me at a distance like a furry shadow, jumping away whenever I glanced back at her. I slung the inn's cream-colored sheets onto my empty curtain rods, afraid of my husband peering in the windows at me, although how would he find me, or even see through the junk piled outside behind the inn? Just in case, I cleared an emergency getaway route by pushing flowerpots from the sill; they split open on the dirt below with birdlike sounds.

I got even more paranoid when the phone rang and nobody spoke on the other end. "Pyrenees Mountain Bed and Breakfast, hello? Hello?" I didn't sleep on those nights, imagining Timor holding the other receiver, the two of us connected by silence and fiber optics. When I told Bigi about the calls, she said, "Some pervs like to call hotels when they're ... you know. It's a thing. At least yours isn't a moaner." I began switching off the ringer at night.

Days were always the same except for the weather. Bigi did it all—morning breakfasts, checkout, laundry and baking, check-in and concierge services, grocery shopping, and instructing the two silent Lysol ladies who cleaned the rooms. No wonder Bigi hated me, the late riser who only walked the dog. During the day, I drove to town and interviewed for jobs I didn't get, shoplifted clothes from the Salvation Army, or took walks along the canyon with Zena, who snapped at anyone who came near us and peed from fright. Bigi said Zena was feral, and she did chew one of her leashes in half, but she never bit me. Bigi hated dogs, which really meant she was afraid of dogs, which really meant she had been bitten once, like Timor had. Besides God, dogs were the only thing he feared.

Something was definitely wrong with Zena. Her tail lived between her legs. The whiteness of her teeth meant she was too young for cataracts, but oystery clouds swam in her eyes nonetheless. Sometimes she toppled over for no reason and lay on her side for a few minutes, simultaneously growling and crying when I tried to approach. I could swear she was saying, "Help me, please, for the love of God." But I don't believe in God, and she wasn't my dog.

I figured Zena was just one of the broken things Jamison and his ex-wife had accumulated without thought, like the two tractors missing the same part, a snowblower that sucked instead of blowing, an industrial vacuum that blew instead of sucking, and an outdoor fishpond with a busted heater. The pond housed four giant koi and a dozen goldfish now suspended in the ice, frozen until spring. Bigi said Jamison's wife had bought the inn early in their marriage, on a handshake and an earnest money check that emptied her trust fund. They spent her parents' retirement money on renovations for twenty years. Once the inn was finished and everyone's money was gone, she left it all behind, including her husband and dog.

Each of the twelve guest rooms in the inn was named after a different ghost town. The rooms flaunted two-person Jacuzzis and antique fireplaces, Tiffany-style stained glass windows and lamps. Whenever the snow melted, my ceiling fan urinated a long, clear stream of water onto my blue carpet. Guests ate eggs Benedict in the morning and Bigi-crafted cherry cheesecake at night under the scrutiny of giant mounted animal heads—wild boar, elk, stag, buffalo. Every now and then, a rattrap snapped on a shelf somewhere and fell to the floor, the rat seizing under the wire, long tail stiff. If guests were around, one of us would distract them while the other tossed the trap into the nearest garbage bin.

One hungry night, I hunted in the larger storage freezer for something to eat. Jamison's wife had baked disaster rations—

frozen cookies, muffins, loaves of sliced banana and zucchini bread. She had baked every kind of bipolar scone an Anglophile could invent, labeled in Sharpie: currant and lemon zest, cranberry and almond, sultana and toffee, and cream and treacle, whatever that was. From the looks of the packed industrial-size freezer, she had been preparing for her departure months in advance. Most of the food was covered in hoarfrost, so I figured Bigi wouldn't notice if I ate some, as nasty as it was, as hungry as I was.

I found something I didn't recognize in an extra-large ziplock. I rubbed the cloudy plastic, trying to see through the frost obscuring what was inside. I dropped the bag on the ground.

It was a cat.

The animal's mouth was frozen in a yelp, eyes closed against its fate. Its long fur had matted against the plastic, and the animal felt brittle, dehydrated, made of frozen fluff and negative space. I stuffed the cat back into the freezer.

The next morning, I found Bigi in the kitchen, baking cowboy cookies with cornflakes in them. I said, "So, you keep a cat in the freezer?"

"That's Thor," Bigi said after a pause. "He died about a year ago. I try to forget it's there."

"Why didn't they cremate it or something?"

Bigi shook her head. "Whatever—they save every fucking thing. Maybe that's why Zena's so weird. She knows she'll end up in the freezer some damn day." What I liked best about Bigi was that she was vulgar behind the kitchen doors, but around guests, she was all propriety: *Yes, ma'am. No, sir. How might I help you today?*

Bigi said, "This happened before. Their Gordon setter died of old age, and Jamison's ex-wife really wanted to bury her on the property, but it was winter and the ground was frozen. So they put the fucking dog corpse in one of those large plastic totes and left it outside in the back. He was a big dog, so they really had to jam him in there. But it was a Colorado winter, freezing and thawing, freezing and thawing. So the dog started oozing this nasty-ass liquid in the daytime, and guests complained about the smell. They ended up taking the dog to the crematorium anyway. That marriage was messed up."

I opened my mouth to say something, but who was I to comment on anyone's marriage?

"What were you doing in the freezer, anyway?" Bigi asked.

"I needed ice."

"For what?"

"A bruise," I said.

"Show it to me," Bigi said, like that was a normal request. Her fists were on her hips. Bigi was on the Atkins diet, and it made her mean, especially in the morning. She frequently dealt disks of pepperoni straight from the bag into her mouth. She had lost five pounds in two weeks but only seemed angrier at the world, personified by me.

"Come on," Bigi said. "I want to see the bruise that was so bad you had to raid our freezer."

For the first time since Timor's divine revelation, I didn't have one to show her, except for my burn. I rolled up my sleeve and pulled off my bandage. It wasn't a bruise, but it looked horrible anyway—red with white ridges, still healing. It hurt and itched, perpetually hovering on the edge of infection.

"You can use the ice, I guess," Bigi said, her gritty voice thinning as she turned away.

Bigi was nicer to me after that, and my burn healed every day. To ingratiate myself, I started doing little chores for her, like sweeping or dishes. Bigi took advantage, saying, "I'm going to do some laundry," and waiting for me to say "I'll do it!" I didn't mind, though. I had applied for at least thirty jobs, but nobody in town would hire me without references. Maybe if I was helpful enough to Bigi, she'd give me one.

To get by in the meantime, I got in the habit of stealing cash. It wasn't that different from stealing muffins. Sometimes there was an unexpected late check-in from a guest who said he'd skip breakfast and check out early. If he paid with cash, it was easy enough to slip the money into my pocket instead of the register. I'd just clean the room after he left and before Bigi arrived in the morning, always exactly at eight. I figured they owed me for all the unpaid work I did there.

One day after a winter windstorm, I was outside cleaning up blown trash, and I found a pile of crushed cigarette butts behind a small boulder down the hill in front. The butts were protected from the wind, unscattered, twenty or more. That was unsafe—fire had already taken the inn once. As I dropped them in my garbage bag, I saw the red WINSTON stamped just above the filter. Timor had smoked Winstons. From the boulder, I had a clear view of the front desk through a window. In the dark, with the inside illuminated, anyone would be able to see in.

My gut said to leave immediately, just fill another trash bag with clothes and find a new town to hide in. But that was ridiculous, fleeing a pile of cigarette butts. Besides, I was out of money and had nowhere to go.

I found Bigi in the kitchen and asked her about it. She said, "Sometimes guests smoke out there, especially during weddings. They go out there to fuck, too." I wanted to kiss her for the bright bouquet of relief she had unknowingly tossed my way. The inn had hosted a small wedding a few weeks before, and I had helped Bigi clean up inside after the ceremony. She said, "Or it could be the ghost, although I doubt she'd want to smoke after she burned up like that."

"The ghost burned up?"

"Haven't you read the article on the wall?" Bigi started on the dishes and raised her voice over the faucet. "When this place burned down in the late 1800s, one of the prostitutes died in the fire. They say her ex-husband tied her up and set fire to the place. He had chased her across four states before finally catching up to her here."

I did not like this story.

Bigi said, "I'm actually a little impressed you're sticking around after the tea incident. That ghost scared off the other night managers. Except for one who stole a hundred dollars from the register and said, 'The ghost did it.'" Bigi snorted, lip curled in contempt. "Like ghosts want *money*."

"I don't believe in ghosts," I said semiapologetically.

"Believe or not," Bigi said. "They exist with or without your approval."

I excused myself and rounded the corner of the kitchen, headed for the door to the lobby. I was stopped short by a champagne glass floating in midair before me.

It bobbed erratically, like the invisible person holding it was startled. I tried to understand what I was seeing. Then the glass rose high into the air and smashed on the ground, as if thrown. Pieces of broken glass skidded everywhere.

"What the fuck?" Bigi rushed around the corner and gawked at the glass shards covering the floor.

"I'm sorry," I said and grabbed the dustpan. She was going to blame me anyway.

So. Ghosts existed. Or, at least, this tea-drinking, champagne-glass-wielding one did. After Timor, it felt remarkably easy to change my entire metaphysical worldview in an instant. Reality was one way, and then it was opposite. It made me strangely hopeful. If ghosts could exist in this world, maybe I could, too.

I read the framed old-timey newspaper articles on the lobby wall. Originally built in 1870, the inn was a brothel/bar for prospectors and men working in the placer mines. They must have been disgusting to have sex with. When the building burned down, a prostitute was locked in her room, unable to escape, likely tied up. She had only been in town for a month, originally from Indiana. Witnesses said she had been pregnant. Law enforcement couldn't find the arsonist.

With the ghost there, I didn't feel quite as safe as I had before, but I didn't have much choice but to hang around the inn all day. Someone had stolen my car from the inn's parking lot, and I couldn't report the theft because the Corolla was under Timor's name. I was sure I had locked it. My first thought was that Timor had reclaimed it, that he had some kind of GPS tracking installed in the car, that he was coming for me next. *You're paranoid*, I told myself, lying awake in the gorgeous bed. *Lots of people steal Toyotas, and lots of people smoke Winstons.*

With my car gone, I had to catch the mountain bus to town on the rare occasions when I scored an interview. I missed a few opportunities because I didn't have a cell phone and Bigi said she was too busy to take messages for me. I was close to getting a job at a fast-food restaurant, if I could only gather courage to ask for a reference from Bigi or Jamison, whom Bigi said was finding himself in the desert "like Jesus," except his desert was the Las Vegas Strip.

One night I was reading at the front desk to escape my creepy room. A guest suddenly appeared right before me. She was a short woman with long hair, bangs, and an old-fashioned high collar open at the throat, maybe Latina. Solid like anyone else. I hadn't heard her approach on the squeaky floorboards. Assuming she wanted the refreshments Bigi had left at the bar next to me, I nodded and returned to my book, squinting in the dim lamplight. I forgot there were no guests checked in that night.

The guest picked up the antique glass pitcher brimming with ice and spring water. Then she tipped it over on its side. Water and ice cascaded across the front desk toward my lap.

I sprang from my chair, away from the waterfall. At that moment, a window in the front of the inn broke, and something struck the wall behind the front desk.

Water, ice, and glass were everywhere. Zena barked and barked. What had just happened? Bewildered, I glanced up at the guest to reassure her somehow, but her whole body collapsed inward to a pinprick in the air. She disappeared.

I leaned over the edge of the bar to see if she had fallen, but nobody was there. I didn't understand. There was nowhere to go, no place to walk that didn't squeak, nothing to hide behind. No matter—the pitcher teetered at the bar's edge, about to fall and shatter on the flagstone next to me. I ducked down behind the bar and caught the pitcher before it hit the floor. This time, I heard the gunshot as another bullet struck the wall behind me.

There weren't any more gunshots before the police arrived and I finally emerged from behind the bar, Zena following me and whining. I showed the cops the two bullet holes in the plaster behind the front desk, exactly where my head would have been each time. Two perfect shots.

In a rush, I told them about Timor and the Winston butts behind the boulder and even the creepy hang-up phone calls back when I first started working at the inn. I couldn't tell them about the missing Corolla without incriminating myself. I showed the cops the water on the floor and the pitcher I had hugged against my chest until they arrived. I could hardly say a ghost saved my life; I told them I had spilled the water myself.

"You're lucky you were clumsy," the cop said. "You think your ex-husband shot the weapon?"

"Current husband," I said. "I didn't divorce him." I hadn't wanted to write my new address on the form.

The cop scribbled down Timor's name and our address in Rexburg while his partner dug the bullets from the wall and dropped them into little baggies. He asked for a picture of Timor, which I didn't have, and said, "I'll put out a BOLO. But without witnesses or the weapon, it may be difficult to prove this is him."

"Who else would it be?"

"A hunter. A past guest. Someone who has a beef with the owner. Or just random violence. Maybe they only wanted to shoot the window, not you."

"What if I file a report on what my husband did to me before?" I showed the cop my arm, still red and puckered with scar tissue.

His face didn't change; he'd seen worse. "If all we've got is a domestic charge, he could be out on bail almost as soon as we put him in. If they go in angry, they come out angrier."

"Restraining order?"

"I once found a dead body with a restraining order stabbed to her chest with a knife. I'm sorry. I do believe you," he said. "But police work isn't fast. We don't know where he is. Or if he did this. Or if we can prove it."

"So what should I do?"

"Stay with a friend, family, someone he doesn't know."

All my friends were Timor's friends. My parents were in the wind, hippies doing hippie things somewhere. "I have no one," I said.

The cop stared at the scratched wooden floor. One shoulder rose and fell.

"Okay," I said. "Say I were your daughter. What would you tell her to do?"

His eyes hardened. "Run," he said.

After the cops left, I turned the lights out and stayed low, crawling to my room with Zena. I locked the door and pulled the bed in front of it. Pressed against me, Zena's body vibrated in the dark. I didn't know if it was from the gunshots or if she always did that—she had never let me touch her before. I stroked her fur, which was shockingly soft. She cried as I murmured into her ears. She studied my face and licked me once on my chin. She had made some kind of decision about me.

I called Timor's cell phone from the hotel's landline extension in my room, now that it didn't matter anymore. He answered on the second ring. "You missed," I said.

He laughed, so I said, "The police are looking for you. You come near me, you're going to jail."

"I'm not worried about the police. I'm worried about you." Timor had his comfortable voice on, the one that used to make me feel safe when he was figuring out how to fix the garbage disposal or deal with a mean neighbor. "If you come back, I can keep you safe from yourself."

"You'll kill me."

"You're killing yourself."

That actually sounded like a good idea. I said, experimentally, "Okay. I will. I'll kill myself."

In the space of his silence, I thought I heard concern. Then God flowed in. "Do you really want to go to hell?"

"No. I want *you* to go to hell."

"God has mercy. Where's yours?"

The future shuffled before me. Mercy as Timor pulls over in the street and shoves me into his Corolla. Timor driving us homeward with his arm around my shoulder, mercifully quiet. Timor stopping the car in an abandoned field, me pleading for mercy as he enacts punishment with his hand, or with my own gun. If I survive by his mercy, he hauls me back to Rexburg for whatever awaits there, mercy, mercy, mercy dragging me down until I buckle and sink, become part of the floor and the beams forever and ever, and he turns me into what he wants—a holy spirit, a living ghost.

"I can't help you unless you cooperate." Timor's voice modulated downward, the way it did during his living room sermons. "God helps those who help themselves."

I knew how to make Timor happy. I just had to say yes, yes, and yes. It's easy to make a narcissist happy. Anyone can do it, for the low, low price of your whole life.

"I *want* to help myself," I said. "Where are you? I'll come to you."

"Right, so you can call the police? I did that for you. I risked my freedom for you." His voice was more tender than it had ever been. "When it's safe again, I'll find you."

"Here?" A wolf spider scurried past Zena, and she ate it.

"Wherever you go, I'll smoke you out." Timor hung up before I did. He was that confident.

I paced the room. Nobody could understand. Except maybe

the ghost, but I didn't know how to summon her. Even the newspaper article didn't know her name. I asked the empty room, "What do I do?"

She didn't answer. I should have thanked her before asking for more.

Or maybe I was asking the wrong person. Her husband murdered her, and she became trapped forever in her thirst. I could see the moments that still held her prisoner: bound to a chair, gagged perhaps by his smelly sock or her own undergarments. Too late for bargaining. She strains against the rope, bucks her chair, tries to make any sound at all. Her husband is gone, and she smells smoke—not familiar smoke, lantern smoke, cigarettes or cigars. This is the smoke of the wrong things burning: curtains, clothing, furniture lacquer. Water will save her, rain, but only heat rises from beneath her feet. Below her on the ground floor, furniture cracks and chandeliers smash, but she hears no human voices. Everyone has already left. She is alone, discarded. The flames crawl the walls of her room. No one will hear her, and even if they do, no one will help her. Her hair is already burning.

I ventured from my room, drew the curtains over the front windows, and got to work. By the light of one dim bulb, I mopped the spilled water in the lobby and swept up the glass, which was everywhere—beneath the windowsill, under the piano, even a shard on a side table doily. I didn't know what to do about the bullet holes.

It was still dark when I jumped at the shucking sound the door's rubber weather stripping made against the floor, but it was just Bigi. She stepped inside and glared at me as her fists rose to her hips.

I started to explain, but Bigi cut me off. "I already talked to the police."

"Then you know it's not my fault."

"No. I don't know that. Jamison and I just spoke. You're fired." Bigi looked a tiny bit sorry, but then frosted over again once I started speaking.

"I'm the victim here," I said.

"I can't endanger the inn if you're bringing these kinds of friends around."

"You think I'm inviting friends over here to *shoot me*?"

"I also know about the stealing." Bigi's gaze was frigid.

"What stealing?" I asked. I thought she was just talking about muffins and cookies until she tossed her purple hair off her shoulders.

Bigi said, "I know you took cash from the register."

I hadn't thought it through. Bigi cared about this place. And she hated thieves. She even hated the guests who took all the soaps and shower caps and little shampoo bottles we left for them. How could she know I had stolen that money? I had been so careful—only cash guests with early checkouts, and I had deleted their reservations. At that moment, Bigi seemed omniscient. I considered telling her everything: Timor, running away, all of it. But if she was omniscient, she already knew. She just didn't care.

Bigi breezed over to the front desk, saying, "You have an hour to clear out. Don't hurt this place any more than you already have." She removed all the cash from the register and tucked it in her shirt pocket, patting it once. "I'm going upstairs to rest before my shift starts; I want you gone by then. I'll be listening if you try anything. Jamison already notified the police about the money you stole from us." My police?

Bigi paused. "I'd like to say it's been a pleasure knowing you. Indeed, I would like to say that." She acted angry, but it wasn't that. I had hurt her feelings. I realized too late that we had become friends.

"Wait." I walked toward her, and she flinched, trying not to look like a scared middle-aged woman. She stiffened when I hugged her. She'd wish she had hugged me back once Timor killed me and I was another ghost, throwing cowboy cookies at her purple hair. But now, Bigi turned away to hide her damp eyes. She mounted the steps slowly to show she wasn't afraid of me, and then the dark green carpet upstairs swallowed her footsteps.

I felt different. No longer hungry, no longer tired. I went into my room to pack. I opened my closet and remembered I had worn those clothes, remembered stealing and washing them, but now they looked far too wrinkled and small for me. I was enormous.

I gathered my wallet and Corolla keys and left everything else. I called the Board of Health and left an after-hours message about the cat in the freezer.

With a whine, a cold nose pushed into my hand. It was Zena, forgotten by everyone else. I fed her one last time and waited for her to finish. Then I slapped my thigh, and she walked next to me, looking up. I didn't pull her chewed leash from the hook. She was free, too.

I passed the old newspaper clippings and the pictures of the inn. Goodbye, goodbye. This place would not burn again. I was the fire, and I was leaving.

As I was about to pass the fireplace, the antique revolver lifted from its hooks and hovered in the air before me.

I searched for the ghost's outline, but she wasn't showing herself to me this time. I reached for the Colt, cool and heavy, no residual warmth from fingers, and it released into my grasp. The mother-of-pearl handle fit my hand perfectly. It was like coming home. I opened the loading gate, angled the cylinder face toward the lamplight, and rotated the cylinder click by click, peering inside. Dusty cartridges plugged two of the six holes.

"What do you want me to do with this?" I asked, but the air in the room was now flat and barren. I was on my own. I pointed the revolver out the window. Up the stairs, where Bigi had gone. At Zena, who side-eyed me. At my own head.

I stuck the revolver down the waistband of my jeans and pushed my way outside, Zena crowding beside me, almost tripping me. She growled low in the dark, but I feared nothing, nobody, no God, wondering if God now feared me.

Outside the inn was wilderness, nothing but mountains and lodgepole pines, interrupted by the occasional cabin or A-frame off the empty road. Zena was a different dog off-leash. In the predawn light, she scampered and grinned for the first time, pausing occasionally to sniff slush. Then she settled into a trot-like walk beside me with her tail high, glancing at me periodically for instructions.

We walked for miles into town as the sun rose and the air dried out. Zena and I checked hotels, one after another, scanning parking lots for the Corolla or Idaho plates. I worried Timor was staying in a different town, or that he had driven off and I had just missed him. I wore a baseball hat and clothes he had never seen before, but I nevertheless worried he would recognize me, that he had somehow become the god he believed in, that he had permeated the air. Because he was nowhere, he was everywhere.

No luck all day. I got so cold, I didn't even feel it anymore. My feet swelled inside my thin sneakers. Every motel, I looked for a warm place to sleep that night, a forgotten nook or unlocked supply closet. With Zena, I couldn't go inside anywhere to warm up, but I was still glad she was with me. Whenever I stopped, she bowed to me and sneezed. When I rested on a bus stop bench, she sat on my feet. When I went inside a gas station to pee, Zena waited for me outside and then inhaled the withered hot dog I bought her, washing it down with bitten snow.

It was the tail end of twilight before I finally circled around to a motel at the edge of town. Goose bumps prickled the scar on my arm. I was thirsty, jittery from lack of sleep and food, and couldn't feel my fingers or toes anymore. I had nowhere to go. I wasn't sure what to do if I didn't find Timor. I also wasn't sure what to do if I did.

And then a Corolla pulled into the motel lot.

Even in the scant light, I recognized the dents from when Timor hydroplaned in a flash flood and hit a pole. When the car door cracked open, I crouched down, flat like a rabbit in the dry grass, pulling Zena down beside me, grateful she matched the dark. Had he seen me? My heart knocked on my knees. I wanted what all rabbits want—to be invisible. I wanted to go home. Not with Timor—a home that was mine alone, with curtains drawn and the smell of food cooking, comfort, safety. But safety is impossible if you have a home. Snakes find burrows. Caves have no back door. Houses have points of ingress that require monitoring, cameras, bulletproof glass, steel bars, retinal security keys. Illuminated by lights inside, you can be trapped in the sights of a gun, the gunman wherever he pleases. The least safe place in the world is home. It is only safe to be the one crouching in the dark.

The motel lights illuminated Timor's pleased expression, but he wasn't looking at me. He was carrying a pizza. He loved pizza. We used to order pepperoni. This was my husband, carrying a pepperoni pizza to eat by himself in a motel room as he waited until it was safe enough to hunt me down. Maybe the old Timor was still in there, buried beneath his newest convictions. But he was broken, and I was too broken to fix him. Timor's God had not helped him as he contorted from love to rage. The belt, his hand, his eyes cut into dark diamonds, me trying to crawl under our bed. The space I had screamed into was uninhabited. What is the character of such a God, to watch and refuse to help either of us?

Timor took the motel stairs two at a time and then walked around the landing to the back of the building. I sprang up and sprinted to follow, hiding from sight under the landing below him. I heard him enter a room on the second floor, and I backed up toward the woods to read the number on the door as it shut, 203. Timor turned on the lights and snapped the curtains closed. The white muslin began to flicker blue; Timor was watching TV as he ate his warm pizza.

My budding plan had been to call the cops and report Timor's location, steal his car once he was safely in jail, and drive somewhere he'd never find me. But none of that made sense anymore. The police had forsaken me. The car could be a trap. Whatever had led him here would lead him to the next place I went, and the next. Death was the period in every sentence. I fingered the car keys, the heavy gun in my pocket.

People think guns are powerful, are power itself. But guns are often named after their calibers—Colt .45, .38 Special, .44 Magnum—because it's the ammunition that kills. And Timor had all the ammo: our money, the car, his convictions, a warm hotel room, a strategy, God on his side. I only had a dog who barked at mirrors and once fell in love with a brick.

If I was going to die anyway, maybe I could decide when and how.

Zena and I climbed the steep, forested hill behind the building. My wet sneakers began to freeze, and I slipped on the snow that hid in the pine trees' daytime shadows. No road, no path back there, nothing but wind tangling itself in the pines and a clear view of room 203, twenty-five yards away.

I told Zena, "We'll just stay here for a bit."

The mountains hovered against the night sky. A hundred million years ago, this was an inland sea. Mollusks, fish, ammonites, even mosasaurs swam, thinking the world was water. But then tectonic forces pushed the land into peaks that drained the sea, and all those animals died. The peaks were already crumbling again, and the land would sink and then rise, water and land and over again. Whatever I did, it would matter for this moment and no more.

Zena thumped her tail. My heart was breaking, but the dog was bored. She stared out at the night, waiting for my answer. I didn't know the answer; I only knew how to decide. The dark, frozen country fell away below me. In my hand I held the Colt. It was almost spring. I wouldn't have to wait a winter to be found. At first light, as Timor stepped out the door to hunt me, I would be the first thing he saw. He'd know he lost.

I held the gun to my temple and thumbed back the hammer. Given her choices, that ghost would have done the same. Anything not to burn alive. The muzzle was freezing; the last thing I would feel was cold. *Timor, I'm helping myself,* I thought. *God help me.*

I squeezed the trigger.

Nothing happened. Of course not—the gun was 150 years old. I pointed it in the air and pressed hard on the stuck trigger while Zena flipped her tail. The frigid metal didn't give.

I pulled the ejector rod to empty the two cartridges into my palm. I knocked the gun against a tree trunk to dislodge any powder and blew into the chambers, wiping the cylinder the best I could with my sleeve. With numb fingers, I half cocked and reloaded the Colt, clicking past the empty chamber and aiming in the air, picking an abandoned paper wasp nest in a lodgepole pine.

The gun kicked back lightly, black powder smoke chafing my eyes and throat, the echo cracking from the canyon walls. The wasp's nest fell a few yards away from my feet, with a soft pat on the snow.

Zena had bolted at the shot, and I thought, *Be free*. But then she circled right back around again, a more opaque black in the long moon shadow of a nearby tree. The wet sound of Zena licking the air.

The lights flickered on in Room 203. Timor opened his door and stepped onto the landing, illuminated by the moon and yellow hotel light. I knew his outline by heart, the curve of his shoulders, the contours of his head. From his posture, I could tell he didn't see me, scanning the tops of trees. He lingered on the landing, a reverse Romeo. Zena growled.

Fear me as you fear God.

One bullet left. All possibilities spread before me and then split like rivers into two. Him or me. Now or never. Only one of us would get saved tonight.

I stepped into the moonlight and waited until Timor saw me. I pointed the gun at his head, to see what it felt like. It felt delicious. It felt like freedom. And that's when I changed my mind, once and forever, about that second bullet.

I Feel Like I Could Stand Here with You All Night and It Would Be the Worst Night of My Life

I work at Ye Olde Curio Shoppe downtown. We consign from the community—people bring in old family items and we split the take. It's a good retirement job. They let me bring my grandbaby to work when my daughter and wife are at their own jobs or need a break, and I have my own section with the things I like to sell. I'm into Americana, especially from here in Ohio: family portraits, war memorabilia, old kitchenware and dental tools. Last weekend I sold a taxidermied great horned owl and a dictionary from an era when "digital" just meant "relating to the fingers." I like to consign that stuff, but I don't take it home. I'm an old man with a cluttery wife and limited space in my life. So the Shoppe feels like it's a little bit mine, even though I just work here.

Shoppe—even now I don't know if I'm supposed to pronounce it "shop-eh." Everyone in town just says "shop." There's a heaviness in here that grounds me. All the items seem to have their own gravity from the years they put in. Beaded lampshades, arrowheads, snake oil bottles, Hardy Boys lunch boxes, antique sweepers from the 1950s, porcelain shoes with presidents and their wives painted on them. Everything means something, or used to. It's different from outside, where I sometimes feel like I'm under a sky too empty and big for me.

So it was a Saturday—that's my shift, Saturday—and I had

my grandbaby behind the counter with me, fed and changed and tucked out of sight in his carrier on the floor. He had just fallen asleep when this African American fella came into the store, in his late twenties or early thirties, good-looking, pretty tall. Taller than me. I thought he'd poke around, but he walked straight to the register without anything in his hands except for a go-cup of coffee. He pushed the coffee across the counter toward me, and two creams and two sugars tumbled onto the counter from his other hand. "I don't know how you take it," he said, glancing around the empty store.

Not even my regulars bought me coffee, and this fella was new to me. I thanked him, not sure what to do with the coffee, which was from the fancy place next door. I usually bring it from home. I like Folgers. But the coffee next door is good, too. I poured it into my thermos, wondering if I should have kept them separate, how the fancy coffee would taste with my can coffee. Oh hell's bells, I could hear my wife saying. It's just coffee.

So the man said, "I was in here last week before one of my classes, looking for a gift for my niece's birthday. I couldn't stop thinking about some of the things you have here in the store." His diction was stiff, with a nervous chug in his throat midsentence, like I get when I'm afraid to talk to someone new. The man was all tucked in and ironed, even though church wasn't until tomorrow. I was wearing my Army cap and a T-shirt with baby spit-up on the shoulder.

"Which items interested you?" I asked.

"More like, they upset me. Like this stuff." He grabbed a ceramic butter container from a display stand. Okay, that was a bad one. Not from my section. That section belonged to James, who retired here from West Virginia, and his section is not to my taste. James just set that item out the day before. The butter container featured the face of another Black fella, but kind of like a cartoon with big eyes and swollen red lips and like that. But fine craftsmanship, perfect condition, dated 1905 and signed on the bottom—probably worth too much to last the weekend before some collector nabbed it. There's a demand. It had been part of a set, and we had one more piece, a ceramic depiction of a heavy Black lady in a headkerchief, fists on her hips, discounted because it was chipped. Meant to hold sugar. I didn't take those consignments, and I told this fella so. Talking about it made my stomach twinge. It's not my area. I resisted the urge to check on my grandson under the blanket.

"The owner's not here," I said, pushing over a pad of paper and a pen. "If you want to write down your—"

But the man interrupted, pointing at the wall next to me. "Also, the Nazi flag. You're just hanging it in your store next to your counter like it's normal."

When we get an original flag in, we hang it low so people can touch it and inspect the stitching. You can't do that in a museum. So he did just that, walked to the wall and touched it, and kind of jumped in a way that made him look like he was a kid, even in his fancy clothing. "Oh my God, is that real? A real Nazi flag? From Nazi Germany?"

His voice made the baby shift around, so I almost whispered, "It's a historical artifact. From the war." I felt proud for a second before I realized I shouldn't.

"But it's disgusting."

Again, it wasn't my section, but it would sound weak to say so. "History consists of good decisions and terrible ones."

"Selling a Nazi flag in your store is a terrible decision." He wasn't angry, just factual in a lecturey way, from a man half my age.

Fear flashed across his face when the bells outside the door jingled, but it was just the wind pushing them around. Who would he even be afraid of in our tiny town? So he'd feel more comfortable, I tried to make my voice friendly, but I overshot and sounded casual instead. "Think of the store as a kind of museum."

"But this isn't a museum. It's essentially a tag sale. You're profiting from this, and propagating it. These objects are really racist."

"America *is* racist," I said. I loved my country, fought for it, but never lied about it. "That's America, too. It's reality. They took down Saddam's statue in Baghdad. They should have left it up so everyone could see what used to happen there. You can't change history by removing the evidence." Like Vietnam. I'm not like other vets. I think we should talk about Mỹ Lai and the war and all of it. I would, but nobody ever asks. Never. So despite the discomfort, it was almost a relief to talk this way now to this younger man, the way I've always been able to talk to my wife since the very first night I met her, despite her lifelong habit of rolling her eyes at me.

The man said, "I think they took down Saddam Hussein's statue because people didn't want to look at the figure of a tyrant who tortured them and butchered their families."

"Should pain be hidden?" Speaking of, my stomach was really hurting now. I didn't know what to do with my hands, which had a strange tremor. They don't usually. I shoved them into my pockets, out of my pockets, in. I wanted the man to leave before he woke up the baby, who had his knuckle in his mouth the way he liked to sleep. His soldier father had died in Afghanistan. I missed the first three months of my grandbaby's life because my own daughter didn't trust me. My Nicky. I changed half her diapers; I didn't miss one softball game or ballet recital, but she still thought I'd reject her child. Now he's all I think about. I gave the car seat a nudge with my foot to get him rocking, and the baby sighed in his sleep.

"What's weirdest for me is this racist crap right next to ordinary stuff, like this carrot peeler," the man said, reaching over and picking it up. "Like, here I am peeling carrots next to my Nazi flag."

"Well, now, I can see about the butter container. I'll give you that one, for sure. It's a bad butter container. But the Nazi flag, that's not your business," I said. "That's for Jewish people to get upset about."

"I am Jewish," the man said.

"You can't be," I said. "You're—"

He waited, his face blank. Was he screwing with me? He didn't look Jewish. Maybe he wasn't even Black. Human, human race, right? That's what my daughter Nicky says. Even though I couldn't categorize him, this fella seemed pretty comfortable categorizing me.

I said, "A Nazi flag is history. A Confederate flag is now."

"You sell those, too?"

I nodded. Not my section. You see them hung on yard flagpoles and painted on cars sometimes around here. Even at a car dealership, waving next to the Ohio flag. My wife said, "Keep driving," and we bought her car in Mansfield instead.

"Don't you ever wonder why people are buying these objects?" the man now asked.

"I don't ask. But we have Black collectors, too," I said.

"Well, even I'm tempted to buy up all this crap so nobody else has it. But I'm an adjunct professor; I don't have the money."

"Young for a professor," I said, trying to flatter him, forgetting that young people hate being told they're young. Most of the professors up at the college look like me. He sighed patiently. I hate that, the sigh through the nose, like I won't notice. My wife does it every damn day.

The fella said, "Okay. So how would you feel if someone bought your Nazi flag and then hung it on their flagpole next to some nice Jewish family's house?"

Now I was able to relax a little. This one I knew the answer to. "That would never happen in a million years."

"Because there aren't any Jewish families in town?"

"No." There was one. "It wouldn't happen because that flag is way too valuable to hang outside."

He flinched. Like I hit him.

It was like I couldn't stop talking. "You know what flag we can't even keep in the store? The KKK flag. As soon as we hang it, it's sold." I didn't know why I was telling him this, except that it was true. I had made one of those sales for James's section, wrapped the flag up in tissue paper and laid it in a garment box for the customer because the owner told me to. That's a survival habit from the military and marriage, doing things people tell me to do without thinking too hard on it.

The man wiped the sweat from his forehead and examined it on his fingers. The thermostat was broken again and it was hot in the store, even though it was brisk outside. My wife's word, "brisk." I wanted to reach for my handkerchief to wipe my own sweaty forehead, but my hands were still weird.

I met my wife at a Rotary Club mixer, right after I recovered and finished my tour, after my second Purple Heart. She was not at all my type, with her smirky pink lips and the way she rolled her blue eyes and talked every time I talked. I like my women quiet. Well, I've been married for forty years to a loud one, so I don't really know how I like them anymore. Anyway, back then, I was pretty sure I wasn't going to see her again, so I just started yakking about the things that matter to me, politics and war and everything that's wrong with America. Then she said, and I'll never forget this, "You know what, Arlon? I feel like I could stand here and talk with you all night. And it would be the worst night of my life." I fell in love with her right then, and told her so. She said, "If *that's* what makes you fall in love with a woman, maybe there's hope for you." I had no idea what she meant, and still don't. She won't tell me.

The baby woke up and started crying, but I didn't want to pick him up around this man. Instead, I dabbed a little honey onto his paci from a plastic honey bear the owner keeps by the register for his morning oatmeal.

"You've got a *baby* back there?" The man got on his tiptoes to try to peer behind the counter, but I had already covered up my grandson's carrier again with the blanket, and he calmed down in the dark with his honey paci. The man said, "You can't give a baby honey! It's unpasteurized."

"It's natural," I said. "From bees."

"Seriously. Honey can make babies really sick."

I didn't believe him, but the man seemed so distressed that I plucked the paci from my grandbaby's mouth, and he wailed again. I felt like he was crying because he didn't like the store, either, that I needed to get him out of there to somewhere he'd be comfortable. But that was ridiculous, and I had four hours left on my shift. I wiped off the paci on my shirt, doctored it with one of the sugar packets the man had given me, and plopped it back in his mouth. He smiled at me, and we both quieted down. I still had honey on my fingers and tried to wipe it off on my pants before I remembered my handkerchief. Now my pants *and* my shirt were sticky.

"I don't know what I can do for you," I said. "Why are you here?"

"I don't know." The man frowned. The front of his shirt had a crease from where the iron had hit it wrong. "Trying to walk my talk, I guess. I teach social change."

I almost laughed. "This is an antique shop. Nothing changes here."

"*You* can change."

Could I? This time I did laugh, nervous as hell. The fella said, "Or at the very least, you could stop consigning this garbage so customers like me stop complaining about it."

"I've worked here ten years, and nobody's ever once complained about it before you." And that was the truth.

"Nobody?" This seemed to bother the man even more than the butter container or the honey.

"So what do you teach up there at the college?"

"Sociology. And I write books."

"What do you write about?"

"Racism," the man said.

"Well," I said. "I hope you have a lot of paper."

By now the baby was crying again and spitting out the pacifier. He wanted a hug. I would have to pick him up. I pulled the blanket off his carrier and picked him up, his body wriggling in my hands, kicking hard against empty air until his strong legs found my belly to push against. I shushed him and he melted into me, cooing. I hoped the man would leave, but instead, when he saw the baby, his brown skin, the fella's mouth fell open. He said, "What the—?" He looked at the two of us, and his gaze flattened. He turned and walked toward the door without saying goodbye.

But now I didn't want him to leave. Not like this, like he was giving up on me. I followed him, carrying my whimpering grandbaby. Even though I didn't know this fella, it felt important that he not leave before he understood me. But I didn't understand me.

"Listen," I said. "Pick out something you like. For your niece, right? Just watch out for lead paint. Anything in the store. It's on me."

But the man just waved at me and my grandbaby with a helpless gesture. His shoulders slouched now, like mine did. I had done that. I had made him feel like me.

The man said, "I know that this"—he waved in the direction of the flag and butter container—"is reality. I teach reality; I understand it. I just don't quite understand *you*."

I said, "I'm sorry you're offended."

"No," he said. "I'm hurt." And then he left.

The man was gone, and the Shoppe was empty. It wasn't a relief. It was the opposite. The weight of the place settled again, but this time it felt too heavy, the air choked with old skin, long dead.

I dug my face into my grandson's shoulder. He smelled like baby. They all smell the same. I patted the springy, coiled hair he inherited from his dead soldier father. I liked the baby's hair. I liked his dark skin, even. I liked everything about him. But just then I realized someday, too soon, he might not like me.

The baby nuzzled into my chest, but even that couldn't save me. I was shaking again, wondering why the hell my eyes were watering. Two Purple Hearts and not one tear, now this? I thought, *I am losing my fucking mind.*

Someone else entered the store, a white lady collector who stops in every couple of weeks to scoop up the valuable stuff for resale. I resettled my grandbaby in his carrier behind the counter and blew my nose good and loud, like I had a cold and that's all. Then I sold her a Monarch typewriter and that infernal butter container and nobody complained about a goddamn thing.

The Blue Hole

In North Texas, there aren't many opportunities to scuba dive, but I signed up for the classes anyway. I didn't like people, so maybe it was the promise of hovering underwater for long periods undisturbed that attracted me so strongly. Everything else about it felt like a series of bad ideas: breathing underwater, taking off my plastic mask so I couldn't see, falling backward off the edge of the pool with forty pounds of junk on my back. I panicked when chlorinated water flooded my sinuses. I lost a contact lens.

For the last skill test before our open water dive, we had to blindly sink to the bottom of the pool and put on all our gear, with no air and no mask. I passed that test, thinking the whole time about the other one, the home pregnancy test I had failed the night before. The fine pink test line had looked at first like a thread, a deep scratch, until I couldn't avoid seeing it for what it really was: a terrible mistake.

I was a junior at UT Dallas, getting my BA in philosophy. I couldn't support a baby. I didn't have money, or friends, or health insurance, or a job. I had been thinking of continuing on for a PhD, living off my student loans and banking on my hypothesis that I'd never have to repay them because world civilization would implode before then. I regretted my nihilistic choices, like booking a scuba certification trip instead of reserving my money for hypothetical hospital bills for a hypothetical birth just in case the condoms Rick had dug out from under his bed were as old as they looked.

I had broken up with Rick a month before. He had been off

buying pot somewhere when I decided to surprise him and let myself into his studio apartment with the key he had given me.

The woman in his bed was naked, eating an orange. She had her red hair knotted up in back, freckled shoulders relaxed into slopes. She said in a chirpy voice, "Bless your heart, you must be Katie."

I lunged for her, and she slapped a foot at my throat without even sitting up. I choked. She said, "I'm a purple belt in tae kwon do. Back off." Then she went back to her orange, sucking a finger where the juice stung a hangnail or something. I left.

It took two days for Rick to call me. "Don't be unreasonable," he said. "We never talked about a commitment. You just make assumptions, Katie. You should communicate more."

"Okay. Fuck you."

"Hey. Hey. I want you in my life, all right? And, um, her."

I said, "You want your cake, and you want your cupcake."

"Which one are you?" he asked.

This was not a good set of genes I was incubating. Rick would never help me. He had already squandered the remainder of his student loan money on drugs, video games, and an electric guitar he didn't know how to play. He was flunking out and would probably return home soon to work in his dad's food truck. I'd only see him again if I wanted a hamburger in Norman, Oklahoma.

And my Baptist parents would never help a daughter pregnant out of wedlock. They loved God more than they loved me. Entire families in our little town had been shunned for much less than incubating a nonsanctioned human. My parents still thought I was a virgin, and they only approved my philosophy degree because I told them we studied the Bible.

In my ethics textbook, I had underlined a quote from Kant: "Act always so that you treat humanity, in your own person or in the person of another, always as an end and never simply as a means." So my "person" shouldn't just be a womb for some baby I didn't want. But the fetus's death shouldn't be a means for me, either, so I could still live my life. If the fetus was a person yet.

I don't know when life begins: conception, birth, or sometime in between. Doctors declare you medically dead once you lose brain function, and they let you starve to death. But I didn't know how that compared to brainless clusters of cells that attach to you like a parasite. Parasites have their own lives, and so do bacteria. Was I a murderer if I fought them off, or was I supposed to sacrifice my body to the lives of so many? I was made of killers: T cells, B cells, antibodies, dendritic cells, natural killer cells, macrophages. I had no control over this rogue genocidal army and couldn't stop them if I tried. But I still couldn't make the mental shift from bacteria to baby, from a future I thought I could plan to a future my body was growing without my permission.

Abortion wasn't legal in Texas anymore, not like New Mexico, where I was headed for my scuba trip. But I didn't have the money for an abortion, if I'd even go through with it. I didn't know how to walk past a crowd of protesters, especially because I didn't know how I felt about abortion myself. Like other sins of the flesh, it wasn't something my family discussed over Frito pie and sweet tea.

I could give it up for adoption. Or could I? I didn't want a baby, but not wanting one wasn't the same thing as saying goodbye forever after housing it in your body for nine months and then pushing it out in a traumatic bloodbath that could possibly kill you. I tried to imagine handing my pink infant, flesh of my flesh, to a waiting mother and making her dreams come true. I wasn't that unselfish, I realized with shuddering certainty, or maybe I wasn't that selfish. I wasn't brave or strong enough, or strong at all.

I only knew I needed options, and options cost money. The dregs of my student loan were tied up in this scuba trip to the Blue

Hole in New Mexico, two days from now. I called the scuba program and asked for a refund, but the lady on the other end said, "It's too late, ma'am. We've already booked the rooms and the facility." I tried to glean from her cool tone if she might judge me for my premarital decisions, if it would be worth telling her how life was suddenly more complicated than I had imagined two months ago when I enrolled. I just thanked her and hung up. Sometimes women are even meaner than men.

But I had a little faith left. In what, I wasn't sure. I was only nine weeks pregnant. I would go to the Blue Hole, and then I'd figure out what to do. At least I'd get underwater once before my life ended for real.

To do your open water test, the last part of scuba certification, the good scuba schools take you to Mexico. The crappy, cheapo ones take you to New Mexico. In the desert town of Santa Rosa, New Mexico, population 2,744, lurks the Blue Hole, a natural sinkhole eighty-one feet deep. It was early March and chilly, but the dive masters told us it didn't matter because the water temperature is 62 degrees Fahrenheit all year round. My body is 98 degrees Fahrenheit all year round. I thought, *Maybe this baby thing will take care of itself.*

I took off right after my last midterm and drove the eight lonely hours to Santa Rosa, worrying my twenty-year-old car would quit as the land rose in altitude. I've lived my whole life in Texas, a state so large it could fit the entire world population inside its borders. I'm used to a long, unbroken view. I wanted to see those New Mexico Rocky Mountains, but the sun set between Childress and Amarillo and then I could have been anywhere in the dark. I drove another three hours before the yellow sign flashed by, WELCOME TO NEW MEXICO, LAND OF ENCHANTMENT.

Land of Abortion, I thought.

Even thinking the word felt wrong, like I might get in trouble with God or my parents. Then I couldn't stop repeating it in my mind, *abortion, abortion, abortion, abortion, abortion*, like a little kid thinking dirty words. The only radio station played Christian rock, DJ'd by a man who called himself Mr. Mercy. I turned it off and just listened to my car's wheels grinding on the empty road, eyes aching from negotiating my cracked windshield.

At eleven at night, I arrived in Santa Rosa, a two-stoplight town of snow globes, Route 66 signs, and Betty Boops. The puffy-faced woman at the motel front desk had clearly been napping. TV laughter gunned from a corner of the lobby, and two preschoolers peeped at me from behind a beige curtain off to the side. It was cold and drafty, but they wore T-shirts with no pants. A bed squeaked, and I realized this family lived in that closet space there, just behind the dirty curtain.

After that, it seemed spoiled to say my room sucked. It was freezing, and I searched for the heater. How could there be no heater? Just an unplugged icebox. The bathtub was spattered with scum and frozen mold, and the sink drain was clogged. A painting of a cowboy staring at a sunset hung above my bed, so ugly nobody would think to steal it, and a lamp was bolted to the nightstand, which was bolted to the wall. Nothing in the drawers, not even a Bible. I lingered on the landing, where it was warmer than inside.

The scuba school had matched me with a random roommate from one of the scuba classes, but she hadn't arrived yet, so I chose the better bed near the door. Shivering, I pulled the polyester blanket over my head. I tried to sleep in the fetal position, my breath warming my knees with each exhale. I considered calling the front desk about the heat, but I didn't want to wake up the lady and her kids again. After an hour, I pulled on three sets of socks and zipped on my leather jacket, then piled everything else in my suitcase over the blanket. I slept precariously, trying not to spill anything over the sides. The top of my head prickled where I had left an air hole. I dreamed of cold, cold water.

In the morning, a stranger lay in the next bed. She sat up and said, "I'm Stacey. Where's the fucking heater?" I didn't recognize her from the classes; everyone looks anonymous in scuba masks. Stacey had light brown hair, brown eyes, and a birthmark on her cheek. She was large-muscled, with long, loose breasts. I admired her T-shirt: a rat caught in a snapped trap, saying, "I regret nothing."

Stacey phoned the front desk and then hung up. "The heater's behind the *desk*," she said. "But of course."

We scurried over to move the elephantine desk and chair, blasted the heater, and sat before the lukewarm stream of air. Stacey's hair blew out in right angles. She was a double major, women's studies and environmental studies, originally from Austin. She said she was getting certified because she was considering a career in marine preservation. She didn't seem tired, even though she said she had rolled in at two in the morning.

Stacey said, "I have breakfast, if you want." From the landing she retrieved a chilly paper bag filled with six sandwiches wrapped in foil. They were labeled PB&J, TuMu, and TuMa.

"What's TuMu?"

"Turkey with mustard."

"And TuMa?"

"Mayonnaise. Not everyone likes mustard. Or mayonnaise," Stacey said.

"Not everyone likes turkey," I said.

"That's why the PB&J. But some folks are allergic to peanuts."

We didn't bother to shower, just pulled sweats over our bathing suits and left for the trailer in the hotel parking lot. Duke, our scuba instructor, was handing out wet suits. The mountains I had craved hovered behind him, but they looked distant, dusty, almost transparent. I knew I should appreciate them, like I knew I should appreciate the half-frozen sandwich Stacey had given me, but both went down cold.

Duke was like an evil grandfather, with eyebrows in peaks and a shark bite down his leg. He was sexy, old, and low-voiced, with indifferent blue eyes and always a white T-shirt over his wrinkled tan. All semester I had peered over the side of the pool at him with stinging eyes as he surveyed me with amusement and demanded the impossible. *Go down and pick up that thumbtack with your eyes closed. Now touch the bottom of the pool with the insides of your elbows. Go down to the bottom, do some quantum physics, and come back up eating a chimichanga.* I hated him, and loved hating him. He still called me Cody, even after I had been insisting for six weeks that my name was Katie.

Duke handed me a men's XXL wet suit, so I ignored him and scrounged for my own. I found a pale-pink-and-black one that fit so tightly I could barely bend my knees. Tight meant warm. Stacey's was also tight, and we thumped stiff-legged down the stairs, like neoprene robots. With some difficulty, we folded ourselves into Stacey's Civic, and she drove us to the Blue Hole.

The Blue Hole was just that: a blue hole off a dirt road, about eighty feet in diameter at the surface. Very blue, though, and perfectly round. Fish swam in it, big orange carp and catfish that looked like bonsai sharks.

"I can see the bottom," Stacey said.

So could I. A ramp was suspended about twenty feet down with buoys tied to it, and another one at thirty-five feet—half the distance down. Stacey reached over the edge and dipped her fingers in the water. When I asked how it felt, she shook her head and frowned. Wind frothed her hair.

We were the first ones in, Team One. "Oh no," I said as I entered the cold water. It stung as it rose above my crotch—like electric voltage, there. The faces above us all wore the same anxious expressions. "Not warm," I told them. A few of the other college girls started whining.

But it wasn't so bad once we started doing the skills they told us to do—submerge yourself, take off your mask, pull your regulator out of your mouth. Fetch a rock from the bottom of a ledge. At twenty feet, strap on a weight belt, take it off, put it back on. I flipped around like a skinny fish until I was done with my drills, hoping I passed the first day of testing. Tomorrow would be harder. After I finished, I kind of hung out underwater, since it was warmer than the air. The water felt dense, puddingy, like a preservative. Weak red wavelengths don't survive at that depth, and the pink in our wet suits turned bluish green.

I suddenly remembered I was pregnant and touched my abdomen, still flat. My white fingers curled up all by themselves. How could those carp live in such cold water all the time—baby carp, even? Only the strong survive. I unzipped my wet suit to let the freezing water flow against my belly. My skin quaked at the contact and then turned numb. I wondered if that was enough, if it would take longer for the cold to penetrate my uterus.

Above me, Stacey was struggling to reattach her weight belt as the weights skidded around the nylon strap. The belt slipped off her waist and through her hands, right past me, and sank down, down, down to the bottom of the Blue Hole, eighty-one feet deep. A cloud erupted as the weights landed, and silt billowed all the way up to us, muddying all the water that had been transparent a moment before.

Duke would be livid. Stacey jerked her head to stare at me through her mask. I could have tried to stop it, but I hadn't even stretched out a hand to grab the belt as it fell. I just watched until my mask clouded with murk. We had to grope for each other's hands and follow our own bubbles upward to swim to the surface.

It took a shower, a bath, and half an hour under the covers before I was warm again. Stacey refused to get into the bathtub with all the slime floating around in there, so she shivered in fits for an hour in a nest of blankets on the floor, her back pressed against the timid heater.

At dinnertime, we walked to Rosalita's for Mexican food and dancing. About thirty scuba students drifted among the locals. They gossiped about a girl who had thrown up into her regulator. Another girl had become hypothermically unreasonable and punched her buddy in the mask. Most divers were university students taking advantage of the cheap elective, as well as a few aging hippies getting certification as an extension class. An older man named Ted gravitated toward Stacey and me, and the three of us formed a dry-erase friendship. Ted narrated his failed marriage—he's Southern Baptist like my parents and his ex-wife is Egyptian and Muslim, and they ultimately broke up over God, or maybe geography. I asked Stacey what her ethnicity was, and she pushed her enchilada remains away. "Don't know. I'm adopted. I know my biological mother was a college student, but that's it."

"How about your biological dad?" I asked. She shook her head.

My mind filled with questions, but instead I just stared at her smooth skin and strong shoulders. Stacey was so solid and upright, and she wouldn't exist if her mother hadn't given birth, given her up, given given given. "How do you feel about being adopted?" I asked.

Stacey looked startled. "Funny, no one's ever asked me. Okay, I guess. Normal. My parents are my parents."

"What if . . ." I didn't know how to finish the question.

Stacey sipped her margarita and said, "You mean, if my biological parents had kept me? I don't know. I might be a drug addict or something. They gave me up for a reason. Some people really shouldn't have babies."

I wanted to ask if she meant that they shouldn't have babies as in give birth to them, or shouldn't have them as in owning them afterward, but before I had the chance, Duke swung by on a Jimmy Buffett tune to tell us we should go to bed early. "And quit drinking. Stick to Coke," he said to me, although I was the only sober one there. The underage college students hid their bottles in their laps.

"What's that thing you have in your hand, Duke?" Ted asked.

Duke smiled at his Corona bottle. "I'm not getting wet tomorrow."

"You might could," I said.

"I'm just going to laugh at you from a folding chair," he said.

"Hey, Duke," Stacey said. "Did you pick out the hotel? Because our room smells like a dead person."

"Probably one of my students from last year," he said. "You're lucky it's not you, after you fucked up the Blue Hole today."

Stacey asked, "Have y'all really had a death here?" We laughed, and Stacey frowned. "No, I mean it. That hole is freezing. It's dangerous. Couldn't we have gone somewhere warmer?"

"Pay me two thousand dollars more and I'll take you to Cozumel. But for now, you get Santa Rosa."

I took a wincing sip from my drink, well gin on the rocks. I remembered stories of bathtub gin curing women's "conditions," but mine was only making me nauseated. Duke said sharply to me, "No more drinking. I mean it. Don't give me the satisfaction of watching you bend over in the bushes tomorrow."

"Oh, honey," I said, "not in front of the kids."

Duke rolled his eyes and left, and I said, "He's got a thing for me." I could tell from their glances that some of the other college girls believed me, thinking it wasn't fair, that I'd get extra chances in the Blue Hole. This is a pretty good example of why nobody ever liked me.

I threw up in the bathroom after dinner—chili con carne. I wiped my mouth with toilet paper. My whole life looked like what was in the toilet bowl. I flushed it.

As I walked out with streaming eyes and a sour mouth, Duke was leaning against the wall, waiting for the men's room. He asked, "You all right?"

I kind of fell forward until my head rested on Duke's chest. His breathing stopped. He smelled like old steak. Then I walked away before he had the chance to touch, to offer comfort, if he would.

On the way out of Rosalita's, we stopped in a darkened gift shop that shared lobby space with the restaurant. Ted shook a Betty Boop snow globe and watched the white particles cover her plastic cleavage. "I have to own this," he said, and shoved it in his pocket. He walked out, and Stacey scrounged in her pockets for spare dollars, which she tucked under the cash register.

"Lauralee," Ted serenaded the empty street. "Oh, Lauralee."

"His wife?" Stacey whispered to me. "His mom?"

"Lauralee. Hey! Get your hansoffame," he said to Stacey, who had started to guide him toward the motel. "I'm gonna do *what I want*!" He staggered off toward a gas station, looking about to tump over every few steps. We watched him go, our nylon windbreakers swishing as we shrugged.

Back in the room, while Stacey took another hot shower, I thought about looking for abortion clinics nearby. Just to ask questions. But instead, I called Rick. When he answered, I said, "Hey. It's me."

"Me who?" he asked.

I almost hung up, but the truth would hurt him more. "I'm pregnant."

A pause. "You have the wrong number."

I did? It sounded like Rick. "Is this Rick Alistair?"

"Katie, sweetheart," Rick said, sounding tired. "You have the wrong number."

I tried to hang up on him, but he beat me to it.

OK, so there weren't going to be any bluebonnet photos with Rick, me, and baby. I wanted to stab something, someone, Rick, myself. How could I have been so stupid, and why was I the only one paying the price for stupidity when that broke bad, all-hat-no-cattle motherfucker was at least twice as stupid as I was with his worthless condoms? He could just walk away, but I had to choose between eternal damnation or a sentence of eighteen years to life. *I'll fucking kill myself*, I thought. *I'll do it right now, and he'll be sorry.*

I was crying and semiseriously scouring the room for sharp objects when Stacey emerged too early from what must have been a shower gone bad because she was goose pimply all over. "What's wrong?" she asked, pulling me down to sit on her bed beside her, still in her towel. "What can I do? What do you need?"

I blurted, half laughing, "You don't happen to have an extra thousand dollars, do you?" And then I started crying again.

"For what?" Stacey sat very still beside me. "Are you in trouble?"

"It's nothing, nothing, it's okay. I'm okay. Thanks." I wiped my face and was about to make up something stupid to laugh it off when a little cramp twinged in my uterus. I held still, listening for it, afraid to screw up a possible miscarriage in progress.

"Katie?" Stacey asked. "What's wrong?"

I scoured my memory for what I had overheard about miscarriages. Pain, and torrential blood. I fled to the bathroom, shut the door, and reached into my underwear. There was no blood, no real pain.

The twinge passed, but it left behind a scrap of hope. Not all life held on when things got hard. Once a sugar maple outside my dorm window up and quit after a late snow barely dusted its buds. Neither Rick nor I were made of tough stuff. Tomorrow was the longer dive, five hours. In and out of the water, over and over. Duke had predicted a freeze.

I left the bathroom. Stacey had changed into red pajamas and was waiting for me, a damp spot on her bedspread from where she had sat in her towel. "Katie, maybe I can help," Stacey said. "Will you trust me with what's troubling you?"

What a ridiculous person. I got into bed and turned away from her.

I felt Stacey's strong gaze on my back for a few minutes more, like a brick wall I could almost lean against. Then she rustled her covers and flicked off the light. I should have thanked her, I realized too late. In the silence, I locked my palms together. I prayed for snow.

We woke at six to the kind of glow that only comes with snow. A fine half inch coated everything in powder—the wet suits we had hung over the chipped railing to dry, the steel grate beneath our feet outside our room. After trying to thaw the wet suits and booties in front of the heater, we waited until the last possible second before dragging our warm bodies into them. Then we stomped to the parking lot and drove to the dive site, our teeth already chattering.

The Blue Hole was no longer blue. The weather had turned it filmy, steamy, soap-scummy. Snow swirled into our hair. The dive masters stood next to the trucks, yelling orders. We found regulators, snorkels, masks, and fins, and strapped on our vests, tanks, and weight belts. I had a sudden glimpse of postapocalyptic life, when we'd all be forced underground or underwater, carrying oxygen to survive.

This dive was much worse. The water had cleared somewhat from Stacey's weight belt accident the day before, but the deeper we dove, the murkier it got. We kept jumping in and out of the water, refilling tanks, waiting our turn for the next round of stupid stuff. Every time we got out of the Hole, water evaporated from our scalps and our hands shook so much we couldn't fasten our belts or adjust our masks. A dive master ordered Stacey to swim across the Hole and back, dragging me. She said, "You do it," and threw her mask at him. Ted bolted from the water to vomit bright yellow bile into the bushes. Stacey and I swam over to help, but Duke blocked our exit from the water.

"Not so bad, is it?" Duke said. He wore a puffy down jacket and fleece hat and gloves, his hands encircling a cup of coffee with steam twirling out the top.

"Imma kick your ass," I said, teeth chattering.

He grinned. "Get 'er done, Cody."

"My name is Katie."

"Guess what?" he asked, so softly I leaned forward. Then he barked in my ear, "I don't care! Stay in there and finish your tests."

"I hate you," Stacey told him.

The next time we stood in line to exchange our tanks, the snow had accumulated a couple of inches under our booties. I couldn't feel my feet or ankles. Even my calves grew numb. I had to use the restroom, but I didn't think I'd be able to get my wet suit back on if I took it off. I didn't care anymore. While Stacey swore next to me, I peed, the heat trailing down my legs and into my booties. It felt so good I almost cried there, soaking in my own urine. Everything was beautiful now, steam encircling the Blue Hole, exhaust rising from Duke's idling pickup like a human offering to some god.

With the relief came instant nausea. I dodged behind the outdoor restrooms and dumped my BCD and weight belt just in time. On my hands and knees, I vomited up the rest of last night's gin and today's TuMu breakfast from some deep place inside me. It went on forever. At one point, I wondered if I was actually throwing up food other people had eaten. After that, I was struck with a headache bigger than Dallas and hurricanes of dizziness so strong I had to lie down in the frozen dirt and snow. I didn't know if I was sweating or freezing, if I had passed out or if no time had passed at all. *A puking, fainting pregnant woman*, I thought. *I'm already a cliché.*

Once I made it back to my feet, I couldn't hold a single thought except that Stacey was waiting for me. I picked up the heavy equipment, now foreign in my numb hands. I stumbled back toward the tank exchange, where Duke was waiting. He whipped around and snarled, "I told you to quit drinking last night! What's wrong with you, Cody? Everyone else is done. Get your gear on and get in the water or you *and* your partner fail your certification."

Next to the Blue Hole, Stacey's face was distorted by the mask, but her shoulders hunched miserably. I tried to strap on my BCD and weight belt, but my hands were in spasms, and I kept swaying under the burden. Another dive master helped me and then gave me a not-so-gentle shove toward the Blue Hole.

The scuba test was almost over. We had passed everything else. All we were expected to do was swim the circumference of the hole three times with our buddy at a depth of thirty-five feet—less than two hundred yards total. Stacey and I submerged ourselves in the cold water, which felt normal now, or maybe I was beyond numb. I felt better once I was in the water, my body alive and working again. We pumped our legs hard, swimming downward, equalizing our ears by blowing through pinched nostrils.

At thirty-five feet below the surface, everything was blue, blue, blue, even our skin, and quiet except for our bubbles and droning breaths. Everyone else had finished their laps already, so it was just me and Stacey down there—crystalline water above us, dark murk below from where the silt from Stacey's weight belt had yet to settle. Underwater, the Blue Hole cinched in before bellying out again at the bottom. I had never seen anything so beautiful as the blue light above refracting and bending color through the long lines of bubbles we made. The first lap was easy, fun even, as we showed each other the life down there. Pointing at a catfish on the second lap, I turned to Stacey as I took a breath through my regulator. But there was no more air.

I tried again, my lungs pulling. Nothing. The air was gone, already cycled through me.

After I threw up behind the restrooms, I had forgotten to change my tank.

I reached for my backup regulator and pressed the purge button. No bubbles came out. I held it in my mouth and tried to breathe, but it was like sucking on a vacuum tube.

Vague hand signals came to me—patting my mouth, drawing a finger across my throat. But I was panicking already. My lungs were aching. *Air*, I thought. *Air. Air. I love air.*

Stacey was looking away. I grabbed at her regulator and tried to rip it from her mouth. Stacey gripped with her teeth, shook her head, pushed at my hands. I shoved her in the face. Stacey's head slid backward until she blurred in the blue. I looked up, but the surface was thirty-five feet above me. I would never make it in time. My body was already out of oxygen.

Involuntarily, I breathed in, one tiny breath of water. I choked on it and then gasped in a lungful. I began to slip down the Blue Hole. I was drowning, me and whatever was inside me. *I'm going to die,* I thought. A catfish clung to the wall of the Blue Hole as I sank past it. *This is my last thought,* I thought. Stacey's dim form faded backward into the blue. I closed my eyes.

Then fingers knotted in my hair, yanking. Stacey pulled me up, toward her suspended body. Her horrified eyes held mine from inside her mask. In that strange moment while I was dying yet still living, Stacey wedged her regulator into my mouth while she reached for her own backup one, air bubbles spilling upward from her lips.

The first breath of air hurt. It was terrible. It was perfect. I struggled to fit air into my lungs above the well of water already there. I was racked with coughing fits and could barely see. Air from Stacey's tank flowed, choking from my mouth in bubbles. Her hand still touched my hair. I cradled the regulator in both hands.

My lungs began to drain. I turned and retched into the water, then bit back on the regulator. Everything was working. Stacey peered into my mask. My body convulsed. But it was still my body, my own.

This is my life, I thought. *My only life. Nobody else can have it.*

The water above us was pale cerulean, snow spotting its outer skin, bubbles rising in a long stream from our mouths below. I looked down to the dark-blue bottom where I had almost gone. I searched for a sign from the depths. Nothing moved. No struggle, no fight. Just water.

Stacey grabbed my jacket and kicked hard. We rose and broke the cold surface. Stacey shouted for help. There was sound. There was light. I cried hard in the open air, inconsolable in this harsh place of reckoning. Because I wanted to live.

Wounds of the
Heart and Great Vessels

My first date with Dr. David Constantino lasted five minutes. It was a windy Friday night in spring, and I was moderately excited to date a doctor, even if he wasn't the kind who saved anybody. "I'm an anesthesiologist. I put people to sleep," he had said over the phone, and on his online profile, and in an article written for an HMO I found on the internet. In the picture, he was giving a patient a handshake with his right hand and a thumbs-up with his left.

David showed up at my doorstep an hour and seven minutes late. He was wearing a white shirt with sweat stains on it. His stubbly jaw wore a smear of ketchup, or maybe it was blood. He was uncommonly tall—even in heels, I was six inches shorter than him—and he had a slight potbelly, but he was good-looking in that douchebag way I can't help but find attractive. It's a problem for me.

"You're late," I said.

"You look nothing like your profile picture," he said.

The exit sign buzzed in the hallway.

He rubbed his eyes. "You want to go, Rachel? Or not?" His face was swollen, like he had been crying.

I considered just shutting up and going on the date. I mean, he was cute and I had at least two hours left on my perfume. I'm a secretary for a flooring company; I hadn't ever had a shot at a doctor before. But an hour and seven minutes late and no apology? I tried again: "You're over an hour late."

He said, "And you're twenty pounds heavier than you said you were, but who's counting."

My hand rose to my neck.

Dr. David Constantino seemed surprised by his own words. A tic began to flutter in his eye. "You ready?"

I gave him a you-first wave, my new nails glittering. He nodded and turned his back. As soon as that schmuck was clear, I slammed the door behind him and locked it.

There was one faint knock, then nothing. The wind pressed against the windows. I ordered a pizza with anchovies and went back online.

A week later, a spring blizzard hit with over a foot of wet snow and counting. I almost spun out driving home from work. It had been a hard week, and I was glad it was Friday. You think answering phones for a flooring company isn't stressful? Guess again. People can do without a lot of things, but show me one person who doesn't need a floor.

So I was in early pajamas, a movie cued up, and a plate of lasagna on my coffee table. It's my famous lasagna, famous to me. It cures menstrual cramps, probably cancer, and definitely heartbreak. My mother used to say, "You want great Italian food, find a Jewish kitchen." Our own food is gross—jellied gefilte fish, chopped liver, p'tcha, kishke, tongue sandwiches. Hence lasagna.

So I was blowing on the first bite when something scratched outside my door. I thought it was my neighbor's rat terrier—they keep him off-leash, and he once pooped on the hallway carpet in front of my apartment. But when I opened the door, a man was crouched over my welcome mat. "Hey," I said, and he straightened up.

I almost didn't recognize him. It was Dr. David Constantino. He was in a snow-covered blue dress shirt and looked ten pounds skinnier than he had been just one week before. His potbelly was nearly gone, and he had that anorexic, wasting-away look behind his ears where the body can't lie. His skin was as mealy as half-dried library paste. Snow encrusted the top of his head in a ring and adorned his shoulders like melting epaulets. The shadows under his eyes were darker than his eyes themselves.

"Rachel. I was just dropping this off." With the toe of his shoe, he nudged something deeper under my welcome mat, which I yanked up. Underneath was a gold watch.

"It was my father's," David said. "He's dead. It's just a token. An apology for hurting your—for my rudeness." His voice was all scratched up, not at all like the smooth, resonant one he had affected on the phone before our aborted date. Snow melted from his hair and down his forehead. He swayed a little and grabbed the doorframe, but he didn't seem drunk.

"You're giving me your dead father's watch?"

"So, we're good?" He seemed relieved, as if checking off the last item on his list. "Goodbye." He started to lurch away, but I grabbed his arm.

"David, where's your coat?"

"I don't need a *coat*." He seemed bewildered at the thought.

"You shouldn't be out driving in this mess." Clods of snow dropped from his shirt to the hallway carpet. I asked, "When's the last time you ate? Or slept?"

"I don't know. The night before I met you."

"You haven't slept or eaten in seven days?" I scanned his face, but he seemed too confused to lie. I felt scared. "Come inside."

He swayed on his feet and shook his head.

"I'm not asking," I said.

He tried to pull himself from my grip, but it turns out I'm stronger than a man who hasn't slept in seven days. I put all my weight into it, and Dr. David Constantino stumbled across my doorstep. I shut the door behind him.

He was crooked and wretched and too tall for my apartment. He smelled like rotten chicken. I nudged him onto my sofa. He clenched his car keys in his fist. "But I have to go do something important," he said. And then he was crying.

I had seen only kids cry like this, not men. His mouth opened, and all his feelings fell out onto my shoulder. He covered me with sobs, erupting from deep inside him. I just hung on. He nearly convulsed, and I didn't feel like I was hugging him so much as trying to hold all his body parts together.

He said, "I killed someone."

"You murdered someone?" Now I didn't dare let go of him; who knew what he'd do.

"A patient. On the table. Last Friday, before I met you. That's why I was late," he slurred into my shoulder.

I continued to hold him while he told me about a teenage patient brought in for emergency surgery, her heart nicked by a knife. "'Wounds of the heart and great vessels' is how they classify it," he said in a suddenly medical-sounding voice before dissolving again. "She was allergic to the anesthesia. I missed the signs. I was flirting with the nurse and wasn't paying close enough attention. She went into shock and died."

"But maybe she would have died anyway. Of the knife wound or whatever."

"Maybe she *would* have," he grieved. "But she died from *me*."

"It's not your fault," I said, but what did I know? "That's why you're not sleeping?"

"I never wanted to be a doctor, even when I was young. Younger." His crying was steady now. "But my parents wanted med school, so that's what I did. Tragic mistake."

I didn't know what to say. I had never heard of anyone settling for medical school. "Do your parents know about all this?"

"No. Yes. I mean, they're dead." He wiped his nose on my shirt and then on his sleeve. "My mom died of cancer two years ago. My dad just gave up and kicked it after that. I'm alone."

I should have said "You're not alone," but instead what came out was "Can't you take some narcotics or something?"

"Don't you think I tried? Nothing makes me sleep." He laughed, semihysterical. "I know. The irony."

By now, both of his hands had wormed onto my breasts, and he started kissing me with his rotten chicken breath. Again, I considered. Go for it? Doctor? But he was just too disgusting, and probably dying at that very moment. I pushed him away and said, "David, we're going to the emergency room."

"No. I work at the hospital." He began to stand up. It took him two tries. "Not your problem. I'll go."

I gave him a little shove, and he splayed backward onto the sofa. Food, I thought. You can always rely on food. I pointed at the lasagna on my coffee table. "Eat that first."

David regarded my lasagna as if it were sewage. He tried to pick up the fork, but his fingers fumbled against it without establishing a grip. His whole body shook. Finally, he managed to raise the fork to his lips, but he spat some lasagna out in an involuntary exhale. "I'm sorry," he said. So I took the fork from him. Fed him. When he opened his mouth, his tongue was trembling.

He swallowed my food. Something changed between us. It was so subtle, only a dog could have smelled it. David looked me full in the eyes for the first time. He grasped my hand and guided another forkful toward his mouth, like a toddler. Then again and again, and soon he had picked up the fork and plate and was gulping great mouthfuls, chewing the noodles and cheese and sausage as if he had never eaten before and damn, isn't this eating thing great? Tomato sauce glistened on his chin, and a small rivulet of saliva ran toward his neck and collar.

When the plate was empty, he gazed at me, stoned. "More," he said.

So I refilled the plate in the kitchen and brought it to him. Again with the gulping and chewing and swallowing. It was deeply satisfying, watching someone eat my lasagna like that. It spoke to my deep talent. Jews know from lasagna.

When David was on his third plate, I remembered an article I had once read about a bulimic model who ate everything in her refrigerator and then dropped dead from the shock. All that cheese couldn't be good for his system. "Perhaps you'd better slow down."

"More," he started to say again, and then a terrified look crossed his face. He ran to my bathroom and slammed the door. I padded behind him and knocked. "David? Are you okay?"

After some time, there was flushing, and more flushing. David stumbled out. "I think I broke your toilet," he said. He fell onto my sofa, his knees buckling beneath him. He panted, "I feel so much better, I don't even want to kill myself anymore."

He closed his eyes. His breath slowed in increments, thickening into snores.

While I hovered over him, a faint cast of pink began to make its way to the surface of his skin, like the slowest dawn in history. Outside, the night snow had turned the world an iridescent blue.

I grabbed the plunger from my utility closet and went to see what he had done to my bathroom.

This is me: I had a baby once. Her name was Leah Sharon Rose. She died of SIDS when she was three weeks old. I had just nursed her. Afterward, I held her upright so she wouldn't get acid reflux.

She slept in my hands with her brow furrowed, like she was working out some complicated math problem in her head. When exhaustion overcame me, too, I pulled her tiny body to my chest, where she nuzzled and stayed. I didn't wake up until she was already cold.

So.

Everyone grieves differently, and my husband's way was to find a different woman, a younger one. He said, "I want a big family, Rachel, with, like, five children and a good mother to protect them from harm." Last I heard, he has a son now, and another baby on the way. Good for him. It's a talent, forgetting. Good for him, good for him.

I read somewhere that doctors can't say "Whoops." Medical schools tell them not to. Instead, when they make a mistake, they have to say "There."

What kind of fuckery is that, a system that doesn't allow for human error? You amputate the wrong leg, you lose a sponge inside a patient, you slip and cut a jugular, and you can only say "There," like you had planned this all along, like your secret mission was to kill the very person you were trying to save.

I spent the weekend watching Cary Grant movies and listening to David snore. He covered the entire length of my sofa and didn't move. I cleaned my apartment, cooked, ate, eventually even vacuumed while he slept. Three or four times a day, I pulled him upright and said, "David. You have to use the bathroom." He groaned and stumbled after me, holding my hand, his eyes still closed. I helped him to my abused toilet, let him do more of his business, and cleaned up. I cleaned him. He never really woke up at all during his toilet breaks. I managed to get a few smoothies inside him with protein powder mixed in. You didn't know a person can drink a smoothie without waking up, did you? I didn't, either.

Then I helped him back to the sofa, where he collapsed again as if he had just climbed a mountain. His eyes rolled from side to side under his thin lids. His jaw was slack and loose, and his lips were cracked. Sometimes I moistened them with lip balm or lifted a lock of dark hair from his eye. While I watched TV, I squeezed water into his mouth from a towel, very gradually so he wouldn't choke.

I knew I should probably call one of his doctor friends, but David's phone wasn't with him—I didn't know who to call. And what could he possibly need beyond what he was getting here on my sofa? I could see the change. It was like his body was an ocean cruiser, still pointed toward death, only beginning its wide arc back toward life.

By Sunday evening, most of the snow had melted, and my apartment smelled like a homeless shelter. It had to be done. I worked off his shirt, pants, and socks and laundered them. I folded his clothes and left them on the coffee table, where he would see them upon waking, if he ever woke. I spread his watch on top. I sponged off his skin with soapy warm water, wiping it dry with a cloth in my other hand so he wouldn't get cold.

His chest heaved under my washcloth. The hairs on his legs straightened and aligned themselves. I washed between his toes. His dark eyebrows relaxed.

After I tucked a fresh cotton blanket around his torso, David's eyes drifted open. "Am I here?" he asked.

"You're here," I said.

"Are you here?" he asked.

"I'm here," I said.

He closed his eyes again. He groped along my arm until he found my hand. I held his hand for a long time with those words thumping in me. Then his grip loosened as his body surrendered to sleep. He let me go.

I found an article Dr. David Constantino wrote for an online medical journal, on anesthesia. He said surgery used to be performed with a bottle of whiskey and four strong men to hold the screaming patient down. People chose suicide instead, preferring death over known pain. And some killed themselves afterward because during the surgery they felt that God had utterly abandoned them.

"But now we just count backward from ten, and our pain vanishes," he wrote. "It's a forgotten taste of death. When we come alive again, we're relieved of our worst memories."

On Monday morning, David was still sleeping and skinny, but his cheeks were pink in the dim light. His chest rose and fell as if he were floating on swells of gentle water. His forehead felt smooth and warm, and he had the beginnings of a beard. He nuzzled his cheek into my touch and resettled into sleep.

I called his office and left a message saying Dr. Constantino was unwell and would not be able to see his patients that day. And I had to go to work, too, right? The phones at A&A Flooring weren't going to yada yada. So I showered, dressed, and left David a note that said, *Dear David, you've been sleeping for three days. If you wake up, please eat something. I'll be home after work. Rachel.*

I worried all day. I dropped a full box of paper clips all over the floor. I fumbled the company's name twice. I ate an entire box of powdered sugar mini doughnuts from a convenience store. When five o'clock finally arrived, I rushed home.

David wasn't on the sofa, or anywhere. His clothes were gone, the blanket folded. My note was flipped over, with doctor-handwriting scrawled across the back of it. It said, *Dear Rachel. You saved my life. David.* Stretched across the top of the note was his father's gold watch.

I picked it up and held it in my hand. I felt . . . I don't know. Maybe it's enough to say I felt.

Here's my lasagna recipe, in case you ever need to save a life.

It's not so much a recipe as an attitude, and all about the balance. Not so many noodles, no more than three layers, and not that oven-ready nonsense. How hard is it to boil water? A whole big container of ricotta with an egg mixed in, and spice up the mixture with a palmful of oregano until you can see specks of green. Use organic Italian sausage; if you keep kosher, skip the meat, but what a shame. No hamburger or onion, or it gets burpy. Expensive spaghetti sauce instead of tomato sauce from a can. Lots of mushrooms, sauté them with garlic beforehand, and drain them so your lasagna doesn't get watery. Speaking of, no spinach, never! Sacrebleu, people! This isn't health food. Sharp cheddar inside with the ricotta, sausage, sauce, and mushrooms, and mozzarella on top. Bake uncovered so the top layer becomes a hard, crunchy shell like a pasta cracker. This is the recipe I inherited from my Jewish relatives, sleeping in the graves they were sent to by cancer and heartbreak.

I called David's clinic a few times to check on him, but he never called back. The nurses sounded surprised at my questions. He seemed just fine; was I a relative? A patient? A friend, I said, the lie evident in my voice. I had no idea what I was.

So one evening after work, I drove to David's clinic to return the watch. I pulled into the parking lot as they were shutting off the lights inside. Deflated office workers and strung-out doctors in blue button-downs seeped from the building. I shut off my car and waited for him. Maybe I was scared again—I don't know. I waited a long time before David pushed through the doors.

It felt strange to see him so far away, like he was anybody else. He looked like he had gained his normal weight back. He strode in my direction. I took a deep breath and reached for the door handle, but David slowed and then stopped. He squinted at the clouds.

It was beginning to rain, great drops splatting on my windshield, thudding against the roof of my car. Everyone scurried to their vehicles, except for David. He raised his face to the sky as it opened up.

Rain ricocheted off every surface. David's clothes stuck to his skin. A few remaining coworkers ran for their cars, their purses or hands shielding their eyes. But Dr. David Constantino didn't move as the rain saturated him with a violence that slicked his hair and darkened his shirt. His shoulders quaked. He was crying. No, he was laughing. He must have been freezing, but he didn't seem to care. I didn't move, David's watch in my lap. The rain began to ease. David lowered his wet face, got into his car, and drove away.

That was a year ago. I still wear the watch. I stopped crying in my sleep. When I think of my daughter, an airy calm settles across my shoulders. I began dating a nice municipal worker who does landscaping in city parks, tending trees and bushes. Last winter on our way to a late movie, a semi almost skidded into his car. Right before it missed us by inches, I thought, *I've done good in this life.*

Someone once told me a legend about an empress who fell off a boat. The river was swift, and she couldn't swim. An empress doesn't learn how. She flailed in the water, and the oarsmen crowded together at the boat's edge.

In whatever culture this was, if you save someone's life, you're responsible for that life forever. The peasants on the boat were poor, powerless. They couldn't bear responsibility for a royal. According to this legend, they all stood on the boat and watched the empress drown to death.

What a load of crap.

Of course they saved her. Because they could. And from then on, through the cracks in their days flowed this constant memory: the relentlessness of the current, a stranger's hand, and that strong pull back to life.

Acknowledgments

I want to express deep appreciation for my supportive family, friends, colleagues, students, and clients, who are all so patient with my foolishness and disorganization. Thank you for keeping me around.

An enormous thank you to my agent, Mary Evans, for making this book possible. Do you know how brilliant you are? I'm grateful to you for your vast experience, wisdom, and integrity, and for always believing in me, no matter what. Thank you to Flatiron executive editor Caroline Bleeke for your outstanding feedback, insight, patience, and expertise. You made these stories so much better than I ever could by myself. Thank you also to the many other people at Flatiron who contributed to this book: assistant editor Sydney Jeon, assistant director of publicity (and genius) Claire McLaughlin, exceptional copy editor Shelly Perron, interior designer Sue Walsh, creative director Keith Hayes, art designer Kelly Gatesman, production editor Morgan Mitchell, senior production manager Eva Diaz, managing editor Emily Walters, marketing manager Erin Kibby, senior vice president and associate publisher Malati Chavali, senior vice president and publisher Megan Lynch, and president and publisher Bob Miller. Few imprints publish short story collections; I'm very grateful to Flatiron Books for your support and hard work on this book.

I'm overwhelmed with gratitude for Murphy's group members Jenny Shank, Rachel Weaver, Paula Younger, Jennifer Sullivan, and Ashley Shires for their editorial help. Ditto for all the other writer friends who provided valuable feedback on individual stories:

Andrea Dupree, Karen Palmer, Tiffany Quay Tyson, Jenny Itell, Rachel Maizes, Ellen Anderman, Ben Whitmer, Laura Mahal, Lynn Schwebach, Amanda Rea, Brad Wetzler, Mariko Tominaga, Steve Caldes, Nana Mizushima, and other people I'm forgetting. Thank you to the many sensitivity/authenticity readers who helped me edit and deepen these stories.

Thank you to Lighthouse Writers Workshop, particularly Andrea Dupree (again) and Mike Henry, who have created a literary home for thousands of writers, including me. The two of you have furthered the art of writing more than anyone I know, and I'm so grateful to be a part of the Lighthouse community. Thank you to Scott Harrison and Ellen Moore, who generously let me write some of these stories in their treehouse. Thank you to Eve and Mike, who have supported my various endeavors so many times over the years.

Thank you to the literary magazine editors who published these stories in their original, rougher forms: Will Allison, Hannah Tinti, and Adina Talve-Goodman at *One Story* for "The Pole of Cold"; Ron Spatz at *Alaska Quarterly Review* for "The Piano"; Linda B. Swanson-Davies and Susan Burmeister-Brown at *Glimmer Train* for "North of Dodge"; Brad Morrow and Pat Sims at *Conjunctions* for "Eat My Moose"; Jessica Rogen at *Boulevard* for "Save Me, Stranger" (originally "Lotus"); Caitlin Horrocks, David Lynn, Abigail Wadsworth Serfass, and John Pickard at the *Kenyon Review* for "When in Bangkok"; Lynne Nugent and Kate Conlow at the *Iowa Review* for "The Standing Man"; Steve Schwartz, Stephanie G'Schwind, Jenny Wortman, and Lauren Furman at *Colorado Review* for "*Jude*"; Sacha Idell and Jessica Faust at the *Southern Review* for "Fear Me as You Fear God"; Genevieve Field at *Glamour* for "The Blue Hole"; and Jonathan Bohr Heinen and Tony Varallo at *Crazyhorse* (now *swamp pink*) for "Wounds of the Heart and

Great Vessels." It's been an honor appearing in your pages, among so many other writers I admire.

Anyone I forgot, please forgive me (see above re: disorganization, etc.).

Most of all, thank you to my chosen family. Everything I do is for you. I love you.

About the Author

Erika Krouse is the author of *Tell Me Everything: The Story of a Private Investigation*, winner of the Edgar Award, the Colorado Book Award, and the Housatonic Book Award; *Come Up and See Me Sometime*, a *New York Times* Notable Book and winner of the Patterson Fiction Prize; and *Contenders*, a finalist for the VCU Cabell First Novelist Award. Krouse's fiction has been published in the *New Yorker*, the *Atlantic*, *Ploughshares, One Story,* and more. She teaches creative writing at Lighthouse Writers Workshop and lives in Colorado.

Recommend *Save Me, Stranger* for your next book club!

Reading Group Guide available at
www.flatironbooks.com/reading-group-guides